# Algebras, Rings and Modules
## Non-commutative Algebras
## and Rings

## Volume 2

# Algebras, Rings and Modules
## Non-commutative Algebras and Rings

## Volume 2

**Michiel Hazewinkel**
Dept. of Pure and Applied Mathematics
Centrum Wiskunde & Informatica
Amsterdam, Netherlands

and

**Nadiya Gubareni**
Institute of Mathematics
Częstochowa University of Technology
Częstochowa, Poland

CRC Press
Taylor & Francis Group
Boca Raton London New York

CRC Press is an imprint of the
Taylor & Francis Group, an **informa** business
A SCIENCE PUBLISHERS BOOK

CRC Press
Taylor & Francis Group
6000 Broken Sound Parkway NW, Suite 300
Boca Raton, FL 33487-2742

First issued in paperback 2021

© 2017 by Taylor & Francis Group, LLC
CRC Press is an imprint of Taylor & Francis Group, an Informa business

No claim to original U.S. Government works

ISBN-13: 978-0-367-78250-4 (pbk)
ISBN-13: 978-1-138-03582-9 (hbk)

**Visit the Taylor & Francis Web site at**
**http://www.taylorandfrancis.com**

**and the CRC Press Web site at**
**http://www.crcpress.com**

# Preface

This the second part of a two-part treatise on (selected parts of) non-commutative algebra and ring theory. The contents of the first part [103] are given below.

Representation theory is a fundamental tool for studying groups, algebras and rings (and many other things). In this volume we consider representation theory for finite posets, finite dimensional algebras and semiperfect rings. The description of modules over some classes of semiperfect rings is reduced to mixed matrix problems over discrete valuation rings and division rings, which are considered in this book.

One of the main goals in ring theory is to reduce in a certain sense the description of large classes of rings to simpler classes by the use of some ring theoretic constructions. The best classical example is the Wedderburn-Artin theorem, however, there are a number of other general results that can be mentioned.

Section 1.1 represents the definition and main properties of the basic main construction of rings which are incidence rings of posets over associative rings.

In section 1.2 we consider some properties of a special class of incidence rings of the form $T(S) = I(S, D)$, where $S$ is a finite partially ordered set and $D$ is a division ring. This class of rings properly contains the class of hereditary serial rings and all Artinian rings with quivers that are trees.

A special class of right hereditary rings $A(S, O)$ of a finite poset $S$ over a family of discrete valuation rings $\{O_i\}_{i \in I}$ with a common skew field of fractions is introduced and studied in section 1.3.

In section 1.4 we introduce and study a special class of incidence rings modulo the radical of the form $I(S, \Lambda, \mathcal{M})$ related to a finite poset $S$ over a local Noetherian domain $\Lambda$ with Jacobson radical $\mathcal{M}$.

Serial and semidistributive rings of the form $I(S, \Lambda, \mathcal{M})$, where $\Lambda$ is a discrete valuation ring with Jacobson radical $\mathcal{M}$ are considered in section 1.5.

The main aim of chapter 2 is to study the properties and structure of different classes of rings whose lattices of submodules are distributive (or semidistributive). Such rings are called distributive (or semidistributive) and they can be considered as a non-commutative generalization of Prüfer domains. The class of distributive rings is very wide and includes, for example the ring of integers and the ring of polynomials $K[x]$ over a field $K$, rings of integral algebraic numbers, and commutative principal ideal rings. More generally, all commutative Dedekind rings and Prüfer domains are examples of distributive rings. The first papers which are devoted to distributive rings appeared in the fifties-sixties of the 20-th century in [26], [27], [29], [155], [156].

The systematic study of these rings began with papers of H. Achkar [2], V. Camillo [39], H.H. Brungs [33] and W. Stephenson [203].

In section 2.1 we consider the main properties of distributive modules and rings. Section 2.2 is devoted to study of semidistributive modules and rings. The structure of Noetherian distributive and semidistributive rings is discussed in section 2.3.

The properties of right hereditary SPSD-rings are studied in section 2.4. It is shown that the structure of all such rings is closely connected with right hereditary rings of the form $A(S, O)$ as described in section 1.3.

Section 2.5 is devoted to the study of the main properties and structure of semihereditary SPSD-rings.

Some of the more fundamental notions and results of the theory of homological algebra were studied in different sections of [100], [101]. Chapter 3 gives some additional facts from homological algebra.

The notions of direct limits and inverse limits for a set of modules were considered in [103, section 1.4] (see also [100, section 4.7]). In the particular case, when this set has only two modules these constructions have their own names, pullback and pushout. They are very important and useful in studying rings and modules. Therefore for the convenience of the reader these module constructions and their main properties are considered in more detail in section 3.1.

In the theory of homological algebra and its applications there is very important the statement which is often known as the "snake lemma" (it is also called "zig-zag lemma", or "serpent lemma"). It is valid in every Abelian category and it is an important tool in the construction of long exact sequences. This lemma consists of two claims: 1) the construction of an exact sequence, which is often called "kernel-cokernel sequence", for any commutative diagram of a special type; and 2) the construction of long exact sequences of homology groups for any given short exact sequence of complexes. The second part of this lemma was proved in [100, theorem 6.1.1]. The proof of the first part of this lemma is given in section 3.2.

The functors $\text{Ext}^n$ as right derived functors for the contravariant left exact functor Hom using projective resolutions were introduced in [100, section 6.4]. This functor is closely related to module extensions.

Let $A$ be a ring, and let $X, Y \in \text{Mod}_r A$. In this chapter the interpretation of the group $\text{Ext}_A^1(Y, X)$ is given in terms of short exact sequences. Section 3.3 is devoted to studying extensions of modules in terms of short exact sequences.

Following R. Baer the addition of extensions of modules, which makes the set of equivalence classes of all extensions an Abelian group, is introduced in section 3.4. Some main properties of the group $\text{Ext}_A^1(Y, X)$ are considered in section 3.5. The results of section 3.6 show that there is an isomorphism between equivalence classes of the group of extensions of $X$ by $Y$ and elements of $\text{Ext}_A^1(Y, X)$ as considered in [100].

In chapter 4 the main results about modules over semiperfect rings are given. Some basic properties of semiperfect rings and modules over them were considered in [100, chapter 10]. There are a number of equivalent definitions of semiperfect rings. One of them is given by H. Bass in terms of projective covers. He proved that a ring is semiperfect if and only if any finitely generated module has a projective cover (see [103, theorem 1.8.2]).

In [100, section 10.4] the structure of finitely generated projective modules over semiperfect rings was discussed. It was proved that any such right module can be uniquely decomposed into a direct sum of principal right modules. The generalization of this statement, the important theorem [103, theorem 7.2.7], states that any projective module over a semiperfect ring is a direct sum of principal modules. In section 4.1 this result is used to study the structure of finitely generated modules over semiperfect rings. The main result of this section, which was proved by R.B. Warfield, Jr. concerns the decompositions of any finitely generated module over a semiperfect ring. This theorem gives a possibility to introduce the notion of stably isomorphic modules.

In section 4.2 we prove that all modules over a semiperfect ring can be divided into the equivalence classes of stably isomorphic modules. Moreover, each stable isomorphism class of finitely generated modules over a semiperfect ring contains a unique (up to isomorphism) minimal element. Most of the results of this section were obtained by R.B. Warfield, Jr. (see [220], [222]).

In [101, section 4.10] we considered the duality in Noetherian rings, which is given by the covariant functor $^* = \text{Hom}_A(-, A)$. For an arbitrary ring $A$ this functor induces a duality between the full subcategories of finitely generated projective right $A$-modules and left $A$-modules. In section 4.3 the main properties of this functor and torsionless modules are studied for the case of modules over semiperfect rings.

Section 4.4 presents an introduction to the duality theory of Auslander and Bridger [13], and yields a connection between finitely presented right modules and finitely presented left modules over semiperfect rings. Some main properties of the Auslander-Bridger transpose, which is closely connected with almost split sequences, are discussed. These sequences were first introduced and studied by M. Auslander and I. Reiten in [15] and [20], and they play an important role in the representation theory of rings and finite dimensional algebras. Section 4.5 is devoted to the study of some main properties of these sequences. In section 4.6 we study almost split sequences over semiperfect rings and prove the existence of these sequences for strongly indecomposable modules. Section 4.7 presents an introduction to the theory of linkage of modules over semiperfect Noetherian rings using two types of functors: syzygy and transpose.

Chapter 5 is devoted to finite partially ordered sets (posets) and their representations, which play an important role in representation theory. They were first introduced and studied by L.A. Nazarova and A.V. Roiter [163] in 1972 in connection with problems of representations of finite dimensional algebras. M.M. Kleiner characterized posets of finite type [129] and described their pairwise non-isomorphic indecomposable representations [130]. He proved a theorem which gives a criterium for posets to be of finite representation type.

Recall that a finite poset $\mathcal{P}$ is called **primitive** if it is a cardinal sum of linearly ordered sets $L_1, \ldots, L_m$. It is then denoted by $\mathcal{P} = L_1 \sqcup \cdots \sqcup L_m$. This chapter gives the proof of the criterium for primitive posets to be of finite representation type following [30].

To prove this criterium there are used only the trichotomy lemma which was proved by P. Gabriel and A.V. Roiter in [82], the Kleiner lemma about the representations of a pair of finite posets proved by M. Kleiner in [129] and the main construction, considered in section 5.5. Note that this construction (in some form) was introduced by L.A. Nazarova and A.V. Roiter in [163].

An important problem in the theory of representations of finite dimensional algebras (or f.d. algebras, in short) is to obtain the full list of different kinds of algebras which are of finite representation type (or finite type, or f.r.t., in short). The first classes of associative f.d. algebras of f.r.t which have been described were the classes of algebras with zero square radical and hereditary algebras over algebraically closed fields.

There are different approaches to study the representations of f.d. algebras. One of them is the approach of P. Gabriel [79], which reduces the study of representations of algebras to the study of representations of quivers. Another approach was first considered by L.A. Nazarova and A.V. Roiter [163]. This approach is to solve "matrix problems", that is, the reducing of some classes of matrices by means of admissible transformations to their simplest form. A third approach is due to M. Auslander and it is connected with the technique of almost split sequences.

Chapter 6 can be considered as an introduction to the theory of representations of quivers and finite dimensional algebras. This chapter gives some main notions and some fundamental results of these representations, most of which are given without proof. In section 6.1 we consider the notions of finite quivers and their representations and give the main results of this theory. Section 6.2 is devoted to species and their representations. In section 6.3 we consider some main notions and results of the representation theory of finite dimensional algebras.

As it turns out the category of representations of finite dimensional algebras is equivalent to the category of representations of special classes of quivers, which are called bound quivers. That is why the quivers play a central role in the theory of finite dimensional associative algebras and their modules.

For right Artinian rings one can also introduce the notion of a ring of finite representation type. As has been shown by D. Eisenbud and P. Griffith [71] this notion is left-right symmetric. They proved this fact using the duality theory of Auslander and Bridger. This result is proved in section 7.1.

For finite dimensional algebras along with the notion of finite representation type there is also considered the notion of bounded representation type. Recall that a finite dimensional algebra $A$ is called of **bounded representation type** if there is a bound on the length of the indecomposable finite dimensional $A$-modules. The first Brauer-Thrall conjecture says that these notions are the same in the case of a finite dimensional algebra $A$ (as was proved by A.V. Roiter [101, theorem 3.5.1]) and in the case of Artinian algebras (as was proved by M. Auslander [15]).

A ring $A$, not necessarily Artinian, is said to be of **finite representation type**, if it has a finite number of non-isomorphic indecomposable finitely presented $A$-modules.

For Artinian rings this definition coincides with the earlier one, because in this case each finitely generated $A$-module is finitely presented as well. The main results concerning Artinian hereditary rings of finite representation type are given in section 7.4.

For Artinian rings along with the notion of finite representation type there is considered the notion of bounded representation type. Recall that a right Artinian ring $A$ is said to be of **bounded representation type** if there is a bound on the length of finitely generated indecomposable right $A$-modules. The first Brauer-Thrall conjecture asserts that these notions are the same. M. Auslander proved that this conjecture is true for right Artinian rings (see theorem 7.2.11).

Following R.B. Warfield, Jr. a ring has **right bounded representation type** if there is an upper bound on the number of generators required for indecomposable finitely presented right $A$-modules. In his paper [222] R.B. Warfield, Jr. puts the following question:

*Question* 4. For what semiperfect rings is there an upper bound on the number of generators required for the indecomposable finitely presented modules?

Chapter 8 describes special classes of semiperfect rings of bounded representation type, which constitute some sort of answer to this question.

As shown by R.B. Warfield, Jr. there is a serious restriction on the structure of rings of bounded representation type connected with modules of finite Goldie dimension. In section 8.1 it is proved that a semiprimary ring of finite right bounded representation type is right Artinian (see [222]).

$O$-species and tensor algebras, which are generalizations of $k$-species as introduced by P. Gabriel, are considered in section 8.2. The connection between right hereditary SPSD-rings and special kinds of $(D, O)$-species is considered in section 8.3.

In section 8.4 we discuss the reduction of representations of $(D, O)$-species to mixed matrix problems over discrete valuation rings and their common skew field of fractions. Some important mixed matrix problems are considered in section 8.5.

Sections 8.6 and 8.7 are devoted to the study of $(D, O)$-species of bounded representation type. There is a theorem which gives the structure of these species in the terms of diagrams which can be considered as generalizations of Dynkin diagrams.

The right hereditary SPSD-rings of finite right bounded representation type are described in section 8.8.

The book is written on a level accessible to advanced students who have some experience with modern algebra. It will be useful for those new to the subject as well for researchers and serves as a reference volume.

While writing this book the second author was in particular supported by FAPEST of Brazil in 2010. The author would like to express cordial thanks to the Institute of Mathematics and Statistics of the University of São Paulo, and especially prof. M. Dokuchaev and prof. V.M. Futorny, for their warm hospitality during her visit in 2010. The author is also grateful to prof. V.V. Kirichenko for useful discussions and fruitful remarks.

# CONTENTS

## Contents of Volume 1

# Contents of Volume 2

# CHAPTER 1

# Rings Related to Finite Posets

One of the main goals in the ring theory is to reduce in a certain sense the description of large classes of rings to simpler classes using some ring theoretic constructions. The most classical example is the Wedderburn-Artin theorem, however one can mention a number of other general results.

Section 1.1 represents the definition and main properties of a basic construction of rings. These are called incidence rings of posets over associative rings.

In section 1.2 we consider some properties of a special class of incidence rings of the form $T(S) = I(S, D)$, where $S$ is a finite partially ordered set and $D$ is a division ring. This class of rings properly contains the class of hereditary serial rings and all Artinian rings with quivers that are trees.

A special class of right hereditary rings $A(S, O)$ of a finite poset $S$ over a family of discrete valuation rings $\{O_i\}_{i \in I}$ with a common skew field of fractions is introduced and studied in section 1.3.

In section 1.4 we introduce and study a special class of incidence rings. These are called incidence rings modulo the radical. They are denoted by $I(S, \Lambda, \mathcal{M})$ and involve a finite poset $S$ over a local Noetherian domain $\Lambda$ with Jacobson radical $\mathcal{M}$.

Serial and semidistributive rings of the form $I(S, \Lambda, \mathcal{M})$, where $\Lambda$ is a discrete valuation ring with Jacobson radical $\mathcal{M}$, are considered in section 1.5.

## 1.1 Incidence Rings

This section covers the definition and some properties of incidence rings of partially ordered sets over associative rings.

Let $S$ be a set with a binary relation denoted by $\leq$. The relation $\leq$ is called a **preorder** if the following properties satisfy:

1. $a \leq a$ for any $a \in S$ (reflexivity).
2. $(a \leq b) \wedge (b \leq c)$ implies $a \leq c$ for any $a, b, c \in S$ (transitivity).
   A set which is equipped with a preorder is called a **preordered set**.
   Note that if a preorder $\leq$ is antisymmetric, that is,

3. $(a \leq b) \wedge (b \leq a)$ implies $a = b$ for each $a, b \in S$ (antisymmetry)
   then $\leq$ is a **partial order**, and a set which is equipped with a partial order is
   called a **partially ordered set** (or **poset**, for short).
   If a preorder $\leq$ is symmetric, that is,
4. $a \leq b$ implies $b \leq a$ for each $a, b \in S$ (symmetry)
   then $\leq$ is called an **equivalence relation**.

Let $S$ be a poset with partial ordering relation $\leq$ (which is transitive, reflexive and antisymmetric). Denote by $x \prec y$ the strict order, i.e. the relation "$x \leq y$ and $x \neq y$".

An element $x \in S$ is called **maximal** if there is no element $y \in S$ satisfying $x \prec y$. Dually, $x$ is **minimal** if there is no element $y \in S$ satisfying $y \prec x$. An element $x \in S$ is said to be a **least element** if $x \leq y$ for all $y \in S$, and $x \in S$ is said to be a **greatest element** if $y \leq x$ for all $y \in S$.

A subset $C$ of a poset $S$ is called a **chain** if it is totally ordered, i.e. for any $x, y \in C$, either $x \leq y$ or $y \leq x$. Denote a chain of cardinality $n$ by $C_n$, or $(n)$. In this case the number $n$ is called the **length** of the chain $C_n$. A subset $A$ of a poset $S$ is called an **antichain** if all its elements are pairwise incomparable in $S$, i.e. for any pair of distinct elements $x, y \in A$, both $x \not\leq y$ and $y \not\leq x$.

In order to visualize a poset $S$ one uses the so called diagram of $S$. Let $x$ and $y$ be distinct elements of $S$. It is said that $y$ **covers** $x$ if $x \prec y$ but there is no element $z$ such that $x \prec z \prec y$. Recall that the **diagram** $H(S)$ of a poset $S$ is the directed graph whose vertex set is $S$ and whose set of edges is given by the set of covering pairs $(x, y)$ of $S$, moreover, there is a (directed) edge from a vertex $x$ up to a vertex $y$ if and only if $y$ covers $x$.

For example, the diagram below

represents the poset $(S, \leq)$ with 3 elements $\{a_1, a_2, a_3\}$ and the one relation $a_2 \prec a_3$.

**Remark 1.1.1.** The diagram $H(S)$ of a poset $S$ is often called its **Hasse diagram**. Usually it is drawn in the plane in such a way that if $y$ covers $x$ then the point representing $y$ is drawn higher than the point representing $x$. In this case the Hasse diagram is drawn without arrows. For example, the Hasse diagram below

represents the same poset $(S, \leq)$ as above, i.e. $S = \{a_1, a_2, a_3\}$ with the one relation $a_2 \prec a_3$.

**Definition 1.1.2.** An **interval** $[x, y]$ of a preordered set $S$ with preorder $\leq$ is defined as the set $[x, y] = \{z \in S \,:\, x \leq z \leq y\}$. A preordered set $S$ is **locally finite** if each its interval is finite.

The idea of incidence algebras had its beginning in the works of R. Dedekind and E.T. Bell. The incidence algebra of a locally finite poset over a field was first introduced by G.-C. Rota in [185]. This notion can be extended to the case of a non-commutative ring $A$.

**Definition 1.1.3.** Let $S$ be a locally finite poset with a partial ordering relation $\leq$, and $A$ an associative (not necessarily commutative) ring with identity. The **incidence ring** $I(S, A)$ of $S$ over $A$ is the set of functions $f : S \times S \to A$ such that $f(x, y) = 0$ if $x \not\leq y$ with operations:

$$(f + g)(x, y) = f(x, y) + g(x, y),$$

$$(fg)(x, y) = \sum_{x \leq z \leq y} f(x, z)g(z, y)$$

$$(af)(x, y) = a(f(x, y))$$

for all $x, y, z \in S$ and $a \in A$.

An important set of elements of $I(S, A)$ is formed by the **characteristic functions** which are defined in the following way. If $X \subseteq S$ then

$$\delta_X(x, z) = \begin{cases} 1 & \text{if } x = z \in X \\ 0 & \text{otherwise.} \end{cases}$$

Let us restrict our attention (for the moment) to the case of finite posets. Let $S = \{\alpha_1, \ldots, \alpha_n\}$ be a finite poset with a partial ordering relation $\leq$, and let $M_n(A)$ be the generalized matrix ring of all $n \times n$-matrices with elements from an associative ring $A$. In this case the incidence ring of $S$ over $A$ is the subring $I(S, A)$ of $M_n(A)$ such that the $(i, j)$-entry of $I(S, A)$ is equal to 0 if $\alpha_i \not\leq \alpha_j$ in $S$. So, letting correspond to any element $\alpha_i \in S$ the matrix unit $e_{ii} \in M_n(A)$, and to any pairs of elements $\alpha_i, \alpha_j \in S$ such that $\alpha_i \leq \alpha_j$ the matrix unit $e_{ij} \in M_n(A)$ one obtain a basis (over $A$) of $I(S, A)$.

**Example 1.1.4.** The poset $\mathcal{R} = \{a, b, c, d \,:\, a < b < d; \, a < c < d\}$ has a Hasse diagram of the following form:

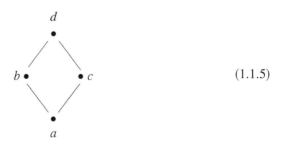

$$(1.1.5)$$

It is called the **rhombus**.

Then

$$I(S, A) = \begin{pmatrix} A & A & A & A \\ 0 & A & 0 & A \\ 0 & 0 & A & A \\ 0 & 0 & 0 & A \end{pmatrix}$$

**Example 1.1.6.** Let $S$ be the poset with Hasse diagram of the following form:

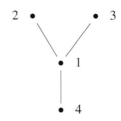

then

$$I(S, A) = \begin{pmatrix} A & A & A & 0 \\ 0 & A & 0 & 0 \\ 0 & 0 & A & 0 \\ A & A & A & A \end{pmatrix}$$

As these examples illustrate, transitivity of the ordering relation has much to do with "$I(S, A)$ is a subring of $M_n(A)$".

The following statement is obvious:

**Lemma 1.1.7.** *Every interval of a finite poset $S$ is a chain if and only if $S$ contains no subposets whose diagrams are rhombuses.*

Denote by $\overline{S}$ the non-oriented graph obtained from the diagram of $S$ by deleting the orientation of all its arrows. Then it is easy to show

**Lemma 1.1.8.** *The non-oriented graph $\overline{S}$ is a disconnected union of trees if and only if $S$ contains no subposets whose diagrams are rhombuses.*

Note also the well-known fact:

**Lemma 1.1.9.** *A finite poset $S$ with partial ordering relation $\leq$ can be labeled as $S = \{\alpha_1, \dots, \alpha_n\}$ where $\alpha_i \leq \alpha_j$ implies $i \leq j$.*

Using this lemma it is easy to show that in the case that $S$ has property just mentioned the incidence ring $I(S, A)$ is isomorphic to a subring of the ring of the upper triangular matrix ring $T_n(A)$. In particular, if a poset $S$ is linearly ordered then $I(S, A) \cong T_n(A)$.

**Proposition 1.1.10.** *Let $S = \{\alpha_1, \dots, \alpha_n\}$ be a finite poset with partial ordering relation $\leq$, and $A$ an associative ring with identity and Jacobson radical $R$. Then*

1. $I(S, A)$ *is an FDI-ring.*
2. $I(S, A)$ *is a semiperfect ring if and only if A is semiperfect.*
3. *The Jacobson radical of $I(S, A)$ is the each set of elements of $I(S, A)$ for which $(i, i)$-entry is in R.*
4. $I(S, A)$ *is a right (left) Noetherian ring if and only if A is right (left) Noetherian.*
5. $I(S, A)$ *is a right (left) Artinian ring if and only if A is right (left) Artinian.*

*Proof.*

1. This is obvious because $1 = e_{11} + \cdots + e_{nn} \in I(S, A) \subset M_n(A)$.
2. This follows from [103, Theorem 1.9.3].
3. Thanks to statement 1 $B = I(S, A)$ is an FDI-ring. Therefore $B$ has a two-sided Peirce decomposition $B = (B_{ij})$. Since for any $i \neq j$ $\alpha_i \not\leq \alpha_j$ or $\alpha_j \not\leq \alpha_i$ in $S$, $B_{ij}B_{ji} = 0$. So the statement follows now from [103, Proposition 2.6.10].

Statements 4 and 5 follows from [103, Theorem 1.1.23]. $\square$

**Remark 1.1.11.** The notion of an incidence ring has been generalized to the case of preordered sets in [75], [204].

Let $S$ be a finite preordered set with a preorder $\leq$, and $A$ an associative (not necessarily commutative) ring with identity. The **incidence ring** $I(S, A)$ of $S$ over $A$ is a ring with the additive structure of a free $A$-module with basis $\{f_{xy} : x \leq y; x, y \in S\}$, where multiplication is given by the linear extension of:

$$f_{xy} \cdot f_{zu} = \begin{cases} f_{xu} & \text{if } y = z \\ 0 & \text{otherwise,} \end{cases} \qquad (1.1.12)$$

for all $x, y, u, v \in V$.

**Example 1.1.13.** Let $\mathfrak{R}$ be a finite preordered set $\{1, 2, \ldots, n\}$ with preordering relation $\leq$. One can consider the corresponding reflexive and transitive Boolean matrix $B = [b_{ij}]$ defined by $b_{ij} = 1$ if and only if $i \leq j$, otherwise $b_{ij} = 0$. To the finite preordered set $\mathfrak{R}$ and an associative ring $A$ with identity one can associate the **structural matrix ring** $M(B, A)$ associated with $B$ and defined by the following way:

$$M(B, A) = \{X = [x_{ij}] \in M_n(A) : b_{ij} = 0 \Longrightarrow x_{ij} = 0\}. \qquad (1.1.14)$$

In fact, $M(B, A)$ is simply the incidence ring of the preordered set $\mathfrak{R}$ over the ring $A$.

In this section we have (so far) considered incidence rings of posets and preordered sets. They are special cases of a general construction - the incidence rings of an arbitrary relation which is not required to be a preorder. Such rings were introduced by G. Abrams in [1] and were called generalized incidence rings.

**Definition 1.1.15.** Let $A$ be an associative ring with unity. Let $\leq$ be a binary relation on a locally finite set $S$. The **generalized incidence ring** $I(S, \leq, A)$ of $S$ with coefficients in $A$ is the free left $A$-module with basis $\{f_{xy} : x \leq y \text{ in } S; x, y \in S\}$, where multiplication is given by the linear extension of:

$$f_{xy} \cdot f_{zu} = \begin{cases} f_{xu} & \text{if } y = z \text{ and } x \leq u \text{ in } S \\ 0 & \text{otherwise,} \end{cases} \tag{1.1.16}$$

for all $x, y, u, v \in S$.

**Remark 1.1.17.** When $\leq$ is a preorder on a set $S$ the definition of a generalized incidence ring given above coincides with the definition of an incidence ring as given in Remark 1.1.11.

For a ring $I(S, \leq, A)$ to be associative it is necessary that the following special property be satisfied.

**Definition 1.1.18.** ([1]). A reflexive relation $\leq$ on a set $S$ is said to be **balanced** if for all $x, y, u, v \in S$ with $x \leq y \leq u \leq v$ and $x \leq v \in S$

$$x \leq u \in S \iff y \leq v \in S.$$

**Remark 1.1.19.** Note that a relation which is both reflexive and transitive is always balanced, but there are balanced relations which are not transitive. Some of these are shown in Example 1.1.23 below.

**Proposition 1.1.20.** [1].

1. *The multiplication in $I(S, \leq, A)$ is associative if and only if the relation $\leq$ on $S$ is balanced.*
2. *If a relation $\leq$ is reflexive then $I(S, \leq, A)$ has multiplicative identity.*

*Proof.*

1. If $x, y, z, w, u, v \in S$ then the expressions $(f_{xy} f_{zw}) f_{uv}$ and $f_{xy} (f_{zw} f_{uv})$ are both zero unless $y = z$, $w = u$ and $x \leq v$. In this case, the first product is equal to $f_{xv}$ precisely when $x \leq w$, while the second product is equal to $f_{xv}$ precisely when $z \leq v$, i.e. this is the case if and only if $\leq$ is a balanced relation.
2. From (1.1.16) it immediately follows that the elements $f_{xx} = e_x$ for all $x \in S$ are idempotents in $I(S, \leq, A)$ and

$$e_x f_{yz} = \begin{cases} f_{yz} & \text{if } x = y \\ 0 & \text{otherwise} \end{cases}, \quad f_{yz} e_x = \begin{cases} f_{yz} & \text{if } x = z \\ 0 & \text{otherwise} \end{cases}$$

for all $x, y, z \in S$. This implies that the element $\sum_{x \in S} e_x$ is a multiplicative identity of $I(S, \leq, A)$. $\square$

The definition of a generalized incidence ring can be given in terms of graphs (see [4]).

**Definition 1.1.21.** Let $S = (V, E)$ be a graph with set of vertices $V = \{1, 2, \ldots, n\}$ and set of edges $E \subseteq V \times V$. Then the associated **generalized incidence ring** $I(S, A)$ of $S$ over an associative ring $A$ is defined as the free left $A$-module with basis consisting of all edges in $E$ where multiplication is defined by the linear extension of

$$(x, y)(u, v) = \begin{cases} (x, v) & \text{if } y = u \text{ and } (x, v) \in E \\ 0 & \text{otherwise,} \end{cases}$$

for all $x, y, u, v \in V$.

**Definition 1.1.22.** A directed graph $S$ contained all loops is said to be **balanced** if for all different $x, y, u, v \in V$ with $(x, y), (y, u), (u, v), (x, v) \in E$

$$(x, u) \in E \iff (y, v) \in E.$$

**Example 1.1.23.** We omit loops in all diagrams represented in these examples. The following directed graphs are balanced and nontransitive.

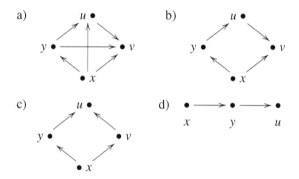

Proposition 1.1.20 and these examples show, in particular, that there are associative generalized incidence rings which are not incidence rings of preordered sets. Note that Example 1.1.23 is in terms of directed graphs but not Hasse diagrams.

**Remarks 1.1.24.**

1. The examples 1.1.23(a,b) may suggest why the terminology 'balanced' is used.
2. In definitions 1.1.18, 1.1.22 we assume that a relation is reflexive (and so its corresponding directed graph has all loops). But some authors (see e.g. [4]) define balanced relation (and balanced graphs) without this assumption.
3. There is also a different notion in graph theory that goes by the name 'balanced directed graph'. The requirement is that the in and out degrees at each vertex are the same. See [50]. Reversing the arrow $(x, v)$ in the picture above turns it into a balanced digraph in this sense.

In [190] V.D. Shmatkov introduced another generalization of an incidence ring.

**Definition 1.1.25.** A set of pairwise orthogonal idempotents $E$ of a ring $A$ is called **maximal** if $\text{r.ann}_A(E) = \text{l.ann}_A(E) = \{0\}$.

**Definition 1.1.26.** Let $S$ be a set with a binary relation $\leq$. A ring $I = I(S)$ with identity is called a **generalized incidence ring** if there exists a maximal set of idempotents $E = \{e_u \; : \; u \in S, e_u \in I\}$ such that:

1. Every idempotent $e_u$ is local, i.e. $eIe/\text{rad}(eIe)$ is a division ring.
2. The set $S$ is locally finite, i.e. each interval $[x, y] = \{z \in S \; : \; x \leq z \leq y\}$ is a finite set.
3. Multiplication and addition in $I$ are defined as follows:
   for any $x, y \in S$ and $f, g \in I$

$$(f + g)(x, y) = f(x, y) + g(x, y),$$

   if $x \nleq y$ then $(fg)(x, y) = 0$, and if $x \leq y$ then

$$(fg)(x, y) = \sum_{x \leq z \leq y} f(x, z)g(z, y)$$

   where $f(x, y) = e_x f e_y$ and $g(x, y) = e_x g e_y$.
4. For any local idempotent $f \in I$ there exist $x, y \in S$ such that $f(x, y) \notin \text{rad}(I)$.
5. For any set $\{c(u, v) \; : \; u, v \in S, c(u, v) \in e_u I e_v\}$ there exists $f \in I$ such that $f(u, v) = e_u f e_v = c(u, v)$ for any $u, v \in S$.

**Example 1.1.27.** Let $A$ be a semiperfect ring. Then there exists a finite set $E = \{e_1, e_2, \ldots, e_n\}$ of pairwise orthogonal local idempotents of $A$ such that $1 = e_1 + e_2 + \cdots + e_n$. Define the binary relation on $E$ such that $e_i \leq e_j$ if and only if $e_i A e_j \neq 0$. Then $A$ with respect to this binary relation is a generalized incidence ring in sense of definition 1.1.26.

**Remark 1.1.28.** Another generalization of these classes of rings is the class of incidence rings of a group automata which is at present studied actively in the literature (see, in particular [120], [121], [122]).

**Remark 1.1.29.** If $S$ and $S'$ are isomorphic then of course the (various) incidence rings $I(S, A)$ and $I(S', A)$ are isomorphic. One may wonder to what extent the reverse is true. Some results and references pertaining to this are given in section 1.6 below.

## 1.2 Incidence Rings $I(S, D)$

In this section we consider the particular example of incidence rings $I(S, D)$ of a finite poset $S$ over a division ring $D$.

Let $S = \{\alpha_1, \alpha_2, \ldots, \alpha_n\}$ be a finite poset with a partial order relation $\leq$, and $D$ a division ring. Denote by $M_n(D)$ the ring of all $(n \times n)$-matrices over $D$ with matrix units $e_{ij}$, where $de_{ij} = e_{ij}d$ for all $d \in D$. Consider the subring $T(S, D) \subset M_n(D)$ which is generated by those $e_{ij} \in M_n(D)$ for which $\alpha_i \leq \alpha_j$ in $S$. This is equivalent to the following: $e_{ii}T(S, D)e_{jj} = D$ if and only if $\alpha_i \leq \alpha_j$ in $S$, otherwise $e_{ii}T(S, D)e_{jj} = 0$.

Obviously, $T(S, D) = I(S, D)$. We will often denote this ring $T(S)$ if a division ring $D$ is fixed and known from the context.

**Example 1.2.1.** Let $S = \{a, b, c, d : a \prec b \prec d;\ a \prec c \prec d\}$ be a rhombus, which has a diagram of the form (1.1.5), and let $D$ be a division ring. Then

$$T(S) = \begin{pmatrix} D & D & D & D \\ 0 & D & 0 & D \\ 0 & 0 & D & D \\ 0 & 0 & 0 & D \end{pmatrix}. \tag{1.2.2}$$

As was shown in [103, Example 9.1.10], the ring $T(S)$ is a piecewise domain but it is neither right nor left hereditary.

**Example 1.2.3.** Let $S = \{\alpha_1, \ldots, \alpha_n\}$ be a linearly ordered set, i.e. $\alpha_i \leq \alpha_j$ if and only if $i \leq j$, and let $D$ be a division ring. In this case $T(S) = T_n(D)$, where $T_n(D)$ is a ring of all upper triangular $(n \times n)$-matrices over $D$. This ring is Artinian serial and hereditary. Moreover, by [100, theorem 13.5.2], any serial Artinian hereditary ring $A$ is Morita equivalent to a finite direct product of rings isomorphic to rings of upper triangular matrices over division rings.

**Proposition 1.2.4.** *There is a labeling of a poset $S$ such that the ring $T(S)$ is a triangular ring, and so is isomorphic to a subring of $T_n(D)$, where $n$ is the number of elements of the poset $S$.*

*Proof.* Since $S = \{\alpha_1, \ldots, \alpha_n\}$ is a finite poset, $S$ can be labeled in such a way that $\alpha_i \leq \alpha_j$ if $i \leq j$. Then for this numbering $T(S)$ is obviously a triangular ring and $T(S) \subseteq T_n(D)$. Moreover, one has a strong equality if (and only if) $S$ is a linearly ordered set. $\square$

**Remark 1.2.5.** In what follows, $T(S, D)$ will be always assumed to be a triangular ring.

From proposition 1.1.10 we immediately obtain the following statement.

**Proposition 1.2.6.**

1. *A ring $T(S, D)$ is a primely triangular Artinian semiperfect ring.*
2. *The Jacobson radical $R$ of $T = T(S, D)$ is equal to the prime radical $N$ of $T(S, D)$, and the two-sided Peirce decomposition of $R$ has the following form:*

$$e_{ii}Re_{ii} = 0 \quad and \quad e_{ii}Re_{jj} = e_{ii}Te_{jj} \quad for\ i, j = 1, 2, \ldots, n;\ i \neq j.$$

Recall that a semiperfect ring $A$ with Jacobson radical is a **basic** ring if $A/R$ is a direct sum of division rings.

**Proposition 1.2.7.** *A ring $T(S, D)$ is a basic ring.*

*Proof.* This follows immediately from proposition 1.2.6. □

**Proposition 1.2.8.** *A ring $T(S, D)$ is Artinian and possesses a classical ring of fractions $\widetilde{T}$ which coincides with $T(S, D)$.*

*Proof.* By proposition 1.2.6, $T(S, D)$ is a primely triangular Artinian semiperfect ring with Jacobson radical $R$ which coincides with the prime radical $N$. Therefore by [103, theorem 8.5.31] $T(S)$ possesses a classical ring of fractions $\widetilde{T}$ which is an Artinian ring.

Now let's show that $T(S) = \widetilde{T}$. From [103, lemma 8.5.26] it follows that any regular element $r \in T(S, D)$ has the form

$$r = \operatorname{diag}(r_1, r_2, \ldots, r_n) + x,$$

where each $r_i$ is a non-zero regular element of $e_{ii}T(S, D)e_{ii} = D$ $(i = 1, 2, \ldots, n)$, and $x \in R$. Therefore any regular element of $T(S, D)$ is invertible in $T(S, D)$, and so $T(S, D) = \widetilde{T}$. □

A finite poset $S$ is said to be **connected** if the Hasse diagram of $S$ is connected. It is obvious that a ring $T(S, D)$ is indecomposable if and only if the poset $S$ is connected. Since $T = T(S, D)$ is an Artinian semiperfect ring it is possible to construct the quiver $Q(T)$ of this ring. Recall that $Q(T) = Q(T/R^2)$. Let $P_i = e_{ii}T$ be a right principal module of $T(S, D)$. If $R$ is the Jacobson radical of $T(S, D)$ then the right quiver of $T(S, D)$ can be constructed in the following way. If

$$P(e_{ii}R) \cong \bigoplus_j P_j^{t_{ij}}$$

or

$$e_{ii}R/e_{ii}R^2 \cong \bigoplus_j U_j^{t_{ij}},$$

where the $U_j$ are right simple $T/R$-modules, then in the quiver $Q(T)$ the vertex $i$ is connected with the vertex $j$ by $t_{ij}$ arrows.

**Proposition 1.2.9.** *The quiver $Q(T)$ of the ring $T = T(S, D)$ coincides with the Hasse diagram $H(S)$ of the poset $S$.*

*Proof.* One can assume that $T = T(S, D)$ is an indecomposable ring. Then by [100, theorem 11.1.9], the quiver $Q(T)$ is connected. Let $\{e_{ij}\}$ be the set of all matrix units of $M_n(D)$ and $1 = e_{11} + e_{22} + \cdots + e_{nn}$ the natural decomposition of the identity of $T(S)$ into a sum of pairwise orthogonal idempotents.

Consider the Hasse diagram $H(S)$ of the poset $S$ and the following cases.

1. Assume that there is an arrow $i \rightarrow j$ in the diagram $H(S)$, which means that $e_{ii} Re_{jj} = D$ and $e_{ii} Re_{kk} = 0$ or $e_{kk} Re_{jj} = 0$ for any integer $1 \leq k \leq n$. Then $e_{ii} R^2 e_{jj} = \sum_{k=1}^{n} e_{ii} Re_{kk} e_{kk} Re_{jj} = 0$. Therefore there is exactly one arrow $i \rightarrow j$ in the quiver $Q(T)$.

2. Assume that $\alpha_i \leq \alpha_j$ and there is no arrow of the form $i \rightarrow j$ in $H(S)$. This means that there is an integer $1 \leq k \leq n$ such that $e_{ii} Re_{kk} = D$ or $e_{kk} Re_{jj} = D$. Then $e_{ii} R^2 e_{jj} = \sum_{k=1}^{n} e_{ii} Re_{kk} e_{kk} Re_{jj} = D$ and therefore $e_{ii} Re_{jj} / e_{ii} R^2 e_{jj} = 0$, i.e. there is no arrow $i \rightarrow j$ in the quiver $Q(T)$.

Conversely, suppose that there is an arrow $i \rightarrow j$ in the quiver $Q(T)$. This means that there is an exact sequence

$$P_j \longrightarrow e_{ii} R \longrightarrow 0$$

Therefore $e_{ii} Re_{jj} \neq 0$ and so $\alpha_i \leq \alpha_j$, and there is an arrow $i \rightarrow j$ in the diagram $H(S)$. $\square$

Recall that the endomorphism ring $\text{End}_A(P)$ of a finitely generated projective $A$-module $P$ is called a **minor** of the ring $A$ (see [100, Section 13.3]). If $P$ can be decomposed into a sum of $n$ indecomposable modules then $\text{End}_A(P)$ is called a minor of order $n$. So that the ring $\text{End}(eA) = eAe$ is a minor for any idempotent $e$ of a ring $A$. If $e$ is a sum of $n$ primitive idempotents then $\text{End}(eA) = eAe$ is a minor of order $n$. It was shown that the minors of right (left) Noetherian rings are right (left) Noetherian, the minors of right (left) Artinian rings are right (left) Artinian (see [100, Theorem 3.6.1]). From [100, Theorem 10.3.8] and [100, Corollary 10.3.11] it follows that any minor of a semiperfect ring is semiperfect. If $e$ is an idempotent of a hereditary (semihereditary) ring then $eAe$ is also hereditary (semihereditary) (see [103, proposition 4.4.6(4)] and [103, theorem 8.2.16]).

Let

$$A = \begin{pmatrix} A_{11} & A_{12} & \cdots & A_{1n} \\ A_{21} & A_{22} & \cdots & A_{2n} \\ \vdots & \vdots & \ddots & \vdots \\ A_{n1} & A_{n2} & \cdots & A_{nn} \end{pmatrix}$$

be a generalized matrix ring, where $A_{ii}$ are rings with identity $1_i$ and $A_{ij}$ are $A_{ii}$-$A_{jj}$-bimodules. Then the identity of the ring $A$ can be decomposed into the sum $1 = e_1 + e_2 + \cdots + e_n$, where $e_i$ is an element of $A$ with entry equal to $1_i$ at $(i, i)$ place and $0$ otherwise. So that $e_i A e_j = A_{ij}$. Obviously, each $e_i$ is an idempotent of $A$. Therefore $eAe$ is a minor of $A$ if $e$ is a sum of some distinct idempotents $e_i$ whose index is in a subset $X \subseteq \{1, 2, \ldots, n\}$.

**Proposition 1.2.10.** *A ring $T(S, D)$ is hereditary if and only if the Hasse diagram $H(S)$ of $S$ is a disconnected union of trees, i.e. the poset $S$ contains no subposet whose diagram is a rhombus.*

*Proof.* Let $T = T(S, D)$ be a hereditary ring. Assume that the Hasse diagram of $S$ contains a subposet whose Hasse diagram is a rhombus. This means that $T$ contains an idempotent $e$ such that $eT(S)e = B$, where

$$B = \begin{pmatrix} D & D & D & D \\ 0 & D & 0 & D \\ 0 & 0 & D & D \\ 0 & 0 & 0 & D \end{pmatrix}.$$

This is impossible, since $T(S)$ is a hereditary ring and for any non-zero idempotent $e \in T(S)$ the ring $B = eT(S)e$ must be also a hereditary ring by [103, proposition 4.4.6(4)], but the ring $B$ is not hereditary.

Conversely, suppose the two-sided Peirce decomposition of $T(S, D)$ does not contain minors of the form $eT(S)e = B$. Then the Hasse diagram of $S$ is a disconnected union of acyclic simply laced quivers with no extra arrows such that its underlying graph $\overline{S}$ (obtained from $S$ by deleting the orientation of the arrows) is a disconnected union of trees. From proposition 1.2.9 it follows that the ring $T(S, D)$ can be considered as the path algebra corresponding to the graph $\overline{S}$ over a division ring $D$. Therefore $T(S, D)$ is a hereditary ring, by [101, theorem 2.3.4]. $\square$

**Proposition 1.2.11.** *A ring $T = T(S, D)$ is an Artinian semidistributive piecewise domain.*

*Proof.* Since for any primitive pairwise orthogonal idempotents $e, f \in T(S, D) = T$ the ring $(e + f)T(e + f)$ is either of the form $\begin{pmatrix} D & D \\ 0 & D \end{pmatrix}$ or $\begin{pmatrix} D & 0 \\ 0 & D \end{pmatrix}$, the ring $T(S, D)$ is semidistributive, by [103, theorem 1.10.9].

Write $P_i = e_{ii}T$ for $i = 1, \ldots, n$. Let $\varphi : P_i \to P_j$ be a non-zero homomorphism. Then $\varphi(e_{ii}a) = \varphi(e_{ii})a = e_{jj}a_0e_{ii}a$, where $a_0, a \in T$ and $e_{jj}a_0e_{ii}$ is a non-zero element from $e_{jj}Te_{ii} = D$. Thus $d_0 = e_{jj}a_0e_{ii}$ defines a monomorphism. Therefore the ring $T(S, D)$ is a piecewise domain. $\square$

**Proposition 1.2.12.** *If the Hasse diagram $H(S)$ of a poset $S$ contains no rhombuses then $T(S, D)$ is an Artinian hereditary semidistributive ring.*

*Proof.* This follows immediately from propositions 1.2.10 and 1.2.11. $\square$

**Proposition 1.2.13.** *A ring $T(S, D)$ is serial if and only if $S$ is a disconnected union of linearly ordered sets.*

*Proof.* Since a ring $T(S, D)$ is indecomposable if and only if the poset $S$ is connected, one can assume that $T(S, D)$ is indecomposable. If $S$ is a chain, then the statement follows from example 1.2.3.

Conversely, suppose that $T(S, D)$ is a serial indecomposable ring. Then by proposition 1.2.9 the quiver of the ring $T(S, D)$ coincides with the diagram of the poset $S$ and by [103, theorem 1.12.2] it is a chain. Thus $S$ is a linearly ordered poset.

$\square$

## 1.3 Right Hereditary Rings $A(S, O)$

Let $O$ be a discrete valuation ring (not necessarily commutative) with division ring of fractions $D$ and Jacobson radical $M = \text{rad}O = \pi O$, where $\pi \in O$ is a generator of $M$. By [100, corollary 10.2.2] $O$ is a local Noetherian hereditary ring which is a right and left principal ideal domain (PID) and $M$ is its unique maximal right and left ideal. Recall (see [100, section 8.4]) that each non-zero element $x \in D$ is uniquely representable in the form $x = \varepsilon \pi^k$, where $\varepsilon$ is an invertible element of the ring $O$ and $k \in \mathbf{Z}$. Each proper $O$-module $X \subset D$ is cyclic and can be represented in the form $X = \pi^k O = O\pi^k$, where $k \in \mathbf{Z}$. Write

$$H_n(O) = \begin{pmatrix} O & O & \cdots & O \\ M & O & \cdots & O \\ \vdots & \vdots & \ddots & \vdots \\ M & M & \cdots & O \end{pmatrix} \tag{1.3.1}$$

which is a subring of the matrix ring $M_n(D)$. Clearly, $H_n(O)$ is a Noetherian serial prime hereditary ring. And so, by the Goldie theorem, it has a classical ring of fractions, which is $M_n(D)$.

Let $\{O_i\}_{i=1,\dots,k}$ be a family of discrete valuation rings (not necessarily commutative) with Jacobson radical $M_i$ and a common skew field of fractions $D$. Let $S = \{\alpha_1, \alpha_2, \dots, \alpha_n\}$ be a finite poset with a partial order $\preceq$, and $S_0 = \{\alpha_1, \alpha_2, \dots, \alpha_k\}$ the subset of minimal points of $S$ ($k \leq n$). Then $S = S_0 \cup S_1$, where $S_1 = \{\alpha_{k+1}, \alpha_{k+2}, \dots, \alpha_n\}$. Corresponding to this partition of $S$ consider a **poset $S$ with weights** so that the point $i$ has the weight $H_{n_i}(O_i)$ for $i = 1, 2, \dots, k$; $n_i \in \mathbf{N}$; and all other points have the weight $D$.

Construct a ring $A = A(S, S_0, S_1; O_1, \dots, O_k; D; n_1, n_2, \dots, n_k)$ (or shortly, $A(S, O)$) as a generalized matrix ring of the form

$$\begin{pmatrix} A_{11} & A_{12} & \cdots & A_{1n} \\ A_{21} & A_{22} & \cdots & A_{2n} \\ \vdots & \vdots & \ddots & \vdots \\ A_{n1} & A_{n2} & \cdots & A_{nn} \end{pmatrix}$$

where $A_{ii} = H_{n_i}(O_i)$ for $i = 1, 2, \dots, k$; $A_{jj} = D$ for $j = k + 1, \dots, n$ and $A_{ij}$ is an $(A_{ii}, A_{jj})$-bimodule for $i, j = 1, 2, \dots, n$. Moreover, if $\alpha_i \not\preceq \alpha_j$ in $S$ then $A_{ij} = 0$,

otherwise $A_{ij} = [D \, D \dots D]^T = [D^{(k_i)}]^T$, where $k_i = \begin{cases} n_i & \text{if } i \in S_0 \\ 1 & \text{otherwise} \end{cases}$.

So this generalized matrix ring $A$ has the following form

$$A = \begin{pmatrix} H_{n_1}(O_1) & \cdots & 0 & M_1 \\ \vdots & \ddots & \vdots & \vdots \\ 0 & \cdots & H_{n_k}(O_k) & M_k \\ \hline 0 & \cdots & 0 & T(S_1, D) \end{pmatrix}, \qquad (1.3.2)$$

where $T(S_1, D)$ is the incidence ring of the poset $S_1$ over the division ring $D$, and $M_i$ is an $(H_{n_i}(O_i), T(S_1, D))$-bimodule for $i = 1, 2, \ldots, k$. Obviously, $A$ is a subring of $M_s(D)$ where $s = n_1 + n_2 + \cdots + n_k + (n - k)$.

The identity of the ring $A$ can be decomposed into the sum $1 = f_1 + f_2 + \cdots + f_n$, where $f_i$ is an element of $A$ with entry equal to identity $1_i$ of the ring $A_{ii}$ at $(i, i)$-th place and $0$ otherwise. Obviously, each $f_i$ is an idempotent of $A$; $f_i A f_i = H_{n_i}(O_i)$ for $i = 1, 2, \ldots, k$ and $f_i A f_i = D$ otherwise; $fAf = T(S_1, D)$ for $f = f_{k+1} + f_{k+2} + \cdots + f_n$; and $f_i A f = M_i$ for $i = 1, 2, \ldots, k$.

**Proposition 1.3.3.** *Let $\{O_i\}_{i=1,\ldots,k}$ be a family of discrete valuation rings with common division ring of fractions $D$. Then $A(S, O)$ is a right Noetherian semiperfect ring.*

*Proof.* By [103, theorem 1.1.23] the ring $A$ is right Noetherian. Since the identity of $A = A(S, O)$ decomposes into a finite sum of pairwise orthogonal local idempotents, $A$ is semiperfect by [103, theorem 1.9.3]. Since all the $H_{n_i}(O_i)$ are Noetherian rings, $T(S_1, D)$ is an Artinian ring, and all $M_i$ are finite dimensional vector spaces over $D$, $A$ is a right Noetherian ring by [103, theorem 1.1.23]. $\square$

Let $N$ be the prime radical of a semiperfect ring $A = A(S, O)$. Then $N$ has the following form

$$N = \begin{pmatrix} I_1 & \cdots & 0 & M_1 \\ \vdots & \ddots & \vdots & \vdots \\ \mathbf{0} & \cdots & I_k & M_k \\ \hline 0 & \cdots & 0 & N_1 \end{pmatrix}$$

where $I_i$ is the prime radical of $H_{n_i}(O_i)$ for $(i = 1, 2, \ldots, k)$, and $N_1$ is the prime radical of $T(S_1, D)$. Since all the $H_{n_i}(O_i)$ are prime rings, $I_i = 0$. By [103, proposition 1.1.25] the prime radical of the Artinian ring $T(S_1, D)$ coincides with its Jacobson radical $R_1 = \mathrm{rad}\, T(S_1, D)$. Therefore

$$N = \begin{pmatrix} 0 & \cdots & 0 & M_1 \\ \vdots & \ddots & \vdots & \vdots \\ 0 & \cdots & 0 & M_k \\ \hline 0 & \cdots & 0 & R_1 \end{pmatrix}$$

Write $A_0 = A/N$ and $W = N/N^2$. Then

$$A = A_0 \oplus N \tag{1.3.4}$$

as a direct sum of two Abelian subgroups and

$$A_0 = \overline{A}_1 \times \overline{A}_2 \times \ldots \times \overline{A}_k \times \overline{T}, \tag{1.3.5}$$

where $\overline{A}_i \simeq H_{n_i}(O_i)$ for $i = 1, \ldots, k$ and $\overline{T} \simeq T(S_1, D)/R_1 = A_{k+1} \times \ldots \times A_n$, $A_j \simeq D$ for $i = k + 1, \ldots, n$. It is obvious that $A_0 \cong A/N$ is a finite direct product of prime rings.

Thus there results the following:

**Proposition 1.3.6.** *The prime radical $N$ of a ring $A = A(S, O)$ is nilpotent, and $A/N$ is a finite direct product of prime rings.*

Let $\overline{1} = \overline{f}_1 + \ldots + \overline{f}_n$ be the corresponding decomposition of the identity of $A_0$ into the sum of pairwise orthogonal idempotents. Set up a correspondence between the idempotents $\overline{f}_1, \ldots, \overline{f}_n$ and vertices $1, \ldots, n$ and connect the vertex $i$ with the vertex $j$ by an arrow if and only if $\overline{f}_i W \overline{f}_j \neq 0$. The thus obtained finite directed graph $PQ(A)$ is the prime quiver of $A$ (see [100, section 11.4]).

Since $A_0$ is a semiprime Noetherian ring, it has a classical ring of fractions which has the following form

$$\widetilde{A}_0 = \widetilde{A}_1 \times \ldots \times \widetilde{A}_n = M_{n_1}(D) \times \ldots \times M_{n_k}(D) \times D \times \ldots \times D \tag{1.3.7}$$

.

**Lemma 1.3.8.** *Let $O$ be a discrete valuation ring with a division ring of fractions $D$. Then a ring*

$$A = A(S, O) = \begin{pmatrix} H_n(O) & M \\ 0 & T(S_1, D) \end{pmatrix},$$

*where $T = T(S_1, D)$ is the incidence ring of the poset $S_1$ over the division ring $D$, and $M$ is a $(H_n(O), T(S_1, D))$-bimodule, has right classical ring of fractions that is an Artinian ring and has the following form*

$$\widetilde{A} = \begin{pmatrix} M_n(D) & \widetilde{M} \\ 0 & T(S_1, D) \end{pmatrix}, \tag{1.3.9}$$

*where $\widetilde{M} = \widetilde{H}M \cong \widetilde{H} \otimes_{H_n(O)} M$, and $\widetilde{H} = M_n(D)$.*

*Proof.* Since $A$ is a primely triangular semiperfect and right Noetherian ring, from [103, theorem 8.5.31] it follows that $A$ has a classical ring of fraction $\tilde{A}$ which is an Artinian ring.

Let $1 = e_1 + e_2$ be the decomposition of the identity of $A$ into a sum of pairwise orthogonal idempotents such that $e_1 A e_1 = A_1 = H_n(O)$ and $e_2 A e_2 = A_2 = T(S_1)$. Then any regular element of $A$ has the following form $r = \begin{pmatrix} r_1 & x \\ 0 & r_2 \end{pmatrix}$, where $r_i$ is a regular element of $A_i$ ($i = 1, 2$), and $x \in M$. Then there exist $r_1^{-1} \in M_n(O)$ and $r_2^{-1} \in T(S_1)$ such that $r^{-1} = \begin{pmatrix} r_1^{-1} & y \\ 0 & r_2^{-1} \end{pmatrix}$, where $y = -r_1^{-1} x r_2^{-1} \in \tilde{M}$, which shows that $\tilde{A}$ has the form (1.3.9). $\square$

**Proposition 1.3.10.** *Let $O_1, O_2, \ldots, O_k$ be a family of discrete valuation rings with a common division ring of fractions $D$. Then the ring $A(S, O) = A(S, S_0, S_1; O_1, \ldots, O_k; D; n_1, n_2, \ldots, n_k)$ possesses a right classical ring of fractions which is an Artinian ring of the form:*

$$\tilde{A} = \left( \begin{array}{ccc|c} M_{n_1}(D) & \cdots & \mathbf{0} & \tilde{M}_1 \\ \vdots & \ddots & \vdots & \vdots \\ \mathbf{0} & \cdots & M_{n_k}(D) & \tilde{M}_k \\ \hline \mathbf{0} & \cdots & \mathbf{0} & T(S_1, D) \end{array} \right) \qquad (1.3.11)$$

*where $T = T(S_1, D)$ is the incidence ring of a poset $S_1$ over a division ring $D$, $\tilde{M}_i = \tilde{H}_i \otimes_{H_{n_i}(O_i)} M_i$, and $\tilde{H}_i = M_{n_i}(D)$ for $i = 1, 2, \ldots, k$.*

*Proof.* Since $A$ is a primely triangular semiperfect and right Noetherian ring, from [103, theorem 8.5.31] it follows that $A$ has a classical ring of fractions $\tilde{A}$ which is an Artinian ring.

Similar to (1.3.2), $A$ can be written in the following form

$$A = \begin{pmatrix} H & M \\ \mathbf{0} & T \end{pmatrix},$$

where

$$H = \begin{pmatrix} H_{n_1}(O_1) & \cdots & \mathbf{0} \\ \vdots & \ddots & \vdots \\ \mathbf{0} & \cdots & H_{n_k}(O_k) \end{pmatrix},$$

$T = T(S_1, D)$, and $M$ is an $(H, T)$-bimodule.

Since $H$ is isomorphic to a finite direct product of serial Noetherian rings, it possesses the classical ring of fractions $\tilde{H}$, by [103, theorem 1.11.5]. Moreover,

$$\tilde{H} = \begin{pmatrix} M_{n_1}(D) & \cdots & \mathbf{0} \\ \vdots & \ddots & \vdots \\ \mathbf{0} & \cdots & M_{n_k}(D) \end{pmatrix}.$$

The ring $T$ now has a classical ring of fractions $\tilde{T} = T$ by proposition 1.2.8.

Let $1 = f_1 + f_2 + \cdots + f_k + f$ be a decomposition of the identity of $A$ into a sum of pairwise orthogonal idempotents such that $e_i A e_i = H_{n_i}(O_i)$ for $i = 1, 2, \ldots, k$ and $f A f = T(S_1, D)$. It is easy to see that any regular element of $A$ has the following form

$$
r = \begin{pmatrix}
r_1 & 0 & \cdots & 0 & x_1 \\
0 & r_2 & \cdots & 0 & x_2 \\
\vdots & \vdots & \ddots & \vdots & \vdots \\
0 & 0 & \cdots & r_k & x_k \\
0 & 0 & \cdots & 0 & r_{k+1}
\end{pmatrix},
\tag{1.3.12}
$$

where each element $r_i$ is regular in $A_i$ for $i = 1, \ldots, k$ and $r_{k+1}$ is regular in $T(S_1, D)$, and $x_i \in M_i$ for $i = 1, 2, \ldots, k$. Therefore taking into account the form of regular elements of $A$ from lemma 1.3.8 it follows that $\widetilde{A}$ has the form (1.3.11). $\square$

The diagram $H(S)$ of a poset $S = \{\alpha_1, \alpha_2, \ldots, \alpha_n\}$ can be considered as a quiver $Q(S)$ with set of vertices $VS = \{1, 2, \ldots, n\}$ and a set of arrows $AS$ which contains an arrow $\sigma_{ij}$ with start at the vertex $i$ and end at the vertex $j$ if and only if $\alpha_i < \alpha_j$ and there is no an element $\alpha_k$ such that $\alpha_i < \alpha_k < \alpha_j$. Now consider the quiver $Q(S)$ provided with weights. Indeed suppose that each vertex $i \in VS$ has a weight corresponding to the element $\alpha_i$. Since $S$ is a poset, the quiver $Q(S)$ is acyclic and has no multiplied arrows, i.e. it is an acyclic simply laced quiver. (Recall that a quiver $Q$ is called **acyclic** if it does not contain oriented cycles and it is called **simply laced** if it does not contain multiple arrows and multiple loops). Write a vertex with a weight $H_{n_i}(O_i)$ as $\odot$ and a vertex with a weight $D$ as $\bullet$.

**Proposition 1.3.13.** *A ring $A(S, O)$ corresponding to a poset $S$ with weights is right hereditary if and only if it does not contain the minors $eAe$ for some idempotent $e$ of $A$ of the following forms:*

$$
B = \begin{pmatrix}
D & D & D & D \\
0 & D & 0 & D \\
0 & 0 & D & D \\
0 & 0 & 0 & D
\end{pmatrix} \quad \text{and} \quad C = \begin{pmatrix}
H_n(O) & C_2 & C_3 & C_4 \\
0 & D & 0 & D \\
0 & 0 & D & D \\
0 & 0 & 0 & D
\end{pmatrix},
\tag{1.3.14}
$$

*where the $C_i$ are uniserial left $H_n(O)$-modules and one-dimensional right $D$-vector spaces for $i = 2, 3, 4$, i.e. any subdiagram of a poset $S$ contains no rhombuses and no diagrams of the following form:*

$$
\tag{1.3.15}
$$

In other words, a ring $A(S, O)$ is a right hereditary indecomposable ring if and only if the diagram $H(S)$ of the poset $S$ is a tree and weights of the form $H_{n_i}(O_i)$ can occur only at minimal points of $S$.

*Proof.*

1. Recall that by [103, proposition 1.9.6] any finitely generated projective module over a semiperfect ring is isomorphic to a finite direct sum of principal modules. Suppose the ring $A = A(S, O)$ is right hereditary and contains for some idempotent $e$ of $A$ the minor $eAe = B$. Since $B$ is a right Noetherian ring, any submodule of a finite generated right $B$-module is finitely generated. Since the right $B$-submodule $(0\ D\ D\ D)$ of the projective right $B$-module $P_1 = e_{11}B$ is not projective over $B$, the ring $B$ is not right hereditary. Analogously the right $C$-module $(0\ C_2\ C_3\ C_4)$ is not projective over $C$, and therefore the ring $C = eAe$ is not right hereditary. Since each minor $eAe$ of a right hereditary ring is right hereditary itself by [103, proposition 4.4.6(4)], $A(S, O)$ contains neither $B$ nor $C$.

2. Conversely, assume the ring $A = A(S, O)$ satisfies the condition of the proposition. For any $i = 1, \ldots, k$ the ring $A_i = H_{n_i}(O_i)$ has a classical ring of quotients $\widetilde{A}_i = M_{n_i}(D)$ which is a simple Artinian ring. Obviously, $A_{ij} \cong \widetilde{A}_i \otimes_{A_i} A_{ij}$, for any $i, j = 1, 2, \ldots, n$. By proposition 1.3.10 $A$ has a classical ring of fractions $\widetilde{A}$ which is an Artinian ring and has the form 1.3.11. For any $i = 1, \ldots, k$ consider the ring

$$B_i = \begin{pmatrix} M_{n_i}(D) & \widetilde{M}_i \\ 0 & T(S_1) \end{pmatrix},$$

which is Morita equivalent to the ring:

$$T_i = \begin{pmatrix} D & X_i \\ 0 & T(S_1) \end{pmatrix},$$

where $X_i$ is a right $T(S_1)$-module and a left one-dimension vector space over $D$. Since $A$ does not contain subrings isomorphic to $B$ or $C$, $B_i$ and $T_i$ are hereditary by proposition 1.2.10, for any $i = 1, \ldots, k$. Then from [103, theorem 8.7.4] it follows that each $\overline{A}_{ij} = A_{ij} / \sum_{i<s<j} A_{is}A_{sj}$ is a projective right $D$-module and all $\mu^0_{isj} : \overline{A}_{is} \otimes_{A_s} A_{sj} \to \overline{A}_{ij}$ induced by the multiplication in $A$ are monomorphisms for all $i, s, j = 1, 2, \ldots, n$.

Since by proposition 1.3.3 $A$ is a right Noetherian ring, all conditions of [103, theorem 8.7.4] are satisfied. So $A$ is also right hereditary. $\square$

**Theorem 1.3.16.** *Let $\{O_1, O_2, \ldots, O_n\}$ be a family of discrete valuation rings with a common division ring of fractions $D$. Suppose that the poset $S$ contains no rhombuses or diagrams of the form (1.3.15) (as a subdiagram). Then $A(S, O)$ is a right Noetherian and right hereditary SPSD-ring.*

*Proof.* The ring $A = A(S, O)$ is semiperfect right Noetherian by proposition 1.3.3 and right hereditary by proposition 1.3.13. Moreover, for any two primitive idempotents $e$

and $f$ the ring $(e+f)A(e+f)$ has one of the following forms: $\begin{pmatrix} O_i & 0 \\ 0 & O_j \end{pmatrix}$, $\begin{pmatrix} O_i & O_i \\ M_i & O_i \end{pmatrix}$, $\begin{pmatrix} O_i & 0 \\ 0 & D \end{pmatrix}$, $\begin{pmatrix} O_i & D \\ 0 & D \end{pmatrix}$, $\begin{pmatrix} D & D \\ 0 & D \end{pmatrix}$ or $\begin{pmatrix} D & 0 \\ 0 & D \end{pmatrix}$, where $D$ is the common skew field of fractions of $O_i$ and $O_j$. It is obvious, that each of these rings is a semidistributive ring. So $A$ is also semidistributive, by [103, theorem 1.10.10]. $\square$

**Corollary 1.3.17.** *The right classical ring of fractions $\widetilde{A}$ of a right hereditary ring $A(S, O)$ is a right Artinian and right hereditary SPSD-ring. The prime radical of $A(S, O)$ coincides with the Jacobson radical of $\widetilde{A}$.*

**Corollary 1.3.18.** *The prime quiver of a right hereditary ring $A(S, O)$ coincides with the quiver of the right classical ring of fractions $\widetilde{A}$ and coincides with the diagram $H(S)$ of the poset $S$.*

## 1.4. Incidence Rings Modulo the Radical

In this section we introduce a new construction of rings which are called incidence rings modulo the radical. Such rings in a more general case[1] were first introduced and studied in [56].

Let $\Lambda$ be a ring with a unique maximal ideal $M$ and let $S = \{\alpha_1, \ldots, \alpha_n\}$ be a finite poset with relation $\leq$. Consider the generalized matrix ring $M_n(\Lambda)$ of all $(n \times n)$-matrices over $\Lambda$ with matrix units $e_{ij}$, where $\alpha e_{ij} = e_{ij}\alpha$ for all $\alpha \in \Lambda$. Denote by $I(S, \Lambda, M) = A$ the following subring of $M_n(\Lambda)$: $e_{ii}Ae_{jj} = e_{ii}M_n(\Lambda)e_{jj} = \Lambda$ if and only if $\alpha_i \leq \alpha_j$, otherwise $e_{ii}Ae_{jj} = M$. That is $I(S, \Lambda, M)$ is like $I(S, \Lambda)$ with the zeros in $I(S, \Lambda)$ replaced by $M$.

**Examples 1.4.1.**

1. If $\Lambda = D$ is a division ring, then we obtain the incidence ring $T(S, D)$ of a poset $S$ over $D$ considered in section 1.2.
2. If $S$ is a linearly ordered set then $I(S, \Lambda, M)$ has the following form:

$$H_n(\Lambda) = \begin{pmatrix} \Lambda & \Lambda & \cdots & \Lambda \\ M & \Lambda & \cdots & \Lambda \\ \vdots & \vdots & \ddots & \vdots \\ M & M & \cdots & \Lambda \end{pmatrix}.$$

In particular, if $\Lambda = O$ is a discrete valuation ring, then $I(S, \Lambda, M) = H_n(O)$.

---

[1]namely, incidence rings modulo an ideal

Recall that a ring $A$ is called a **tiled order** if it is a prime Noetherian semiperfect semidistributive ring with non-zero Jacobson radical.

According to [100, theorem 14.5.2] each tiled order is isomorphic to a ring of the following form

$$A = \begin{pmatrix} O & \pi^{\alpha_{12}}O & \cdots & \pi^{\alpha_{1n}}O \\ \pi^{\alpha_{21}}O & O & \cdots & \pi^{\alpha_{2n}}O \\ \vdots & \vdots & \ddots & \vdots \\ \pi^{\alpha_{n1}}O & \pi^{\alpha_{n2}}O & \cdots & O \end{pmatrix},$$

where $n \geq 1$, $O$ is a discrete valuation ring with prime element $\pi$, and the $\alpha_{ij}$ are integers such that $\alpha_{ij} + \alpha_{jk} \geq \alpha_{ik}$ for all $i, j, k$ (and $\alpha_{ii} = 0$ for any $i$).

Denote by $\mathcal{E}(A) = (\alpha_{ij})$ the (so called) exponent matrix of $A$. The matrix $\mathcal{E}$ is called **reduced** if $\alpha_{ij} + \alpha_{ji} > 0$ for $i, j = 1, \dots, n, i \neq j$. If a tiled order is a basic ring, then $\alpha_{ij} + \alpha_{ji} > 0$ for $i, j = 1, \dots, n, i \neq j$, i.e. $\mathcal{E}(A)$ is a reduced matrix.

A tiled order $A = \{O, \mathcal{E}(A)\}$ is called a $(0, 1)$-**order** if $\mathcal{E}(A)$ is a $(0, 1)$-matrix. With a basic $(0, 1)$-order $A$ we can associate a finite partially ordered set

$$S = \{p_1, \dots, p_m\}$$

with the partial ordering relation $\leq$ defined by $p_i \leq p_j \iff \alpha_{ij} = 0$.

Conversely, to any finite poset $S = \{p_1, \dots, p_m\}$ assign a reduced $(0, 1)$-matrix $\mathcal{E}_S = (\alpha_{ij})$ in the following way: $\alpha_{ij} = 0 \iff p_i \leq p_j$, otherwise $\alpha_{ij} = 1$. Then $A(S) = \{O, \mathcal{E}_S\}$ is a $(0, 1)$-order which is a basic ring. We have that $A(S) = I(S, O, M)$.

From [103, theorems 1.1.23, 1.9.3] and [100, proposition 10.7.8] there immediately follows the result.

**Proposition 1.4.2.** *Let $\Lambda$ be a Noetherian semiperfect ring with Jacobson radical $M$. Then a ring $I(S, \Lambda, M)$ is a Noetherian semiperfect ring. Further, if $\Lambda$ is a local domain which is not a division ring then $I(S, \Lambda, M)$ is a piecewise domain.*

**Proposition 1.4.3.** *If the ring $\Lambda$ is semiperfect then $I(S, \Lambda, M)$ is a basic ring.*

*Proof.* Due to proposition 1.4.2 the ring $T = I(S, \Lambda, M)$ is semiperfect. We write $P_i = e_{ii}T$. If $P_i \neq P_j$ for $i \neq j$, then it follows from [103, proposition 1.9.9] that $T$ is a basic ring. Suppose that $P_i \simeq P_j$, then $e_{ij} \in T$ and $e_{ji} \in T$. Consequently, $\alpha_i \leq \alpha_j$ and $\alpha_j \leq \alpha_i$. So, $\alpha_i = \alpha_j$. A contradiction. $\square$

Let $\Lambda$ be a semiperfect ring with Jacobson radical $M$. Then $T = I(S, \Lambda, M)$ is also a semiperfect ring with Jacobson radical $R$. By [103, proposition 1.9.9] the two-sided Peirce decomposition of $R$ has the following form:

$$e_{ii}Re_{ii} = M \text{ for } i = 1, \dots, n \text{ and } e_{ii}Re_{jj} = e_{ii}Te_{jj} \text{ for } i \neq j. \tag{1.4.4}$$

Let $\Lambda$ be a Noetherian local domain, which is not a division ring, with unique maximal ideal $\mathcal{M}$. Suppose that $\mathcal{M}/\mathcal{M}^2 = U^m$, where $\Lambda/\mathcal{M} = U$ is the unique simple $\Lambda$-module.

Denote by $\widetilde{Q}(S, m)$ the quiver obtained from the Hasse diagram $H(S)$ by adding $m$ arrows from $i$ to $j$ for all $(\alpha_i, \alpha_j) \in S_{max} \times S_{min}$.

The following theorem can be considered as a generalization of [100, theorem 14.6.3].

**Theorem 1.4.5.** *Let $\Lambda$ be a Noetherian local domain, which is not a division ring, with unique maximal ideal $\mathcal{M}$. Suppose that $\mathcal{M}/\mathcal{M}^2 = U^m$, where $\Lambda/\mathcal{M} = U$ is the unique simple $\Lambda$-module. If $T = I(S, \Lambda, \mathcal{M})$ then the quiver $Q(T)$ coincides with the quiver $\widetilde{Q}(S, m)$.*

*Proof.* Let $T = I(S, \Lambda, \mathcal{M})$ with Jacobson radical $R$. By proposition 1.4.2 $T$ is semiperfect and the two-sided Pierce decomposition of $R$ has the form (1.4.4). Suppose that in the diagram $H(S)$ there is an arrow from the vertex $s$ to the vertex $t$. This means that $e_{ss}Te_{tt} = \Lambda$ and there is no positive integer $m$ ($m \neq s, t$) such that $e_{ss}Te_{mm} = \Lambda$ and $e_{mm}Te_{tt} = \Lambda$. Then $e_{ss}R^2e_{tt} = \sum\limits_{i=1}^{n} e_{ss}Re_{ii}Re_{tt} = \mathcal{M}$. So, $e_{ss}Re_{tt} = \Lambda$ strongly contains $e_{ss}R^2e_{tt} = \mathcal{M}$ and by [103, lemma 1.9.12] there is only one arrow from $s$ to $t$.

Suppose that $p \in S_{max}$. This means that $e_{pp}Te_{mm} = \mathcal{M}$ for $m \neq p$. Therefore the $p$-th row of the Peirce decomposition of $R$ is $(\mathcal{M}, \mathcal{M}, \ldots, \mathcal{M})$. Similarly, if $q \in S_{min}$, the $q$-th column of $R$ is $(\mathcal{M}, \mathcal{M}, \ldots, \mathcal{M})^T$ (here $T$ is the transposition sign). So, $e_{pp}R^2e_{qq} = \mathcal{M}^2$ and $e_{pp}Re_{qq} = \mathcal{M}$. By assumption $\mathcal{M}/\mathcal{M}^2 = U^m$. Consequently, in $Q(T)$ there exist $m$ arrows from $p$ to $q$, and this proves that $\widetilde{Q}(S, m)$ is a subquiver of $Q(T(S, \Lambda))$.

We now prove the converse inclusion. Suppose that $e_{pp}R^2e_{qq} = \mathcal{M}^2$. Then it is obvious that

$$e_{pp}R = (\mathcal{M}, \mathcal{M}, \ldots, \mathcal{M})$$

and

$$Re_{qq} = (\mathcal{M}, \mathcal{M}, \ldots, \mathcal{M})^T.$$

Therefore $p \in S_{max}$, $q \in S_{min}$ and there are $m$ arrows from $p$ to $q$.

Suppose $e_{pp}R^2e_{qq} = \mathcal{M}$ and $e_{pp}Re_{qq} = \Lambda$. Consequently, $p \neq q$ and $e_{pp}Te_{qq} = \Lambda$. So, $\alpha_p \leq \alpha_q$. Since $e_{pp}R^2e_{qq} = \sum\limits_{k=1}^{n} e_{pp}Re_{kk}Re_{qq}$, it follows that $e_{pp}Re_{kk}Re_{qq} \subseteq \mathcal{M}$ for $k = 1, \ldots, n$. Thus $e_{pp}Te_{kk}Te_{qq} \subset \mathcal{M}$ for $k \neq p, q$. This is a contradiction. Therefore, there is no positive integer $k$ ($k \neq p, q$) such that $\alpha_p \leq \alpha_k$ and $\alpha_k \leq \alpha_q$. This means that there is an arrow from $p$ to $q$ in $Q(T)$ and the opposite inclusion is proved. $\square$

**Corollary 1.4.6.** *Let $\Lambda$ be a Noetherian local domain, which is not a division ring, with unique maximal ideal $M$. Then the quiver $Q(T(S, \Lambda, M))$ is strongly connected.*

Let $S = \{\alpha_1, \ldots, \alpha_n\}$ be a finite poset. Write $S^{op} = \{\beta_1, \ldots, \beta_n\}$, the finite poset defined by the following rule:

$$\beta_i \leq \beta_j \text{ if and only if } \alpha_j \leq \alpha_i.$$

Consider $\widetilde{Q}(S^{op}, m)$. Let $\Lambda$ be a Noetherian local domain with unique maximal ideal $M$. Suppose that $M/M^2 = V^m$ as a left $\Lambda$-module, where $V = \Lambda/M$ is the unique left simple $\Lambda$-module.

The next theorem uses the concept of the left quiver of a ring. For a first definition and first properties of left (and right) quivers of (Artinian) rings (see [100, Chapter 11] and [101, Section 5.2]).

**Theorem 1.4.7.** *If $\Lambda = D$ is a division ring, then the left quiver $Q'(T)$ of $T = I(S, D)$ coincides with the diagram $H(S^{op})$ of the finite poset $S^{op}$. If $\Lambda$ is a local Noetherian domain, which is not a division ring, with the unique maximal ideal $M$ and $M/M^2 = V^m$, then the left quiver $Q'(I(S, \Lambda, M))$ coincides with the quiver $\widetilde{Q}(S^{op}, m)$.*

*Proof.* The proof is analogous to the proof of theorem 1.4.5. $\square$

**Corollary 1.4.8.** *Let $\Lambda$ be a local Noetherian domain, which is not a division ring, with unique maximal ideal $M$. A ring $A = I(S, \Lambda, M)$ is a tiled $(0, 1)$-order if and only if the left quiver $Q'(A)$ and the right quiver $Q(A)$ of $A$ are strongly connected and simply laced.*

*Proof.* The right and left quivers of a tiled $(0, 1)$-order are strongly connected and simply laced by [100, theorem 14.6.3] and [100, corollary 14.3.4].

Let $\Lambda$ be a Noetherian local domain and $M/M^2 = U^k$, where $k \geq 1$. In this case the right quiver $Q(I(S, \Lambda, M))$ coincides with $\widetilde{Q}(S, k)$, by theorem 1.4.5. If $k \geq 2$, then there exist multiple arrows ($k$ arrows from a maximal element to a minimal element). Consequently, $k = 1$ and $\Lambda/M^2$ is a right uniserial ring. Analogously, the left quiver $Q'(T(S, O, M))$ coincides with $\widetilde{Q}(S^{op}, m)$, and $\widetilde{Q}(S^{op}, m)$ is simply laced if and only if $m = 1$. Therefore $M/M^2 = V$ as a left $\Lambda$-module and $\Lambda/M^2$ is a left uniserial ring. Therefore the ring $B = \Lambda/M^2$ is a uniserial ring. Since the Jacobson radical $J$ of $B$ is equal to $M/M^2$, $J^2 = 0$ and $B/J \cong \Lambda/M$. Consequently, $B$ is a semiprimary Noetherian ring. So, by the Hopkins-Levitzki theorem [103, theorem 1.1.22], $B$ is an Artinian ring. Thus, $B = \Lambda/M^2$ is an Artinian uniserial ring. Thanks to [100, theorem 12.3.10] this implies that the ring $\Lambda$ is serial. So $\Lambda$ is a Noetherian local serial ring. Proposition [100, proposition 13.3.1] states that in this case $\Lambda$ is either a discrete valuation ring or an Artinian uniserial ring. Since $\Lambda$ is a domain, and so its Jacobson radical cannot be nilpotent, $\Lambda$ is a discrete valuation ring. In this case $A = I(S, \Lambda, M)$ is a tiled $(0, 1)$-order. $\square$

## 1.5 Serial and Semidistributive Rings $I(S, \Lambda, \mathcal{M})$

**Definition 1.5.1.** (See also [100, p.347]). Let $A$ be a semiperfect ring with Jacobson radical $R$ such that $A/R^2$ is an Artinian ring. The ring $A$ is called $Q$-**symmetric** if the left quiver $Q'(A)$ can be obtained from the right quiver $Q(A)$ by reversing all arrows.

**Examples 1.5.2.**

1. Every *SPSD*-ring is $Q$-symmetric. This follows from [100, corollary 14.3.4].
2. If $O = D$ is a division ring then a ring $I(S, D, 0) = T(S, D)$ is $Q$-symmetric. This follows from proposition 1.2.9.

**Proposition 1.5.3.** *Let $O$ be a Noetherian local domain with Jacobson radical $\mathcal{M}$. A ring $A = I(S, O, \mathcal{M})$ is $Q$-symmetric if and only if $\mathcal{M}/\mathcal{M}^2 = U^m$ as a right $O$-module and $\mathcal{M}/\mathcal{M}^2 = V^m$ as a left $O$-module, where $U$ is a simple right $O$-module and $V$ is a simple left $O$-module.*

The proof follows from theorems 1.4.5 and 1.4.7.

**Proposition 1.5.4.** *Let $O$ be a Noetherian local domain with Jacobson radical $\mathcal{M}$. A ring $I(S, O, \mathcal{M})$ is semidistributive if and only if $O$ is either a division ring or a discrete valuation ring.*

*Proof.* It follows from [100, proposition 14.4.10] that a minor of the first order of a Noetherian *SPSD*-ring is uniserial and it is either a discrete valuation ring or an Artinian uniserial ring.

In the case of a ring $T(S, O, \mathcal{M})$ there results that $O$ is either a division ring or a discrete valuation ring and every indecomposable minor of the second order of $T = I(S, O, \mathcal{M})$ is isomorphic to one of the following rings:

$$\text{a. } T_2(D) = \begin{pmatrix} D & D \\ 0 & D \end{pmatrix}; \quad \text{b. } H_2(O) = \begin{pmatrix} O & O \\ \mathcal{M} & O \end{pmatrix}; \quad \text{c. } \begin{pmatrix} O & \mathcal{M} \\ \mathcal{M} & O \end{pmatrix}.$$

All these rings are *SPSD*-rings. By the reduction theorem for *SPSD*-rings [103, theorem 1.10.10], a ring $T = I(S, O, \mathcal{M})$ is semidistributive. $\square$

**Corollary 1.5.5.** *Let $O$ be a Noetherian local domain with Jacobson radical $\mathcal{M}$. A ring $I(S, O, \mathcal{M})$ is an SPSD-ring if and only if the quiver $Q(T(S, O, \mathcal{M}))$ is simply laced and the ring $T(S, O, \mathcal{M})$ is $Q$-symmetric.*

The proof follows from theorems 1.4.5 and 1.4.7.

**Theorem 1.5.6.** *Let $O$ be a Noetherian local domain with Jacobson radical $\mathcal{M}$. An indecomposable ring $T = I(S, O, \mathcal{M})$ is serial if and only if the poset $S$ is linearly ordered set and $O$ is either a division ring or a discrete valuation ring. In the first case $T \simeq T_n(D)$, and in the second case $T \simeq H_n(O)$.*

*Proof.* By proposition 1.4.2 $T = I(S, O, M)$ is a Noetherian semiperfect ring. If $O = D$ is a division ring the proof follows from proposition 1.2.13. Since every serial ring is semidistributive, the local domain $O$ is a discrete valuation ring, by proposition 1.5.4. In this case a ring $T$ is an indecomposable serial non-Artinian ring.

By [100, theorem 12.3.11], the quiver $Q(T)$ of $T$ is a simple cycle and by [100, theorem 12.3.8] and [100, corollary 12.3.7], $T$ is isomorphic to a ring $H_n(O)$. The theorem is proved. $\square$

**Remark 1.5.7.** By [100, theorem 13.5.1] a serial hereditary ring is Noetherian. It follows from [100, theorem 13.5.2] that a serial hereditary ring is Morita equivalent to a finite direct product of the rings isomorphic to rings $T_n(D)$ of the upper triangular matrices over division rings $D$ and rings of the form $H_n(O)$, where $O$ is a discrete valuation ring. So, all hereditary serial basic rings can be obtained from either a division ring or a discrete valuation ring using the construction $T(S, O, M)$ for a linearly ordered set $S$.


## 1.6 Notes and References

The idea of incidence algebras goes back to R. Dedekind and E.T. Bell. Incidence rings of locally finite posets were first defined by G.C. Rota in [185], who showed that this class of rings is a useful instrument for solving enumeration combinatorial problems on posets.

W. Belding [24] and [158] extended the definition of incidence rings to the case of locally finite preordered sets.

The algebraic properties of incidence rings were studied by D.R. Farkas [73], R.B. Feinberg [74], P. Leroux, J. Sarraillé [143], N.A. Nacev [158], [159], [160], C.J. O'Donnell [170], E. Spiegel [201] and others. Most available information about incidence algebras over commutative rings is presented in the book of E. Spiegel, Ch.J. O'Donnell [200].

Isomorphism and automorphism problems for incidence rings have been studied by of many authors [24], [95], [158], [202], [218]. J.K. Haack in [96] proved that if $A$ is a semiperfect ring and $X$ is a poset then $I(X, A) \cong I(X, B)$ implies $A \cong B$. W. Belding, [24] and [158], proved that if $A$ is a simple Artinian ring and $X$ is a locally finite preordered set then $I(X, A) \cong I(Y, A)$ implies that $X$ is order isomorphic to $Y$ as preordered sets.

E. Voss in [218] generalized the Stanley-Belding-Nachev results and proved the following statement:

*If $X$ and $Y$ are locally finite preordered sets and $A$ is an indecomposable semiperfect ring, then $I(X, A) \cong I(Y, A)$ implies that $X$ and $Y$ are order isomorphic.*

The notions of generalized incidence rings and balanced relations were introduced by G. Abrams in [1], where he investigated the structure of these rings. Note that in his paper G. Abrams considered a balanced relation as a reflexive relation.

Later R. Alfaro and A.V. Kelarev in [4] considered a balanced relation without the assumption of reflexivity and they considered incidence rings of graphs with a balanced relation. Such rings have applications in coding theory.

Definition 1.1.26 of generalized incidence rings was introduced by V.D. Shmatkov [190], who studied the isomorphism problem for such class of rings. In particular he proved that if two generalized incidence rings $I(S_1, \rho_1)$ and $I(S_2, \rho_2)$ are isomorphic then the corresponding sets $(S_1, \rho_1)$ and $(S_2, \rho_2)$ are isomorphic. The problem of elementary equivalence for such rings was studied by E.I. Bunina and A.S. Dobrokhotova-Maykova in [37]. They proved a similar result, namely, if two generalized incidence rings $I(S_1, \rho_1)$ and $I(S_2, \rho_2)$ are elementarily equivalent then the corresponding sets $(S_1, \rho_1)$ and $(S_2, \rho_2)$ are elementarily equivalent.

Questions connected with simply connectedness of incidence algebras were studied in [9], where there was given a combinatorial condition on an incidence algebra of a finite poset to be simply connected. Simply connected algebras have played an important role in representation theory.

A general construction of an incidence algebra $I(A, C)$ of a decomposition-finite category $C$ over an associative ring $A$ was considered in [143].

Incidence algebras and rings have found numerous applications in combinatorics [202].

By analogy with incidence rings of graphs A.V. Kelarev in [120] introduced the concept of an incidence ring of group automata. Recall that a **group automaton** is an algebraic system $S = (X, G, \delta)$ where

1. $X$ is a nonempty set of states.
2. $G$ is a group of input symbols.
3. $\delta : X \times G \longrightarrow X$ given by $\delta(x, g) = xg$ is a transition function satisfying the equality $x(gh) = (xg)h$ for all $x \in X$ and $g, h \in G$.

An **incidence ring of a group automaton** $S = (X, G, \delta)$ over a ring $A$ with identity is the ring $I(S, A)$ that is spanned as a free left $A$-module by the set of all triples $< x, g, xg >$, with $x \in X$, $g \in G$. Multiplication is defined by the distributive law and the rules

i. $\quad < x, g, xg > \cdot < y, h, yh > = \begin{cases} < x, gh, xgh > & \text{if } xg = y, \\ 0 & \text{otherwise} \end{cases}$

ii. $\quad < x, g, xg > \cdot a = a \cdot < x, g, xg >.$

Incidence rings of group automata were studied in [121], [120], [122]. They have an application in coding theory [4].

Structural matrix rings were studied by K.C. Smith and L. van Wyk in [199], [224], [225], [188].

The general construction and properties of incidence rings modulo an ideal of the form $I(S, A, R, \mathcal{I})$, where $A$ is an associative ring with Jacobson radical $R$ and $\mathcal{I} \subseteq R$ is a two-sided ideal of $A$, were first introduced and considered in the paper [56]. Particular cases, rings of the form $T(S, D)$ and $A(S, O)$ connected with a finite poset $S$, were studied in [87], [88], [90], [55].

# CHAPTER 2

# Distributive and Semidistributive Rings

The main aim of this chapter is to represent the properties and structure of various classes of rings whose lattices of submodule are distributive (or semidistributive). Such rings are called distributive (or semidistributive) and they can be considered to be noncommutative generalization of Prüfer domains. The class of distributive rings is very wide and includes, for example, the ring of integers, the ring of polynomials $K[x]$ over a field $K$, rings of integral algebraic numbers, and commutative principal ideal rings. More generally, all commutative Dedekind rings and Prüfer domains are examples of distributive rings.

This chapter presents properties of and the structure results for semiperfect semidistributive rings (SPSD-rings, in short). In section 2.1 we consider some important properties of distributive modules and rings. Section 2.2 is devoted to the study of semidistributive modules and rings. The structure of Noetherian distributive and semidistributive rings is discussed in section 2.3.

Some important properties of right hereditary SPSD-rings are studied in section 2.4. It is shown that the structure of all such rings is closely connected with right hereditary rings of the form $A(S, O)$ as described in section 1.3.

Section 2.5 is devoted to the study of some important properties and of the structure of semihereditary SPSD-rings.

## 2.1 Distributive Modules and Rings

**Proposition 2.1.1.** *The following statements are equivalent*:

1. *For all submodules $K, L, M$ of a module $M$,*

$$K \cap (L + N) = (K \cap L) + (K \cap N).$$

2. *For all submodules $K, L, M$ of a module $M$,*

$$K + (L \cap N) = (K + L) \cap (K + N).$$

*Proof.*

$1 \Longrightarrow 2$. Since $K \subseteq K + N$, applying the modular law [103, theorem 1.1.2] we obtain $(K + L) \cap (K + N) = K + (K + N) \cap L$. Since statement 1 satisfies for any $K, L, N \subseteq M$ we obtain $(K + N) \cap L = (K \cap L) + (N \cap L)$. Therefore

$$(K + L) \cap (K \cap N) = K + (K + N) \cap L = K + (K \cap L) + (N \cap L) = K + (N \cap L).$$

$2 \Longrightarrow 1$. Using the modular law [103, theorem 1.1.2] we have

$$(K \cap L) + (K \cap N) = K \cap (K \cap L + N) \subseteq K \cap (L + N),$$

and $K \cap (L + K \cap N) = K \cap N + K \cap L$.

On the other hand by statement 2:

$$K + K \cap L = (K + K) \cap (K + L) = K \cap (K + L)$$

and $(K + L) \cap (L + N) = L + K \cap N$. Therefore

$$K \cap (L + N) \subseteq (K + K \cap L) \cap (L + N) = [K \cap (K + L)] \cap (L + N) =$$

$$= K \cap [(K + L) \cap (L + N)] = K \cap (L + K \cap N) = K \cap N + K \cap L.$$

So $K \cap (L + N) = (K \cap L) + (K \cap N)$. $\square$

**Definition 2.1.2.** A module $M$ is called **distributive** if it satisfies the equivalents statements of proposition 2.1.1. A ring $A$ is called **right (left) distributive** if the right (left) regular module $A_A$ ($_A A$) is distributive. A ring is **distributive** if it both right and left distributive.

Obviously, any submodule of a distributive module is distributive. The class of distributive modules includes the class of all simple and all uniserial modules.

**Lemma 2.1.3.** *A local commutative distributive ring is uniserial.*

*Proof.* Let $A$ be a local commutative distributive ring. We now prove that for any two elements $a, b \in A$ at least one of the statements $a|b$ or $b|a$ is true.

Since the ideals of $A$ form a distributive lattice,

$$(a) = (a) \cap [(b) + (b - a)] = (a) \cap (b) + (a) \cap (b - a).$$

So that $a = t + (a - b)c$ for some elements $t, c \in A$ with $t \in (a) \cap (b)$ and $bc \in (a)$. If $c$ is a unit, $b \in (a)$. If $c$ is not a unit, $1 - c$ is a unit, since $A$ is local. Therefore $a(1 - c) = a - ac = t - bc \in (b)$, and so $a \in (b)$. This means that either $a|b$ or $b|a$.

Let $I, J$ be two ideals of the ring $A$. Suppose that both $I \not\subseteq J$ and $J \not\subseteq I$. Then there exist two elements $a, b \in A$ such that $a \in I \setminus J$ and $b \in J \setminus I$. By the proving above $a|b$ or $b|a$. Suppose $a|b$, then $b \in (a) = aA \subseteq I$. A contradiction. So either $I \subseteq J$ or $J \subseteq I$, i.e. $A$ is a uniserial ring, as required. $\square$

Recall that a commutative ring $A$ is **arithmetical** if the ideals of the local ring $A_M$ are totally ordered by inclusion for all maximal ideals $M$ of $A$. Below we give some further characterization of distributive rings in terms of ideals.

**Theorem 2.1.4.** (Ghr. U. Jensen [113]). *A commutative ring $A$ is distributive if and only if $A$ is an arithmetical ring.*

*Proof.* Let $M$ be a maximal ideal of a commutative ring $A$ and $A_M$ be a localization of $A$ at $M$. Then any ideal $I$ of $A$ is uniquely determined by its local components $I_M = IA_M$ for all maximal ideals $M$ of $A$. Moreover, for any ideals $I, J$ of $A$ we have that:
1. $I \cap J = K$ if and only if $I_M \cap J_M = K_M$.
2. $I + J = K$ if and only if $I_M + J_M = K_M$.

So that it suffices to prove the theorem for local rings. If $A$ is a local distributive ring then $A$ is uniserial, and so arithmetical, by lemma 2.1.3. Suppose that $A$ is a local arithmetical ring, i.e. $A$ is an uniserial ring. Then $A$ is obviously a distributive ring. $\square$

Recall that a Prüfer domain is a semihereditary integral domain. Another equivalent condition on an integral domain is that any finitely generated ideal is invertible. There are many other equivalent definitions for Prüfer domains. [103, Theorem 3.1.18] gives an equivalent definition of these rings in terms of valuation domains. From [103, theorems 3.1.18, 11.1.4] we immediately obtain a characterization of Prüfer domains in terms of ideals.

**Proposition 2.1.5.** *Let $A$ be a commutative domain. Then the following conditions are equivalent:*

1. *$A$ is a Prüfer domain.*
2. *$A$ is a distributive ring.*
3. *$A$ is an arithmetical ring.*

In view of proposition 2.1.5 the class of distributive rings may be considered as an extension of Püfer domains and arithmetical rings.

**Proposition 2.1.6.** *Let $e$ be a non-zero idempotent of a ring $A$. Then*

1. *If $M$ is a distributive right $A$-module then $Me$ is a right distributive $eAe$-module.*
2. *If $eA$ is a distributive right $A$-module then the ring $eAe$ is right distributive.*
3. *If $A$ is a right distributive ring then the ring $eAe$ is right distributive.*

*Proof.*
1. This follows immediately from [103, proposition 5.7.23(1)].
$1 \implies 2 \implies 3$. These are trivial. $\square$

Let $M$ be an $A$-module. Given two elements $m, n \in M$ we set

$$(m : n) = \{a \in A \mid na \in mA\}. \tag{2.1.7}$$

From this place on we will deal with noncommutative distributive rings. The next theorem gives some equivalent characterizations of distributive modules.

**Theorem 2.1.8.** (V. Camillo [39], W. Stephenson [203]). *Let $M$ be an $A$-module. Then the following statements are equivalent.*

1. *$M$ is a distributive module.*
2. *All submodules of $M$ with two generators are distributive modules.*
3. *Let $M$ be a distributive module over a ring $A$. Then for any $m, n \in M$ there exist $a, b \in A$ such that $1 = a + b$ and $maA + nbA \subseteq mA \cap nA$.*
4. *For any $m, n \in M$ there exist four elements $a, b, c, d \in A$ such that $1 = a + b$ and $ma = nc$, $nb = md$.*
5.

$$(m : n) + (n : m) = A$$

*for all $m, n \in M$.*

*Proof.* The proof of this theorem follows immediately from [100, proposition 14.1.1, lemma 14.1.2, lemma 14.1.3 and theorem 14.1.4]. □

From this theorem we obtain the following equivalent definition for distributive rings.

**Proposition 2.1.9.** (W. Stephenson [203]). *A ring $A$ is distributive if and only if*

$$(a : b) + (b : a) = A$$

*for any two elements $a, b \in A$.*

**Proposition 2.1.10.** *Let $M_i$ be a distributive module over a ring $A_i$ for all $i \in I$, and let $A = \prod_{i \in I} A_i$. Then the $A$-modules $\prod_{i \in I} M_i$ and $\bigoplus_{i \in I} M_i$ are distributive.*

*Proof.* Denote by $e_i$ the identity of $A_i$ and write $M = \prod_{i \in I} M_i$. Let $\{m_i\}_{i \in I}$ and $\{n_i\}_{i \in I}$ be arbitrary elements of $M$. By theorem 2.1.8(4) there exist elements $a_i, b_i, c_i, d_i$ such that $e_i = a_i + b_i$ and $m_i a_i = n_i c_i$, $n_i b_i = m_i d_i$. Let $a = \{a_i\}_{i \in I} \in A$, $b = \{b_i\}_{i \in I} \in A$, $c = \{c_i\}_{i \in I} \in A$, and $d = \{d_i\}_{i \in I} \in A$. Then $1 = \{e_i\}_{i \in I} = a + b$, is the identity of $A$, $ma = nc$, and $nb = md$. So, again by theorem 2.1.8(4) $M$ is a distributive $A$-module. Since $\bigoplus_{i \in I} M_i$ is a submodule of $M$, it is also distributive. □

**Proposition 2.1.11.** (W. Stephenson [203]). *Let $M$ be a right distributive $A$-module. Then $\mathrm{Hom}_A(X, Y) = 0$ for any submodules $X, Y$ of $M$ such that $X \cap Y = 0$.*

*Proof.* Let $f \in \mathrm{Hom}_A(X, Y)$ and $x \in X$. Write $y = f(x) \in Y$. Then by theorem 2.1.8(3) there exists elements $a, b \in A$ such that $1 = a + b$, and $xa + yb = 0$, since

$xA \cap yA \subseteq X \cap Y = 0$. Therefore $xa = -yb = y(a-1) = 0$. So $y = ya = f(x)a = f(xa) = f(0) = 0$, i.e. $\text{Hom}_A(X, Y) = 0$, as required. $\square$

**Corollary 2.1.12.** (W. Stephenson [203]). *Every idempotent of a right (left) distributive ring is central*[1].

*Proof.* Let $e \in A$ be an idempotent of a right distributive ring $A$. Then $A = eA \oplus (1-e)A$. Since $eA \cap (1-e)A = 0$, from proposition 2.1.11 it follows that $\text{Hom}(eA, (1-e)A) = 0$ and $\text{Hom}((1-e)A, eA) = 0$. Therefore $eA(1-e) = (1-e)Ae = 0$, i.e. $e$ is a central idempotent.

The proof for a left distributive ring $A$ is similar. $\square$

Recall that a module $U$ is called **uniform** if the intersection of any two non-zero submodules of $U$ is non-zero. A ring is called **right (left) uniform** if it is uniform as a right (left) module over itself.

**Lemma 2.1.13.** *Let $M$ be a non-zero right ideal of a ring $A$ such that the $A$-module $M$ is distributive. Then the following statements hold*:

1. *If $m, n \in M$ and $mA \cap nA = 0$ then $mAn = 0$.*
2. *If $A$ is a prime ring then $M$ is a uniform $A$-module.*

*Proof.*

1. Let $a \in A$. Consider the homomorphism $f : nA \to maA$ defined by the rule $f(nb) = mab$ for any $b \in A$. By proposition 2.1.11, $f \equiv 0$. Therefore, in particular, $man = 0$ for any $a \in A$, i.e. $mAn = 0$.
2. This follows immediately from the previous statement. $\square$

**Lemma 2.1.14.** *Let $M$ be a non-zero distributive right module over a ring $A$. Then the following statements hold*:

1. *If $m$, $n$ are non-zero elements of $M$ and $mA \cap nA = 0$ then $mA \oplus nA = (m+n)A$ and $\text{r.ann}_A(m+n) = \text{r.ann}_A(m) \cap \text{r.ann}_A(n) \neq \text{r.ann}_A(m)$.*
2. *If $A$ has d.c.c. on right annihilators of elements of $M$ then $M$ is finite dimensional (in the sense of [103, Definition 5.1.13]).*

*Proof.*

1. Let $m, n \in M$ and $m, n \neq 0$. Then by theorem 2.1.8(3) there exist elements $a, b \in A$ such that $1 = a+b$ and $maA + nbA \subseteq mA \cap nA = 0$. Hence $ma = nb = n(1-a) = 0$. Therefore $(m+n)a = n \neq 0$ and $(m+n)(1-a) = m \neq 0$, which means that $mA \oplus nA = (m+n)A$ and $a \in \text{r.ann}_A(m)$ and $a \notin \text{r.ann}_A(m+n)$. So $\text{r.ann}_A(m) \setminus \text{r.ann}_A(m+n) \neq 0$. Let $x \in \text{r.ann}_A(m+n)$, i.e. $(m+n)x = 0$.

---

[1]Note that a ring whose every idempotent is central is sometimes called **normal** (see e.g. [100]), or **Abelian**).

Then $mx = -nx \in mA \cap nA = 0$, i.e. $x \in$ r.ann$_A(m) \cap$ r.ann$_A(n)$. Therefore r.ann$_A(m + n) =$ r.ann$_A(m) \cap$ r.ann$_A(n) \neq$ r.ann$_A(m)$ and r.ann$_A(m + n) \subsetneq$ r.ann$_A(m)$.

2. Suppose that $M$ is not finite dimensional. Then $M$ contains an infinite direct sum $\bigoplus_{i=1}^{\infty} M_i$ of submodules of $M$. Therefore there is a set $\{m_i\}_{i=1}^{\infty}$ of non-zero elements $m_i \in M_i$ such that the sum of all the cyclic modules $m_i A$ forms an infinite direct sum $\bigoplus_{i=1}^{\infty} m_i A$. Hence the cyclic module $s_{n+1} A$ properly contains the cyclic module $s_n A$, where $s_n = (\sum_{i=1}^{n} m_i)A$, for any $n$. Then from the previous statement r.ann$_A(s_{n+1}) \subset$ r.ann$_A(s_n)$, which is not the case. So $M$ is finite dimensional. $\square$

**Proposition 2.1.15.** *Let $A$ be a right distributive ring. Suppose that one of the following conditions hold*:

1. *$A$ has d.c.c. on right annihilators.*
2. *$A$ has a.c.c. on left annihilators.*

*Then $A$ is a right finite dimensional ring (in the sense of* [103, Definition 5.1.13]).

*Proof.* If $A$ satisfies condition 1 then the statement follows from lemma 2.1.14. Condition 2 is equivalent to condition 1, by [103, proposition 5.3.23]. $\square$

Recall that a right ideal $I$ of $A$ is called **essentially closed** in $A$ if for any right ideal $J$ of $A$ such that $I \subseteq J \subset A$ and $I \subseteq J$ is an essential extension of $I$ it follows that $I = J$.

**Lemma 2.1.16.** (A.A. Tuganbaev [213]). *Any essentially closed right ideal of a right distributive ring $A$ is a two-sided ideal.*

*Proof.* Let $I$ be an essentially closed right ideal in $A$. By [103, proposition 5.1.37] $I$ is a complement in $A$, i.e. there is a right ideal $N$ in $A$ such that $I \cap N = 0$ and $J = I \oplus N$ is an essential right ideal in $A$. Since $A$ is a right distributive ring, $J$ is distributive as a right $A$-module. Therefore $B = \text{End}_A(J)$ is a right distributive ring and so $\text{Hom}_A(I, N) = 0$ and all idempotents of $B$ are central, by proposition 2.1.11 and corollary 2.1.12. Therefore $I$ is a two-sided ideal. $\square$

Recall that a ring is **reduced** if it does not have non-zero nilpotent elements. Bellow we give some important properties of reduced rings.

**Lemma 2.1.17.** *If $A$ is a reduced ring then a right annihilator of any subset of $A$ coincides with its left annihilator.*

*Proof.* Let $A$ be a reduced ring and let $S$ be a subset of $A$. Assume that $x \in$ r.ann$_A(S)$, i.e. $ax = 0$ for any non-zero element $a \in S$. Then $(xa)^2 = x(ax)a = 0$, hence

$xa = 0$, since $A$ is reduced. This means that $\mathrm{r.ann}_A(S) \subseteq \mathrm{l.ann}_A(S)$. Analogously $\mathrm{l.ann}_A(S) \subseteq \mathrm{r.ann}_A(S)$. So $\mathrm{r.ann}_A(S) = \mathrm{l.ann}_A(S)$. $\square$

**Lemma 2.1.18.** *Let $A$ be a reduced ring and $a, b \in A$. Then*

1. *If $ab = 0$ then $aAb = bAa = 0$.*
2. *$\mathrm{r.ann}_A(a) = \mathrm{r.ann}_A(a^n)$ for all $n \in N$.*

*Proof.*

1. Let $ab = 0$ and let $x = bca \in bAa$. Then $x^2 = (bca)(bca) = 0$. Hence $x = 0$ since $A$ is reduced, so $bAa = 0$. Let $y = ba$, then $y^2 = (ba)(ba) = 0$ which implies $y = 0$. Therefore as above $aAb = 0$.
2. Let $x \in \mathrm{r.ann}_A(a^n)$, i.e. $a^n x = 0$. Then $a(a^{n-1}x) = 0$. Therefore by the previous statement $aAa^{n-1}x = 0$, and so $ax(a^{n-1}x) = (axa)(a^{n-2}x) = 0$. Again using this statement we obtain $(axa)A(a^{n-2}x) = 0$ and so $(axax)(a^{n-2}x) = 0$. Continuing this process we see that $(ax)^n = 0$. Since $A$ is reduced this means that $ax = 0$, i.e. $x \in \mathrm{r.ann}_A(a)$. Hence $\mathrm{r.ann}_A(a) = \mathrm{r.ann}_A(a^n)$. $\square$

**Lemma 2.1.19.** *If $A$ is a right order in a right classical ring of fractions $Q$ then $A$ is a reduced ring if and only if $Q$ is a reduced ring.*

*Proof.* Since $A$ is a subring in $Q$, then if $Q$ is reduced then $A$ is also reduced.

Suppose that $A$ is a reduced ring and $Q$ is not reduced. Then there exists $q \in Q$ such that $q^2 = 0$. Let $q = ab^{-1}$ where $a \in A$ and $b \in C_A(0)$, where $C_A(0)$ is the set of all regular elements in $A$ (see [103, section 1.11]). Since $A$ is a right order in $Q$, there exist elements $u \in C_A(0)$ and $v \in A$ such that $au = bv$. Then

$$av = a(b^{-1}au) = qau = q(qbu) = q^2bu = 0,$$

so $v \in \mathrm{r.ann}_A(a)$. By lemma 2.1.17, $\mathrm{r.ann}_A(a)$ is a two-sided ideal in $A$. Therefore $bv \in \mathrm{r.ann}_A(a)$, i.e. $abv = a(au) = a^2u = 0$, which implies that $a^2 = 0$ since $u \in C_A(0)$. Hence $a = 0$ and $q = 0$. $\square$

**Proposition 2.1.20.** *Any reduced ring is nonsingular.*

*Proof.* Assume that $\mathcal{Z}(A_A) \neq 0$. Here $\mathcal{Z}(A_A)$ is the right singular ideal of $A$ (see [103, Definition 5.3.5]). Let $0 \neq x \in \mathcal{Z}(A_A)$. This means that $\mathrm{r.ann}_A(x)$ is an essential right ideal in $A$, i.e. $\mathrm{r.ann}_A(x) \cap J \neq 0$ for any right ideal $J$ of $A$. Let $J = xA$ and $I = \mathrm{r.ann}_A(x) \cap J \neq 0$. So there is a non-zero element $y = xa \in I$ such that $xy = 0$. Then $(yx)^2 = y(xy)x = 0$ which means that $yx = 0$ since $A$ is reduced. Therefore $y^2 = y(xa) = (yx)a = 0$. A contradiction. So $A$ is a right nonsingular ring. In a similar way one can show that $A$ is a left nonsingular ring. So the $A$ is nonsingular. $\square$

Note that not every nonsingular ring is reduced. For example, the simple Artinian ring $M_n(D)$, where $D$ is a division ring, is nonsingular, but it is not reduced for $n > 1$. However if a nonsingular ring is distributive then it is reduced.

**Proposition 2.1.21.** *A right distributive and right nonsingular ring is a reduced ring.*

*Proof.* Let $A$ be a right distributive and right nonsingular ring. Suppose that $A$ is not reduced. Then there is a non-zero element $a \in A$ such that $a^2 = 0$, i.e. $a \in$ r.ann$_A(a)$. Since $A$ is right nonsingular ring, r.ann$_A(a)$ is not an essential ideal. By [103, corollary 5.1.28] for a right non-zero ideal $aA$ there exists an essentially closed right ideal $H$ such that $aA \cap H = 0$ and $aA \oplus H$ is an essential ideal in $A$. Since $A$ is a right distributive ring, $H$ is a two-sided ideal, by lemma 2.1.16. Then $aH \subseteq aA \cap H = 0$. Since $a^2 A = 0$, we obtain that $aA \oplus H \subseteq$ r.ann$_A(a)$, i.e. r.ann$_A(a)$ is an essential ideal of $A$. A contradiction.

**Lemma 2.1.22.** *If $A$ is a distributive ring then*

$$\text{r.ann}_A(a^n) \subseteq_e \text{r.ann}_A(a^{n+1})$$

*for any $a \in A$ and any $n \in N$.*

(Here the symbol $\subseteq_e$ means 'essential extension').

*Proof.* We write $N = $ r.ann$_A(a^n)$ and $M = $ r.ann$_A(a^{n+1})$. Then $N \subseteq M$. We now have to show that $N \subseteq_e M$, i.e. $N \cap T \neq 0$ for any non-zero right ideal $T \subseteq M$.

Assume that there exists a non-zero right ideal $T \subseteq M$ such that $N \cap T = 0$. Then $a^{n+1}x = 0$ for any element $x \in T$. So $a^n(ax) = 0$, which means that $ax \in$ r.ann$_A(a^n) = N$. Therefore $aT \subseteq N$ and $aT \cap N = 0$ since $N \cap T = 0$. Then Hom$_A(T, aT) = 0$ by proposition 2.1.11. In particular the homomorphism $f \in$ Hom$_A(T, aT)$ given by $f(t) = at$ for any $t \in T$ is zero, i.e. $at = 0$ for any $t \in T$. But $T \cap N = 0$, which means that $at \neq 0$. A contradiction. $\square$

**Lemma 2.1.23.** *If $A$ is a right distributive ring then all nilpotent elements of $A$ belong to $\mathcal{Z}(A_A)$, the right singular ideal of $A$.*

*Proof.* Let $A$ be a right distributive ring, and $0 \neq a \in A$, $a^n = 0$. Consider the sequence

$$\text{r.ann}_A(a) \subseteq_e \text{r.ann}_A(a^2) \subseteq_e \cdots \subseteq_e \text{r.ann}_A(a^{n-1}) \subseteq_e \text{r.ann}_A(a^n) = A$$

Then by [100, Lemma 5.3.2] r.ann$_A(a) \subseteq_e A$, which means that $a \in \mathcal{Z}(A_A)$, the right singular ideal of $A$. $\square$

**Proposition 2.1.24.** *Let $A$ be a right distributive ring with prime radical $N$. If $A$ has a.c.c. on right annihilators then $N$ is nilpotent and $N = \mathcal{Z}(A_A)$, the right singular ideal of $A$.*

*Proof.* If $A$ has a.c.c. on right annihilators then the right singular ideal $\mathcal{Z}(A_A)$ of $A$ is nilpotent, by [103, proposition 5.3.21]. Let $X$ be a set of all nilpotent elements of $A$. Then $X \subseteq \mathcal{Z}(A_A)$ by lemma 2.1.23. On the other hand $\mathcal{Z}(A_A) \subseteq N$, since by [100, Proposition 11.2.2] $N$ contains all nilpotent ideals, and $N \subseteq X$ since $N$ is a nil-ideal by [100, Corollary 11.2.7]. So $X \subseteq \mathcal{Z}(A_A) \subseteq N \subseteq X$, which implies that $\mathcal{Z}(A_A) = N$, and so $N$ is nilpotent. $\square$

The concept of a prime ideal has played an important role in the theory of commutative rings. In a commutative ring $A$ an ideal $P \neq A$ is prime if and only if $ab \in$ implies $a \in P$ or $b \in P$ for all $a, b \in A$, or equivalently that the quotient ring $A/P$ is a domain. Recall that in a noncommutative ring an ideal $P$ is prime if $IJ \subseteq P$ implies $I \subseteq P$ or $J \subseteq P$ for all ideals $I, J$ of $A$. The definition of prime ideals as in the case of commutative rings defines completely prime ideals in the case of noncommutative ring, as has been pointed out presumably first by H. Fitting [76]. Completely prime ideals have been studied in the case of associative rings by V.A. Andrunakievich and Yu.M. Ryabukhin [5], [6] and also by N.H. McCoy [153].

**Definition 2.1.25.** An ideal $I$ in $A$ is called **completely prime** if $ab \in I$ implies $a \in I$ or $b \in I$ for all $a, b \in A$, or equivalently that the quotient ring $A/I$ is a domain.

From this definition it follows immediately that any maximal ideal is completely prime.

Note that a ring $A$ is a domain if and only if 0 is a completely prime ideal in $A$. So $A/I$ is a domain if and only if $I$ is a completely prime ideal of $A$.

Obviously, completely prime ideals are prime, but the converse is not true. For example, the zero ideal in the ring $M_n(D)$, where $D$ is a division ring, is a prime ideal, but it is not completely prime for $n > 1$.

**Lemma 2.1.26.** *Let $A$ be a right distributive ring with prime radical $N$, and let $N$ be a completely prime ideal. Then the following statements hold:*

1. $N = mN$ for any $m \in A \setminus N$, and $A$ has no nontrivial idempotents.
2. $IN = N$ for any right ideal $I$ of $A$ which is not contained in $N$.

*Proof.*

1. Let $m \in A \setminus N$ and $n \in N$. Then by theorem 2.1.8(4) there exist elements $a, b, c, d \in A$ such that $1 = a + b$ and $ma = nc$, $nb = md$. Since $N$ is a completely prime ideal and $m \notin N$, from $ma \in N$ and $md \in N$ it follows that $a, d \in N$. Since $N \subseteq \mathrm{rad}(A)$, $b = 1 - a$ is a unit in $A$. Therefore $n = mdb^{-1} \in mN$, i.e $N \subseteq mN \subseteq N$. Hence $N = mN$. Since $A/N$ is a domain and $N$ is a nil-ideal, $N$ does not contain nontrivial idempotents. Therefore $A$ has no nontrivial idempotents as well.
2. This follows from the first statement. $\square$

**Definition 2.1.27.** Two ideals $I$ and $J$ of a ring $A$ are called **incomparable** if neither $I \subseteq J$ nor $J \subseteq I$.

**Lemma 2.1.28.** *Let A be a right distributive ring. If I and J are incomparable completely prime ideals of A then I + J = A.*

*Proof.* Since $I$ and $J$ are incomparable there exist $m \in I \setminus J$ and $n \in J \setminus I$. By theorem 2.1.8(4) there exist elements $a, b, c, d$ such that $1 = a + b$ and $ma = nc$, $nb = md$. Therefore $ma \in J$ and $nb \in I$. Since $I, J$ are completely prime ideals, $a \in J$ and $b \in I$, i.e. $Aa \subseteq J$ and $Ab \subseteq I$. Therefore $A = Aa + Ab \subseteq I + J \subseteq A$, hence $A = Aa + Ab = I + J$. $\square$

**Proposition 2.1.29.** (A.A. Tuganbaev [213].) *Let A be a right distributive indecomposable ring with prime radical N. Then the following conditions are equivalent*:

1. *$A/N$ is a uniform domain.*
2. *$N$ is a completely prime ideal in A.*
3. *$A/N$ has a.c.c on right annihilators or on left annihilators.*

*Proof.*
$1 \Longrightarrow 2$ and $2 \Longrightarrow 3$ are trivial.

$3 \Longrightarrow 1$. Suppose that $A/N$ has a.c.c on left annihilators. Since $N$ is the prime radical of $A$, $A/N$ is a semiprime ring. It is right distributive and indecomposable, since $N$ is a nil-ideal and all idempotents of $A$ are central by corollary 2.1.12, and so idempotents can be lifted modulo $N$. From proposition 2.1.15 it follows that $A/N$ is right finitely dimensional. By [103, proposition 5.3.25] $A/N$ is right and left nonsingular. So from proposition 2.1.21 it follows that $A/N$ is a reduced ring. Therefore $A/N$ has also a.c.c. on right annihilators, by lemma 2.1.17. Thus, $A/N$ is a semiprime right Goldie ring. Therefore it has a right classical ring of fractions which is semisimple. Since $A/N$ is indecomposable and reduced, $A/N$ is an order in a division ring by proposition 2.1.21. So $A/N$ is a domain, and $N$ is completely prime. Since $A/N$ is a finitely dimensional domain, $A/N$ is a uniform domain, by [103, theorem 8.2.6].

Suppose that $A/N$ has a.c.c on right annihilators. Then just as in the previous case $A/N$ is a semiprime right distributive indecomposable right and left nonsingular ring. So it is reduced by proposition 2.1.21, and it has a.c.c. on left annihilators and the previous argument completes the proof. $\square$

The following statement can be considered as a generalization of lemma 2.1.3.

**Lemma 2.1.30.** (W. Stephenson [203]). *A right A-module M over a local ring A is right distributive if and only if M is a right uniserial module.*

*Proof.* Let $A$ be a local ring. If $M$ is a right uniserial $A$-module then it is obviously distributive. Conversely, suppose that $M$ is a right distributive $A$-module. Let $K, L$ be two submodules of $M$. It suffices to treat the case where $K = xA$ and $L = yA$ are cyclic $A$-modules. Then by theorem 2.1.8 there exist elements $a, b, c, d \in A$ such that $1 = a + b$, $xa = yc$, $yb = xd$. If $A$ is a unit then $x = yca^{-1} \in yA = L$ and so $K \subseteq L$.

If $A$ is not a unit, then $b = 1 - a$ is a unit, since $A$ is local. In this case $y \in xA$ and so $L \subseteq K$. $\square$

**Corollary 2.1.31.** (W. Stephenson [203]). *A local ring $A$ is right distributive if and only if $A$ is right uniserial.*

**Theorem 2.1.32.** (W. Stephenson [203]).

1. *A ring $A$ is semiperfect and right distributive if and only if $A$ is a finite direct product of right uniserial rings.*
2. *A ring $A$ is right (or left) perfect and right distributive if and only if $A$ is a finite direct product of right uniserial Artinian rings.*

*Proof.*

1. Let $A$ be a semiperfect ring. Then the identity of $A$ decomposes into a finite sum of pairwise orthogonal local idempotents, i.e. $1 = e_1 + e_2 + \cdots + e_n$, and all the $e_i A e_i$ are local rings. Since $A$ is a right distributive ring, all idempotents are central, by corollary 2.1.12. Therefore $A = A_1 \times A_2 \times \cdots \times A_n$ is a direct product of local right distributive rings $A_i$, and these are right uniserial, by corollary 2.1.31.

   Since a finite direct product of right uniserial rings is semiperfect, and a finite direct product of right distributive rings is right distributive, by proposition 2.1.10, we obtain the converse statement.

2. Suppose that $A$ is a right perfect and right distributive ring. Since any right perfect ring is semiperfect, $A = A_1 \times A_2 \times \cdots \times A_n$ is a finite direct product of right uniserial rings. By [101, corollary 4.7.4] $A$ is a right Artinian ring. So, by [103, theorem 1.1.25], each $A_i = e_i A e_i$ is also right Artinian.

   If $A$ is a left perfect and right distributive ring, then as in the previous case $A = A_1 \times A_2 \times \cdots \times A_n$ is a finite direct product of right uniserial and left perfect rings $A_i$. By the Bass theorem (see also [100, theorem 10.5.5]), each $A_i$ has d.c.c. on principal right ideals. Suppose that we have a strictly descending chain $I_1 \supset I_2 \supset \cdots \supset I_n \supset \cdots$ of right ideals of $A_i$. For any $n$ we can choose an element $a_n \in I_n \setminus I_{n+1}$. Since $A_i$ is a uniserial ring for any $i$, $I_n \supseteq a_n A \supset I_{n+1}$, and we obtain a strictly descending chain of principal right ideals in $A$. This contradiction shows that each $A_i$ is right Artinian. $\square$

## 2.2 Semiprime Semidistributive Rings

This section considers some main properties of semiprime semidistributive rings.

Recall that a module is called **semidistributive** if it is a direct sum of distributive modules. A ring $A$ is called **right (left) semidistributive** if the right (left) regular module $A_A$ ($_A A$) is semidistributive. A right and left semidistributive ring is called **semidistributive**.

Obviously, the class of semidistributive modules includes the class of semisimple and serial modules.

**Proposition 2.2.1.** *Let e be a non-zero idempotent of a ring A. Then*

1. *If M is a semidistributive right A-module then Me is a right semidistributive eAe-module.*
2. *If eA is a semidistributive right A-module then the ring eAe is right semidistributive.*
3. *If A is a right semidistributive ring then the ring eAe is right semidistributive.*

*Proof.*
1. This follows immediately from [103, proposition 5.7.23(1)].
$1 \Longrightarrow 2 \Longrightarrow 3$. These implications are trivial. $\square$

**Lemma 2.2.2.** *Let A be a semiprime ring. Then the right and left annihilators of any two-sided ideal of A coincide.*

*Proof.* Let $I$ be a two-sided ideal of $A$. Then $J_1 = \mathrm{r.ann}_A(I)$ and $J_2 = \mathrm{l.ann}_A(I)$ are also two-sided ideals of $A$. Since $A$ is semiprime and $(J_1 I)^2 = 0$, we have that $J_1 I = 0$, i.e. $J_1 \subseteq J_2$. Analogously $J_2 \subseteq J_1$, which means that $J_1 = J_2$. $\square$

**Lemma 2.2.3.**

1. *Let P be a proper prime ideal of a ring A. Then ePe is a proper prime ideal of the ring eAe for any idempotent e of A such that $e \notin P$.*
2. *Let P, S be incomparable proper prime ideals of a ring A and let e be an idempotent of A which is not contained in $P + S$. Then ideals ePe, eSe are incomparable prime ideals in a ring eAe and $ePe + eSe \neq eAe$.*

*Proof.*

1. Let $I$ be a prime ideal of $A$ and let $e \in A \setminus P$ be an idempotent of $A$. Then $ePe$ is an ideal of $eAe$ and $ePe \neq eAe$, i.e. $ePe$ is a proper ideal of $eAe$. Assume that there exist ideals $eMe$ and $eNe$ of $eAe$ such that $(eMe)(eNe) \subseteq ePe \subseteq P$, where $M, N$ are ideals of $A$. Then $(AeA)M(eAe)N(eAe) \subseteq P$. Since $P$ is a proper prime ideal of $A$, $M \subseteq P$ or $N \subseteq P$. Hence $eMe \subseteq ePe$ or $eNe \subseteq ePe$, i.e. $ePe$ is a proper prime ideal of $eAe$.
2. By statement 1 $ePe$ and $eSe$ are prime ideals of $eAe$. Since $e \in A \setminus (P + S)$ we obtain that $eAe \neq ePe + eSe$. Suppose that $eSe \subseteq ePe \subseteq P$, then $(AeA)S(AeA) \subseteq P$. Since $P$ is prime, $S \subseteq P$. A contradiction. $\square$

**Proposition 2.2.4.** (A.A. Tuganbaev). *Let* $1 = e_1 + e_2 + \cdots + e_n$ *be a decomposition of the identity of a ring A into a sum of pairwise orthogonal idempotents, and let all right A-modules $e_i A$ be distributive. Then the following statements hold.*

1. *If A is a right nonsingular ring then all the rings $e_i A e_i$ are reduced for $i = 1, 2, \ldots, n$.*
2. *If A is a prime right nonsingular ring then A is a right finite dimensional ring.*
3. *If A is a prime right nonsingular ring then all the rings $e_i A e_i$ are domains.*
4. *Let P be an ideal of a ring A such that $A/P$ is a prime right nonsingular ring. Then ePe is a completely prime ideal of a ring eAe for any idempotent e of A such that $e \notin P$.*

*Proof.* Let $1 = e_1 + e_2 + \cdots + e_n$ be a decomposition of the identity $1 \in A$ into a sum of pairwise orthogonal idempotents and let $e_i A$ be a distributive right $A$-module for all $i = 1, \ldots, n$.

1. By proposition 2.1.6(2) all the rings $e_i A e_i$ are right distributive. Since all these rings are also right nonsingular by [103, proposition 5.3.13(2)], they are reduced by proposition 2.1.21.
2. Since $A$ is a prime ring and $e_i A$ is a distributive right $A$-module, also $e_i A$ is a uniform module, by lemma 2.1.13(2). So $A = e_1 A \oplus e_2 A \oplus \cdots \oplus e_n A$, where each $e_i A$ is uniform. Therefore, u.dim$(A_A) = n$, i.e. $A$ is a right finite dimensional ring.
3. Since $A$ is a prime right nonsingular and right finite dimensional ring, it is a right Goldie ring by [103, theorem 5.4.10]. Hence $A$ is a right order in a simple Artinian ring. By proposition 2.1.6, $B_i = e_i A e_i$ is a right distributive ring, for each $i = 1, 2, \ldots, n$. Since $A$ is a prime ring, $B_i$ is also prime. Since $A$ is right nonsingular, $B_i$ is a reduced ring, by statement 1. So, $B_i$ is a reduced prime ring which is subring of a simple Artinian ring, which means that $B_i$ is a domain. $\square$
4. Since $eAe \neq ePe$ and $eAe/ePe \simeq (e + P)(A/P)(e + P)$ we can assume that $P = 0$, i.e. that $A$ is a prime right nonsingular ring. Then by statement 3 $eAe$ is a domain. $\square$

**Lemma 2.2.5.** *Let A be a semiprime right semidistributive and right Goldie ring. If P and S are distinct minimal prime ideals of A, then $P + S = A$.*

*Proof.* By [103, proposition 5.4.21] $A/P$ and $A/S$ are prime right Goldie rings, and $P \cap S = 0$. Since $A$ is a right semidistributive ring, there exists a decomposition of $1 = e_1 + e_2 + \cdots + e_n$ into a sum of pairwise orthogonal idempotents such that all the $e_i A$ are distributive modules. So all the $e_i A e_i$ are right distributive rings, by proposition 2.1.6(2).

Suppose $P + S \neq A$. Then there exists an idempotent $e_i \in A \setminus (P + S)$. Then $e_i P e_i$, $e_i S e_i$ are incomparable proper completely prime ideals of $e_i A e_i$ by proposition 2.2.4(4), and $e_iPe_i + e_iSe_i \neq e_iAe_i$, by lemma 2.2.3(2). But this contradicts lemma 2.1.28. □

**Theorem 2.2.6.** (A.A. Tuganbaev [211], [213, proposition 5.15]). *Let A be a semiprime right semidisributive ring. Then the following conditions are equivalent*:

1. *A is a ring with a.c.c. on right annihilators.*
2. *A is a ring with a.c.c. on left annihilators.*
3. *A is right nonsingular and does not contain infinite direct sums of non-zero ideals.*
4. *A is a right Goldie ring.*
5. *A is a finite direct product of prime right Goldie rings which are direct sums of uniform distributive right ideals.*

*Proof.*

$1 \Longleftrightarrow 2$. Since $A$ is a semiprime ring, statement follows from [103, lemma 5.3.30].

$2 \Longrightarrow 3$. Since $A$ is semiprime, $A$ is right and left nonsingular, by [103, proposition 5.3.25]. By proposition 2.1.15 the ring $A$ is right finite dimensional.

$4 \Longleftrightarrow 3$. This follows from [103, theorem 5.4.9].

$4 \Longrightarrow 1$. This follows from the definition of a right Goldie ring.

$3 \Longrightarrow 2$. By [103, proposition 5.3.28] $A$ is a ring with d.c.c. for right annihilators. Therefore $A$ is a ring with a.c.c. on left annihilators by [103, proposition 5.3.23].

$5 \Longrightarrow 4$. Let $A = \prod_{i=1}^{n} A_i$, where each $A_i$ is a prime right Goldie ring. Since the direct product of nonsingular rings $A_i$ is nonsingular by [103, proposition 5.3.13], $A$ is nonsingular. By [103, corollary 5.1.23(2)] $\text{u.dim}(A_A) = \sum_{i=1}^{n} \text{u.dim}(A_i) < \infty$. So, $A$ is a right Goldie ring.

$4 \Longrightarrow 5$. Since $A$ is a semiprime right Goldie ring, it has a right classical ring of fractions $Q$. From [103, proposition 5.4.21] it follows that there exists a finite number of distinct prime ideals $P_1, P_2, \ldots, P_n$ such that $P_i \cap P_j = 0$ and the $A_i = A/P_i$ are prime right Goldie rings for all $i, j$ and $i \neq j$. Then by lemma 2.2.5 $P_i + P_j = A$ for all $i \neq j$. Therefore by the Chinese Remainder theorem[2] for noncommutative rings $A$ is isomorphic to the direct product of prime right Goldie rings $A_1, A_2, \ldots, A_n$, since $P_1 \cap P_2 \cap \cdots \cap P_n = 0$. Since $A$ is a right semidistributive ring, each $A_i$ is a prime right semidistributive ring which decomposes into a direct sum of uniform distributive right ideals by lemma 2.1.13(2). □

---

[2]The Chinese Reminder theorem states that if $I_1, I_2, \ldots, I_n$ are two-sided ideals of a ring $A$ (not necessarily commutative) such that $I_i + I_j = A$ whenever $i \neq j$ then

$$A/(I_1 \cap I_2 \cap \cdots \cap I_n) \simeq A/I_1 \times A/I_2 \times \cdots \times A/I_n$$

as $A$-modules. The proof of this theorem is done for commutative case in many books, e.g. [141], [100, theorem 7.6.2]. For noncommutative case the proof is almost the same.

**Theorem 2.2.7.** (A.A. Tuganbaev [211]). *Let A be a semiprime right semidistributive indecomposable ring with a.c.c. on right annihilators. Then*

1. *A is a prime right Goldie ring.*
2. *A is a finite direct sum of distributive uniform right ideals $e_1 A, e_2 A, \ldots, e_n A$ such that $e_i^2 = e_i$, and all $e_i A e_i$ are right distributive domains.*

*Proof.*

1. By theorem 2.2.6 $A$ is a direct product of prime right Goldie rings. Since $A$ is indecomposable, $A$ is a prime ring.
2. By theorem 2.2.6 all $e_i A$ are uniform distributive right ideals, and so the $e_i A e_i$ are domains by proposition 2.2.4(3). □

**Theorem 2.2.8.** (A.A. Tuganbaev [211]). *Let A be a semiprime right hereditary and right semidistributive ring such that A contains no infinite direct sums of non-zero ideals. Then A is a finite direct product of prime right Noetherian rings.*

*Proof.* Assume that $A$ is a semiprime right hereditary and right semidistributive ring such that $A$ contains no infinite direct sums of non-zero ideals. Since any right hereditary ring is nonsingular, $A$ is a finite direct product $\prod_{i=1}^{n} A_i$ of prime right Goldie rings $A_i$, by theorem 2.2.6. Therefore $\mathrm{u.dim}(A_i) < \infty$ and $A_i$ is a right hereditary ring for each $i = 1, 2, \ldots, n$. Then according to [103, theorem 8.2.5] each $A_i$ is right Noetherian. □

**Lemma 2.2.9.** *A right uniform and right nonsingular ring is a domain.*

*Proof.* Let $A$ be a right uniform and right nonsingular ring, and $0 \neq x \in A$. Consider $I = \mathrm{r.ann}_A(x)$ which is a right ideal of $A$. Since $A$ is a right nonsingular ring, $I$ is not an essential right ideal of $A$. On the other hand, by [103, proposition 5.1.3], a ring $A$ is right uniform if and only if any right non-zero ideal is essential in $A$. This contradiction shows that $I = 0$, and so $A$ is a domain. □

**Theorem 2.2.10.** (A.A. Tuganbaev [206], [207], [208]). *Let A be a Noetherian right distributive indecomposable ring. Then it is either a Noetherian right distributive and right uniform domain or an Artinian right uniserial ring.*

*Proof.* Let $A$ be a Noetherian right distributive indecomposable ring with prime radical $N$.

If $N = 0$ then $A$ is a semiprime ring. Therefore $A$ is a semiprime Noetherian right distributive ring, and so it is a Goldie ring. Since $A$ is indecomposable, $A$ is a prime Noetherian right Goldie ring which decomposes into a direct sum of uniform distributive right ideals $A = I_1 \oplus I_2 \oplus \cdots \oplus I_n$ by proposition 2.2.6. Since $A$ is a right distributive ring, $\mathrm{Hom}_A(I_i, I_j) = 0$ whenever $i \neq j$ by proposition 2.1.11. Taking into account that $A \simeq \mathrm{End}_A(A)$ is a right distributive ring and all its idempotents are central by corollary 2.1.12, there results that all of these right ideals $I_i$ are two-sided. Therefore $A$ decomposes into a finite direct product of right distributive right uniform right nonsingular rings and all of them are domains by lemma 2.2.9. Since $A$ is an indecomposable ring, it is a right uniform domain. Therefore $A$ is a Noetherian right distributive right uniform domain.

Suppose that $N \neq 0$ and $R$ is the Jacobson radical of $A$. Then $N \subseteq R$. Since $A$ is a Noetherian right distributive, $N$ is nilpotent by proposition 2.1.24. Since $A$ is a Noetherian ring, $A/N$ is also right Noetherian. Therefore $A/N$ is a uniform domain and $N$ is a completely prime ideal in $A$, by proposition 2.1.29. Moreover, $A/N$ is an order in a division ring $D$. Therefore $A/N = D$.

Suppose $N \neq R$. Then $RN = N$ by lemma 2.1.26(2), since $A$ is a right distributive ring and $N$ is a completely prime ideal. Using Nakayama's lemma we find that $N = 0$, since $A$ is a Noetherian ring and so $N$ is a finitely generated ideal. This contradiction shows that $N = R$. Therefore $A/R = D$ and $R$ is nilpotent, i.e. $A$ is a local and semiprimary ring. Since $A$ is a Noetherian ring, $A$ is Artinian by [103, theorem 1.1.22]. Since $A$ is a local and right distributive ring, $A$ is right uniserial, by corollary 2.1.31. Hence $A$ is an Artinian right uniserial ring. $\square$

Theorem 2.2.10 in fact comes in a stronger form that follows from the next theorem which we state without proof.

Recall that a ring is called a **right (left) duo ring** when all its right (left) ideals are two-sided. A ring is said to be a **duo ring**[3] if it is both a right and left duo ring.

**Theorem 2.2.11.** (A.A. Tuganbaev [206], [207], [208]). *Let $A$ be a Noetherian right distributive indecomposable ring. Then it is a Noetherian hereditary duo domain or an Artinian right uniserial ring.*

The next results are immediate consequences of this theorem.

**Proposition 2.2.12.** *If $A$ is a Noetherian right distributive ring then* K.dim$(A_A) \leq 1$.

*Proof.* The proof follows from theorem 2.2.11 and [103, proposition 8.2.14] taking into account that an Artinian ring has Krull dimension 0. $\square$

**Proposition 2.2.13.** *If $A$ is a Noetherian ring and Noetherian rings $e_i A e_i$ are right distributive, for $i = 1, 2, \ldots, n$, then* K.dim$(A_A) \leq 1$.

*Proof.* This follows from [103, proposition 5.7.24(4)] and proposition 2.2.12. $\square$

**Theorem 2.2.14.** (A.A. Tuganbaev [213, Theorem 11.51]). *If $A$ is a Noetherian right semidistributive ring then* K.dim$(A_A) \leq 1$.

---

[3]Note that in the literature these rings are also called **invariant** (see e.g. [39], [213], [214]) or **subcommutative** (see e.g. [203]).

*Proof.* From proposition 2.1.6 it follows that the identity of $A$ can be decomposed into a finite sum of orthogonal idempotents $e_1, \ldots, e_n$ such that all $e_i A e_i$ are distributive rings. Now one can apply proposition 2.2.12. $\square$

**Theorem 2.2.15.** (A.A. Tuganbaev [213, Theorem 11.51]). *Let $A$ be a Noetherian right semidistributive ring with Jacobson radical $R$. Then*

$$\bigcap_{n \geq 0} R^n = 0.$$

*Proof.* By theorem 2.2.14 K.dim$(A_A) \leq 1$. So the theorem follows from [103, theorem 9.5.14]. $\square$

From theorem 2.2.15 there immediately follows the next statement:

**Corollary 2.2.16.** [125, Theorem 2.4] *The intersection of all natural powers of the Jacobson radical of a Noetherian SPSD-ring is zero.*

**Remark 2.2.17.** Note that this corollary is exactly [103, theorem 9.5.5], which was proved earlier by means of other arguments.

## 2.3 Semiperfect Semidistributive Rings

In this section we consider a special class of semihereditary rings; namely semiperfect semidistributive rings (abbreviated **SPSD-rings**). For these rings one can obtain some interesting characterizations. Note that Chapter 14 of [100] is also all about SPSD rings.

**Theorem 2.3.1.** (A.A. Tuganbaev [213, Proposition 11.27(3)]). *Any semiperfect semiprime right semihereditary right finite dimensional ring is right serial.*

*Proof.* Since $A$ is a right finite dimensional ring, there is an essential right ideal $I = \bigoplus_{i=1}^{n} I_i$, where each $I_i$ is a uniform right ideal. Since $A$ is a right semihereditary, it is right nonsingular by [103, corollary 8.1.4]. Now, from [103, lemma 5.4.3] it follows that any essential right ideal of $A$ contains a regular element of $A$. Therefore $I$ contains a regular element $b \in A$, i.e. $bA \cong A_A$. We now prove that $N = bA$ is a direct sum of uniform ideals. Suppose that u.dim$_A(N) = m \leq n$. We will proceed by induction on $m$. If $m = 1$ then $U \subseteq N = bA \subseteq \bigoplus_{i=1}^{n} I_i$, where $U$ is a uniform right ideal, and so $N = U$. Now let $1 < m \leq n$. Consider the natural projection $\pi_i : I \to I_i$ and the exact sequence $0 \to N \cap (\bigoplus_{j \neq i} I_j) \to N \to \pi_i(N) \to 0$. Since $N$ is a principal ideal, $\pi_i(N)$ is also a principal ideal, and so it is projective. Therefore this sequence is split and so $N \cong N \cap (\bigoplus_{j \neq i} I_j) \oplus \pi_i(N)$. Since $\pi_i(N)$ is a uniform ideal, u.dim$(N \cap (\bigoplus_{j \neq i} I_j)) < m$ and therefore by the induction hypothesis $N$ is a direct sum of uniform ideals. Therefore $A_A$ is also a direct sum of uniform right $A$-modules. Let $U$ be a uniform direct summand of $A_A$. It will now be shown that $U$ is a uniserial right $A$-modules. Assume that $U$ is not uniserial, i.e. there exist two cyclic incomparable submodules $X, Y \subset U$ of $A$. Then $X + Y$ is a finitely generated nonlocal right $A$-module, and so $X + Y$ is projective since $A$ is right semihereditary. Since $U$ is uniform, $X \cap Y \neq 0$, i.e. $X + Y$ is indecomposable. Therefore $X + Y$ is an indecomposable projective right $A$-module, where $A$ is a semiperfect ring. So, $X + Y$ is isomorphic to a principal right $A$-module which is local. A contradiction. $\square$

Since any right Noetherian ring is right finite dimensional, we immediately obtain the follows.

**Corollary 2.3.2.** *Any semiperfect semiprime semihereditary right Noetherian ring is right serial.*

**Theorem 2.3.3.** *A semiperfect right hereditary and right distributive ring is right Noetherian.*

*Proof.* Since $A$ is a semiperfect ring, it has only a finite number of principal right $A$-modules. Let $I$ be a right ideal of $A$. Since $A$ is a right hereditary ring, $I$ is a projective right $A$-module and so, by [103, theorem 7.2.16], $I$ decomposes into a direct sum of principal right $A$-modules. If this sum is infinite, then in this decomposition there exist at least two identical principal modules. In this case the socle of $I/IR$ is not square-free, where $R$ is the Jacobson radical of $A$. But this impossible by [100, theorem 14.1.5], since $I$ is a semidistributive $A$-module. So any right ideal of $A$ is finitely generated, and $A$ is a right Noetherian ring. $\square$

**Corollary 2.3.4.** *A right hereditary local ring $A$ is semidistributive if and only if it is either a discrete valuation ring or a division ring.*

*Proof.* By the previous theorem 2.3.3, $A$ is a right Noetherian ring, and so from [103, proposition 8.3.12] it follows that $A$ is a principal right ideal domain. The statement follows now from [103, proposition 3.5.5]. $\square$

**Theorem 2.3.5.** (A.A. Tuganbaev [209], [212], [213, Proposition 11.28]). *Let $A$ be a semiperfect semiprime right hereditary ring. Then the following conditions are equivalent:*

1. *$A$ is right Noetherian.*
2. *$A$ is right serial.*
3. *$A$ is right semidistributive.*

*Proof.* $1 \implies 2$. This is corollary 2.3.2. The implication $2 \implies 3$ is trivial. The implication $3 \implies 2$ is theorem 2.3.3. $\square$

**Proposition 2.3.6.** *If A is a right hereditary (resp. semihereditary) SPSD-ring and* $e \in A$ *is a non-zero idempotent then* $eAe$ *is a right hereditary (resp. semihereditary) SPSD-ring.*

The proof follows from [103, theorems 4.4.6, 8.2.16] and [103, proposition 1.10.12]. □

**Proposition 2.3.7.** *A semiperfect right semidistributive ring A with Jacobson radical R is right Artinian if and only if R is nilpotent.*

*Proof.* If $A$ is right Artinian, then by the Hopkins theorem [103, Theorem 1.1.21] the Jacobson radical $R$ of $A$ is nilpotent.

Conversely, if $R$ is nilpotent, then $A$ is a semiprimary right semidistributive ring. By [100, theorem 14.1.6] $A$ is a right Artinian ring. □

**Corollary 2.3.8.** *An SPSD-ring A with Jacobson radical R is Artinian if and only if R is nilpotent.*

## 2.4 Right Hereditary SPSD-Rings

This section is devoted to the study of special classes of semiperfect semidistributive rings, namely right hereditary SPSD-rings. We will show that the structure of all such rings is closely related to right hereditary rings of the form $A(S, O)$ as described in section 1.3.

**Theorem 2.4.1.** (V.V. Kirichenko, M.A. Khibina [125, Theorem 2.5]). *The endomorphism rings of simple modules over a Noetherian indecomposable SPSD-ring are division rings. And they are all isomorphic.*

*Proof.* Let $A$ be an indecomposable Noetherian SPSD-ring. Then the quiver $Q(A)$ is connected by [100, theorem 11.1.9]. Since $Q(A) = Q(A/R^2)$ and every simple $A$-module is an $A/R^2$-module, one can assume that $R^2 = 0$. Let $A = P_1 \oplus \ldots \oplus P_s$ be a decomposition of the ring $A$ into a direct sum of projective indecomposable modules. One can assume that the ring $A$ is a basic ring[4] and therefore that the modules $P_1, \ldots, P_s$ are pairwise non-isomorphic. Let $1 = e_1 + \ldots + e_s$ be the corresponding decomposition of $1 \in A$ into a sum of mutually orthogonal local idempotents, and let $A_{ij} = e_i A e_j$ ($i, j = 1, \ldots, s$). Every simple module $U_i = P_i/P_i R$ is of the form: $U_i \cong A_{ii}/R_i = D_i$, where $D_i$ is a division ring ($i = 1, \ldots, s$). By the Schur lemma, the endomorphism ring of $U_i$ is isomorphic to $D_i$. Since $A$ is a Noetherian SPSD-ring and $R^2 = 0$, $A_{ij}R_j = R_iA_{ij} = 0$ for $i \neq j$, by [103, theorem 1.10.11]. If $A_{ij} \neq 0$ then, by [103, theorem 1.10.9], $A_{ij}$ is a right uniserial $A_{jj}$-module and a

---

[4]Note that a semiperfect ring $A$ with the Jacobson radical $R$ is **basic** if $A/R$ is a direct sum of division rings.

left uniserial $A_{ii}$-module. Therefore if $A_{ij} \neq 0$, there exist an element $x$ such that $A_{ij} = D_i x = x D_j$. So the correspondence $\sigma : D_i \to D_j$, $\alpha x = x \alpha^\sigma$ is an isomorphism between the division rings $D_i$ and $D_j$. Note that if $A_{ij} \neq 0$ $(i \neq j)$ then there is an arrow from $i$ to $j$ in $Q(A)$. Conversely, if there is an arrow from $i$ to $j$ then $A_{ij} \neq 0$. Thus if the points $i$ and $j$ are connected by an arrow then the endomorphism rings of the simple modules $U_i$ and $U_j$ are isomorphic division rings. If the vertex 1 is not connected with any other vertex, then $Q(A)$ is not connected. Hence, one can assume that the vertex 1 is connected with the vertex 2. Denote all points which are connected with the vertex 1 by $2, \ldots, m$, where $m \geq 2$. Consider the set $\{m + 1, \ldots, s\}$. Since the quiver $Q(A)$ is connected there exists a vertex $j \in \{m + 1, \ldots, s\}$ which is connected by an arrow with one of the vertices $2, \ldots, m$. One can assume that $j = m + 1$. Thus the endomorphism rings of the modules $U_1, \ldots, U_{m+1}$ are isomorphic division rings. Considering the sets $\{1, \ldots, m + 1\}$ and $\{m + 2, \ldots, s\}$, one can again assume that the vertex $m + 2$ is connected with one of the vertices $1, \ldots, m + 1$. Continuing this process we obtain the statement of the theorem. $\square$

**Remark 2.4.2.** Let $\mathbf{Q}$ be the field of rational numbers, $p$ a prime integer, $\mathbf{Z}_p = \{\frac{m}{n} \in \mathbf{Q} | (n, p) = 1\}$. Consider the Herstein-Small ring

$$A = \begin{pmatrix} \mathbf{Z}_p & \mathbf{Q} \\ 0 & \mathbf{Q} \end{pmatrix}$$

(see [100], section 5.6) which shows that theorem 2.4.1 is not true if one just assume right Noetherian and serial. Indeed, the Jacobson radical $R$ of $A$ is

$$\begin{pmatrix} p\mathbf{Z}_p & \mathbf{Q} \\ 0 & 0 \end{pmatrix}$$

and the ring $\bar{A} = A/R$ decomposes into the direct product of rings:

$$\bar{A} = \mathbf{Z}_p / p\mathbf{Z}_p \times \mathbf{Q}.$$

Therefore, $A$ has two non-isomorphic simple modules $\mathbf{Z}_p / p\mathbf{Z}_p$ and $\mathbf{Q}$, whose endomorphism rings are non-isomorphic division rings.

Denote by $M_n(D)$ the ring of $n \times n$ matrices over a division ring $D$ with matrix units $e_{ij}$. Let $S = \{\alpha_1, \ldots, \alpha_n\}$ be a finite poset, which is labeled by natural numbers in such a way that $i \leqslant j$ if $\alpha_i \leq \alpha_j$. Let $\Omega = (\omega_{ij})$ be a set of automorphisms of the division ring $D$, such that $\omega_{ij} \in \Omega$ if and only if $\alpha_i \leq \alpha_j$; $\omega_{ii} = \mathrm{id}$, for $i = 1, 2, \ldots, n$. Consider the ring $T = T(S, D, \Omega)$, which is generated by those matrix units $e_{ij} \in T$ for which $\alpha_i \leq \alpha_j$ in $S$, and with relations $e_{ij}e_{jk} = e_{ik}$ if $e_{ij}, e_{jk} \in T$; and $d e_{ij} = e_{ij} d^{\omega_{ij}}$ for any $d \in D$. Thus, the ring $T(S, D, \Omega)$ is defined by a finite poset $S$, the basis of matrix units $\{e_{ij}\}$ related to the poset $S$, and the relations

$$d e_{ij} = e_{ij} d^{\omega_{ij}} \tag{2.4.3}$$

for any $d \in D$.

In the case that all automorphisms $\omega_{ij}$ are identity automorphisms the ring $T = T(S, D, \Omega)$ is simply the ring $T(S, D)$ considered in section 1.2.

**Lemma 2.4.4.** *Let S be a finite poset, and D a division ring. The ring $T = T(S, D, \Omega)$ is an Artinian semidistributive piecewise domain, and the quiver $Q(T)$ coincides with the Hasse diagram $H(S)$ of the poset S.*

*Proof.* From the construction of the ring $T$ it follows that $T$ is an Artinian ring, by [103, theorem 1.1.23].

For any $i, j$ either $e_{ii}Te_{jj} \simeq D$ or $e_{ii}Te_{jj} = 0$. Therefore $T$ is a semidistributive ring, by [103, theorem 1.10.9].

Since $e_{ii}Te_{jj} = D$, $e_{jj}Te_{kk} = D$ and $e_{ii}Te_{kk} = D$ if $e_{ij}, e_{jk} \in T$, $T$ is a piecewise domain.

It remains only to prove that the quiver $Q(T)$ coincides with the diagram $H(S)$ of the poset $S$. Let $J = \mathrm{rad}(T)$ be the Jacobson radical of $T$. Note that a right and left $D$-basis of $J$ are given by the matrix units $e_{ij}$ such that $\alpha_i < \alpha_j$. Suppose that $\alpha_j$ do not cover $\alpha_i$. Then there is an element $\alpha_k \in S$ such that $\alpha_i < \alpha_k < \alpha_j$. Therefore $e_{ik}, e_{kj} \in J$ and $e_{ij} \in J^2$, since $e_{ik} \cdot e_{kj} = e_{ij}$. Therefore $e_{pq} \in J \setminus J^2$ if and only if when $\alpha_q$ covers $\alpha_p$. From the annihilation lemma [103, Lemma 1.9.11] and the $Q$-Lemma [103, Lemma 1.9.12] it follows that in $Q(T)$ there exists an arrow from $p$ to $q$ if and only if when there exists an arrow from $p$ to $q$ in the diagram $H(S)$ of the poset $S$.

From [100, Theorem 14.3.1] it follows, that the quiver $Q(T)$ of the ring $T$ is simply laced. Therefore, $Q(T)$ coincides with $H(S)$. The lemma is proved. $\square$

**Lemma 2.4.5.** (V.V. Kirichenko [127, Lemma 3.18]). *Given a ring $A = \begin{pmatrix} A_1 & X \\ 0 & A_2 \end{pmatrix}$, and isomorphisms of rings $\varphi : A_1 \to C_1$ and $\psi : A_2 \to C_2$, the ring $A$ is isomorphic to the ring $C = \begin{pmatrix} C_1 & X \\ 0 & C_2 \end{pmatrix}$ with multiplication defined by the following rule:*

$$\begin{pmatrix} c_1 & x \\ 0 & c_2 \end{pmatrix} \cdot \begin{pmatrix} c_1' & x' \\ 0 & c_2' \end{pmatrix} = \begin{pmatrix} c_1 c_1' & c_1^{\varphi^{-1}} x' + x c_2'^{\psi^{-1}} \\ 0 & c_2 c_2' \end{pmatrix}.$$

*Proof.* It is easy to show that the map $\Psi : A \to C$ given by the rule:

$$\Psi\left(\begin{pmatrix} a_1 & x \\ 0 & a_2 \end{pmatrix}\right) = \begin{pmatrix} a_1^\varphi & x \\ 0 & a_2^\psi \end{pmatrix} \tag{2.4.6}$$

defines an isomorphism of rings. $\square$

**Theorem 2.4.7.** (V.V. Kirichenko [127, Theorem 3.19]). *Every hereditary semidistributive Artinian ring $A$ is Morita equivalent to a finite direct product of indecomposable rings of the form $T(S, D)$, where $D$ is a division ring and the diagram of the finite poset $S$ contains no rhombuses. Conversely, every ring of this form is a hereditary semidistributive Artinian ring.*

*Proof.* Let $A$ be an indecomposable hereditary semidistributive Artinian ring. Without loss of generality one can assume that $A$ is a basic ring. By [103, corollary 8.5.7] this ring is primely triangular. By theorem 2.4.1 all endomorphism rings of simple modules over $A$ are isomorphic division rings. From [103, theorem 1.10.9] it follows that every indecomposable Artinian semidistributive hereditary ring is Morita equivalent to a ring $T$ whose two-sided Peirce decomposition has the following form:

$$B = \begin{pmatrix} D & T_{12} & \cdots & T_{1m} \\ 0 & D & \cdots & T_{2m} \\ \vdots & \vdots & \ddots & \vdots \\ 0 & 0 & \cdots & D \end{pmatrix}, \tag{2.4.8}$$

where $T_{ij}$ is either zero or a one-dimensional left and right vector space over the division ring $D$, and if $T_{ij} \neq 0$ and $T_{jk} \neq 0$ then $T_{ik} \neq 0$ as well. Denote by $S$ the finite poset whose diagram is $Q(T)$. Obviously, one can assume that $S = \{1, 2, \ldots, m\}$ and $i \prec j$ if and only if $T_{ij} \neq 0$. If the diagram $H(S)$ of the poset $S$ contains rhombuses, then there exists an idempotent $e$ of the ring $T$ such that the ring $eTe$ is of the form:

$$B = \begin{pmatrix} D & B_{12} & B_{13} & B_{14} \\ 0 & D & 0 & B_{24} \\ 0 & 0 & D & B_{34} \\ 0 & 0 & 0 & D \end{pmatrix},$$

where $B_{12}, B_{13}, B_{14}, B_{24}, B_{34}$ are non-zero one-dimensional right and left vector spaces over $D$. Clearly, the right ideal $\begin{pmatrix} 0 & B_{12} & B_{13} & B_{14} \end{pmatrix}$ is not projective. Thus the poset $S$ contains no rhombuses.

Let $T_{ij}$ be non-zero and $x_{ij} \in T_{ij}$, $x_{ij} \neq 0$. Then the correspondence $\omega_{ij} : D \to D$ given by the formula $dx_{ij} = x_{ij} d^{\omega_{ij}}$ ($d \in D$) is an automorphism of the division ring $D$. Clearly, in this case the ring $T$ is isomorphic to the ring $T(S, D, \Omega)$, where $\Omega$ is the set of all automorphisms $\omega_{ij}$ of $D$. Applying lemma 2.4.5 to the case when the diagram of the poset $S$ is a tree, one can assume that all the $\omega_{ij}$ are identities. In this case $T(S, D, \Omega) = T(S, D)$.

The converse statement follows from proposition 1.2.12. $\square$

**Theorem 2.4.9.** *A prime right hereditary SPSD-ring is right and left Noetherian. Every such ring is right and left hereditary, and it is Morita-equivalent to either a division ring or a ring of the form $H_m(O)$, where $O$ is a discrete valuation ring.*

*Proof.* Let $A$ be a right hereditary SPSD-ring. Then, by theorem 2.4.1, $A$ is a right Noetherian ring. By [100, theorem 14.5.1], a right Noetherian prime SPSD-ring is

Morita equivalent to either a division ring $D$ or a semimaximal prime ring of the form

$$A = \begin{pmatrix} O & \pi^{\alpha_{12}}O & \cdots & \pi^{\alpha_{1n}}O \\ \pi^{\alpha_{21}}O & O & \cdots & \pi^{\alpha_{2n}}O \\ \vdots & \vdots & \ddots & \vdots \\ \pi^{\alpha_{n1}}O & \pi^{\alpha_{n2}}O & \cdots & O \end{pmatrix},$$

where $n \geq 1$, $O$ is a discrete valuation ring with a prime element $\pi$, and the $\alpha_{ij}$ are integers such that $\alpha_{ij} + \alpha_{jk} \geq \alpha_{ik}$ for all $i, j, k$ ($\alpha_{ii} = 0$ for all $i$). In both cases $A$ is a Noetherian ring. Since $A$ is a right hereditary ring, it is a left hereditary as well, by the Auslander theorem [103, Theorem 4.1.20].

So $A$ is a prime hereditary Noetherian semiperfect ring. Therefore, by the Michler theorem [103, theorem 8.3.9], $A$ is Morita equivalent to either to a division ring or a ring of the form $H_n(O)$, where $O$ is a discrete valuation ring. $\square$

**Lemma 2.4.10.** *Let $O$ be a discrete valuation ring with unique maximal ideal $M$, and let $X$ be a serial left (resp. right) $O$-module such that the inclusion $MX \subset X$ (resp. $XM \subset X$) is strong. Then there is an element $x \in X$ such that $X = Ox$ (resp. $X = xO$).*

*Proof.* Let $x \in X \setminus MX$. The submodule $Ox \subseteq X$ is not contained in $MX$. Since the module $X$ is serial, the quotient module $X/MX$ is simple. Therefore $Ox = X$. $\square$

**Lemma 2.4.11.** *([126, Lemma 3.11]). Any indecomposable right hereditary SPSD-ring with a two-vertex quiver is a serial ring and is Morita equivalent to one of the following rings:*

(a) $T_2(D) = \begin{pmatrix} D & D \\ 0 & D \end{pmatrix}$.

(b) $H_2(O) = \begin{pmatrix} O & O \\ M & O \end{pmatrix}$.

(c) $H(O, 1, 1) = \begin{pmatrix} O & D \\ 0 & D \end{pmatrix}$,

*where $O$ is a discrete valuation ring with Jacobson radical $M$ and a classical ring of fractions which is a division ring $D$.*

*Proof.* Without loss of generality one can assume that $A$ is a basic ring. If $A$ is an indecomposable prime ring with a two-vertex quiver, then it is Morita equivalent to the ring $H_2(O)$, by theorem 2.4.9. Therefore one can assume that $A$ is not prime. Let $1 = e_1 + e_2$ be a decomposition of the identity into a sum of pairwise orthogonal local idempotents. Suppose $e_1 A e_2 \neq 0$ and $e_2 A e_1 \neq 0$. Then, by [101, theorem 6.9.2], $A$ is a weakly prime ring[5]. Therefore, by [103, corollary 8.5.10], $A$ is a prime ring.

---

[5]Recall that a ring $A$ is called **weakly prime** if the product of any two non-zero ideals not contained in the Jacobson radical of $A$ is non-zero.

This contradiction shows that $e_1Ae_2 = 0$ or $e_2Ae_1 = 0$. Without loss of generality one can assume that $e_2Ae_1 = 0$. Write $A_i = e_iAe_i$, $R_i = \operatorname{rad}A_i$, for $i = 1, 2$, and $X = e_1Ae_2 \neq 0$. By corollary 2.3.4, $A_i$ is either a division ring or a discrete valuation ring. Thus, there are only four cases.

1. $A_1$ and $A_2$ are division rings. Then, by [103, theorem 1.10.9], $X$ is a one dimensional right $A_2$-vector space and a one-dimensional left $A_1$-vector space. Therefore $A \simeq T_2(D)$.

2. $A_1 = D$ is a division ring and $A_2 = O$ is a discrete valuation ring. Then, by [103, theorem 1.10.9], $X$ is a one dimensional left $D$-vector space and a uniserial right $O$-module. By theorem 2.3.3 $A$ is a right Noetherian ring, so $X$ is a finitely generated right $O$-module, by [103, theorem 1.1.23]. Since $A$ is a right hereditary ring, $X$ is a projective right $O$-module by [103, theorem 4.7.3]. So $X \simeq O$. Then there is an element $x \in X$ such that $X = Dx = xO$. Since $O$ is a discrete valuation ring, $X$ is a torsion-free and faithful right $O$-module by [103, proposition 5.4.18], and so $D \simeq O$ which is impossible.

3. $A_1 = O$ is a discrete valuation ring and $A_2 = D$ is a division ring. Then, as before, one sees that $X$ is a one dimensional right $D$-vector space and a uniserial left $O$-module. Therefore, by [100, lemma 13.3.4], $X$ is isomorphic to a division ring of fractions $D_1$ of $O$. So $X = D_1x = xD$ for some element $x \in X$. The map $\sigma : D_1 \to D$ given by $\alpha x = x\alpha^\sigma$ for any $\alpha \in D_1$ is an isomorphism of division rings. Therefore the map $\tau : H(O, 1, 1) \to A$ given by

$$\tau\left(\begin{pmatrix} \alpha & \beta \\ 0 & \gamma \end{pmatrix}\right) = \begin{pmatrix} \alpha & \beta x \\ 0 & \gamma^\sigma \end{pmatrix}$$

for any $\begin{pmatrix} \alpha & \beta \\ 0 & \gamma \end{pmatrix} \in H(O, 1, 1)$ is an isomorphism between the rings $H(O, 1, 1)$ and $A$.

4. $A_1 = O_1$ and $A_2 = O_2$ are discrete valuation rings. Then, by [103, theorem 1.10.9], $X$ is a uniserial left $A_1$-module and a uniserial right $A_2$-module. Then

$$A = \begin{pmatrix} O_1 & X \\ 0 & O_2 \end{pmatrix}.$$

Consider the right $A$-module $I = \begin{pmatrix} 0 & X \end{pmatrix}$. Since $A$ is a right hereditary semiperfect ring, $I$ is a direct sum of indecomposable projective modules. But $X$ is uniserial, so the projective right $A$-module $I$ is indecomposable, i.e. $I$ is isomorphic to either $P_1 = e_1A$ or $P_2 = e_2A$. But $P_1e_1 \neq 0$ and $Ie_1 = 0$. Consequently, $I \simeq P_2$. Let $\Theta : P_2 \to I$ be an isomorphism. Then $\Theta(e_2^2) = \Theta(e_2) = \Theta(e_2)e_2$. Write $\Theta(e_2) = x_0$. Then

$$\begin{pmatrix} 0 & x_0 \\ 0 & 0 \end{pmatrix}\begin{pmatrix} a_1 & x \\ 0 & a_2 \end{pmatrix} = \begin{pmatrix} 0 & x_0a_2 \\ 0 & 0 \end{pmatrix},$$

i.e. $X = x_0O_2$.

Let $M_1 = \pi_1 O_1 = O_1 \pi_1$ be the unique maximal ideal of $O_1$. Let's show that $M_1 X$ is strongly contained in $X$. Assume on the contrary that $M_1 X = X$, i.e. $\pi_1 X = X$ and so $\pi_1^m X = X$ for all $m \geq 1$. Therefore $X = \pi_1^{-m} X$ for all $m \geq 1$. Let $D_1$ be the classical division ring of fractions of $O_1$. Then every element $d_1 \in D_1$ has the following form: $d_1 = \pi_1^t \varepsilon$, where $t$ is an integer and $\varepsilon$ is a unit of $O_1$. So

$$d_1 X = \pi_1^t \varepsilon X = \pi_1^t X \subset X,$$

which implies that $X$ is a left $D_1$-vector space. Denote by $[X : D_1]$ the dimension of a left $D_1$-vector space $X$ over $D_1$. Suppose $[X : D_1] \geq 2$. Let $x_1$ and $x_2$ be two linearly independent elements in $X$, i.e. $X$ contains the two-dimensional left $D_1$-vector space:

$$D_1 x_1 \oplus D_1 x_2.$$

In this case we obtain a contradiction: the left uniserial $O_1$-module $X$ contains the left decomposable $O_1$-module $O_1 x_1 \oplus O_1 x_2$. Consequently,

$$[X : D_1] = 1 \quad \text{and} \quad X = D_1 x_0 = x_0 O_2.$$

We now show that $D_1$ is isomorphic to $O_2$. Let $d_1 \in D_1$ and $d_1 \neq 0$. Since $X$ is a faithful right $O_2$-module and a faithful left $O_1$-module, $d_1 x_0 \neq 0$ and $d_1 x_0 \in X$. So $d_1 x_0 = x_0 d_1^\varphi \neq 0$ and $d_1^\varphi \neq 0$. Therefore, $\varphi : D_1 \to O_2$ is an injective map.

Let $a_2 \in O_2$ be an arbitrary non-zero element of $O_2$. Then $x_0 a_2 \neq 0$ and $x_0 a_2 \in X$. But $x_0 a_2 = d_1 x_0$. Therefore, $a_2 = d_1^\varphi$ and $\varphi$ is a surjective map.

Moreover, $(d_1 d_2) x_0 = d_1 (d_2 x_0) = d_1 (x_0 d_2^\varphi) = (d_1 x_0) d_2^\varphi = (x_0 d_1^\varphi) d_2^\varphi = x_0 (d_1 d_2)^\varphi$ and $(d_1 + d_2) x_0 = x_0 (d_1 + d_2)^\varphi$. So $(d_1 + d_2) x_0 = d_1 x_0 + d_2 x_0 = x_0 d_1^\varphi + x_0 d_2^\varphi = x_0 (d_1^\varphi + d_2^\varphi)$. Therefore,

$$(d_1 d_2)^\varphi = d_1^\varphi d_2^\varphi, \tag{2.4.12}$$

$$(d_1 + d_2)^\varphi = d_1^\varphi + d_2^\varphi, \tag{2.4.13}$$

and so $\varphi$ is an isomorphism. This is a contradiction: the discrete valuation ring $O_2$ can not be isomorphic to the division ring $D_1$. Consequently, $M_1 X$ is strongly contained in $X$ and, by lemma 2.4.10, it follows that $X = O_1 x$. Therefore $M_1^{k+1} X$ is strongly contained in $M_1^k X$.

Consider the right ideal

$$L_k = \begin{pmatrix} M_1^k & X \\ 0 & 0 \end{pmatrix}.$$

From the equalities:

$$\begin{pmatrix} \pi_1^k & 0 \\ 0 & 0 \end{pmatrix} \begin{pmatrix} O_1 & X \\ 0 & O_2 \end{pmatrix} = \begin{pmatrix} M_1^k & M_1^k X \\ 0 & 0 \end{pmatrix}$$

and

$$\begin{pmatrix} 0 & x_0 \\ 0 & 0 \end{pmatrix} \begin{pmatrix} O_1 & X \\ 0 & O_2 \end{pmatrix} = \begin{pmatrix} 0 & X \\ 0 & 0 \end{pmatrix},$$

it follows that $L_k$ is a two-generated ideal for $k \geq 1$. Since the ring $A$ is right hereditary and semiperfect, $L_k$ is a finite direct sum of indecomposable projective modules. Let

$$R = \begin{pmatrix} M_1 & X \\ 0 & M_2 \end{pmatrix}$$

be the Jacobson radical of $A$, then

$$L_k R = \begin{pmatrix} M_1^k & X \\ 0 & 0 \end{pmatrix} \begin{pmatrix} M_1 & X \\ 0 & M_2 \end{pmatrix} = \begin{pmatrix} M_1^{k+1} & XM_2 \\ 0 & 0 \end{pmatrix}.$$

Therefore, $L_k / L_k R \simeq U_1 \oplus U_2$, where $U_1 \simeq P_1 / P_1 R$ and $U_2 \simeq P_2 / P_2 R$ ($P_1 = e_1 A$, $P_2 = e_2 A$). So, the projective cover $P(L_k) \simeq P_1 \oplus P_2$. Since $A$ is a right hereditary semiperfect ring, $L_k \simeq P(L_k) \simeq P_1 \oplus P_2$. But if $\psi : P_1 \oplus P_2 \to L_k$ is an epimorphism then

$$\psi(P_1) = \begin{pmatrix} \pi_1^k O_1 & \pi_1^k X \\ 0 & 0 \end{pmatrix}$$

and

$$\psi(P_2) = \begin{pmatrix} 0 & X \\ 0 & 0 \end{pmatrix}.$$

Consequently,

$$\psi(P_1) \cap \psi(P_2) = \begin{pmatrix} 0 & \pi_1^k X \\ 0 & 0 \end{pmatrix}.$$

This is a contradiction, since $A$ is a right hereditary ring. So case 4 is impossible.

Note that the rings $T_2(D)$, $H_2(O)$ and $H(O, 1, 1)$ are serial, by [103, theorem 1.10.7]. □

**Remark 2.4.14.** The case c of lemma 2.4.11 shows that the Chatter theorem [103, theorem 8.7.6] does not hold in general for right hereditary right Noetherian rings.

**Corollary 2.4.15.** *Any indecomposable hereditary SPSD-ring with a two-pointed quiver is a serial Noetherian ring and is Morita equivalent to either an Artinian ring $T_2(D)$ or a tiled order $H_2(O)$.*

**Theorem 2.4.16.** (V.V. Kirichenko [127, Theorem 3.20]). *Every hereditary SPSD-ring A decomposes into a finite direct product of rings each of which is Morita equivalent to either a ring of the form $H_m(O)$ or an Artinian ring of the form $T(S, D)$, where D is a division ring, and the diagram of S contains no rhombuses. Conversely, such rings exhaust all hereditary SPSD-rings up to Morita equivalence.*

*Proof.* Without loss of generality one can assume that $A$ is an indecomposable basic ring. So, one can assume that $A_A = P_1 \oplus \ldots \oplus P_n$ is a decomposition of $A_A$ into a direct sum of the indecomposable right projective $A$-modules, where $n \geq 2$. If $A$ is local, then by [100, corollary 14.4.11] this ring is either a division ring or a discrete valuation ring. Let $1 = e_1 + \ldots + e_n$ be the corresponding decomposition of $1 \in A$ into a sum of pairwise orthogonal idempotents, i.e. $P_i = e_i A$ for $i = 1, \ldots, n$.

Let $A = \bigoplus_{i,j}^{n} A_{ij}$ be the corresponding two-sided Peirce decomposition, i.e.
$e_i A e_j = A_{ij}$ $(i, j = 1, \ldots, n)$. Since $A$ is a right Noetherian semiperfect hereditary ring, it is primely triangular. So the idempotents $e_i$ can be labeled in such a way that $e_i A e_j = 0$ for $i > j$. Since all the $e_i A e_i$ are local rings, there are the following cases.

1. All $A_{ii}$ are division rings, for $i = 1, \ldots, n$. Then $R = \operatorname{rad} A$ is equal to the prime radical and so it is nilpotent, by [103, proposition 8.5.30]. Therefore $A$ is a semiprimary ring. By [103, theorem 1.10.13], $A$ is Artinian and by theorem 2.4.7 $A$ is Morita equivalent to a ring of the form $T(S, D)$, where the diagram of the finite poset $S$ contains no rhombuses.
2. All $A_{ii}$ are discrete valuation rings, $i = 1, \ldots, n$. Then from the indecomposability of $A$ and corollary 2.4.15 it follows that $A_{ij} \neq 0$. Therefore $A$ is a weakly prime ring, by [101, theorem 6.9.2]. Then, by [103, corollary 8.5.10] and theorem 2.4.9, $A$ is isomorphic to a ring of the form $H_m(O)$, where $O$ is a discrete valuation ring.
3. After a suitable renumbering (if needed) there exists a $k$ $(1 \leq k < n)$ such that all rings $e_i A e_i$ are discrete valuation rings for $1 \leq i \leq k$ and the $e_j A e_j$ are division rings for $k + 1 \leq i \leq n$. We now show that $A_{ji} = 0$ for all $k + 1 \leq j \leq n$ and $1 \leq i \leq n$. Suppose that there exists an $s$ such that $k + 1 \leq s \leq n$ and an $t$ such that $1 \leq t \leq k$ for which $A_{st} \neq 0$. Consider the following minor of the second degree of the ring $A$:

$$B_{ts} = \begin{pmatrix} A_{tt} & A_{ts} \\ A_{st} & A_{ss} \end{pmatrix}. \tag{2.4.17}$$

Since $B_{ts}$ is a hereditary $SPSD$-ring by proposition 2.3.6, the form (2.4.17) contradicts corollary 2.4.15. The same arguments show that all $A_{ji}$ for $k + 1 \leq j \leq s$ and $1 \leq i \leq k$ are equal to zero. This contradicts the indecomposablity of the ring $A$. $\square$

**Corollary 2.4.18.** ([126, Corollary 3.12]). *The two-sided Peirce decomposition of an indecomposable right hereditary basic SPSD-ring A has the following form*:

$$A = \begin{pmatrix} H_{n_1}(O_1) & \ldots & O & M_1 \\ \vdots & \ddots & \vdots & \vdots \\ O & \ldots & H_{n_k}(O_k) & M_k \\ \hline O & \ldots & O & B \end{pmatrix} \tag{2.4.19}$$

*where $M_i$ is a $(H_{n_i}(O_i), B)$-bimodule, and the $O_i$ are discrete valuations rings with a common classical division ring of fractions $D$ for $i = 1, 2, \ldots, k$. Moreover, $B$ is an indecomposable hereditary Artinian ring of the form $T(S, D)$, where $S$ is a finite poset whose diagram does not contain rhombuses.*

*This two-sided Peirce decomposition of $A$ is formed via a decomposition of the identity of $A$ into a sum of pairwise orthogonal idempotents $1 = f_1 + \ldots + f_k + f_{k+1} + \ldots + f_{k+r}$ which correspond to the vertices of the prime quiver $PQ(A)$.*

*Moreover, for any local idempotent e among the $f_i$ ($i = 1, \ldots, k$) $eAf_j$ ($j = k+1, \ldots, k+r$) is either 0 or D.*

*Proof.* Since any right hereditary SPSD-ring $A$ is an FDD-ring, the prime quiver of $A$ is connected if and only if $A$ is an indecomposable ring, by [100, theorem 11.7.3].

Suppose that in the prime quiver $PQ(A)$ there is an arrow which connects the vertex $i$ corresponding to a ring $H_{n_i}(O_i)$ with the vertex $j$ corresponding to $H_{n_j}(O_j)$. Then the two-sided Peirce decomposition of $A$ has a minor of the form

$$\begin{pmatrix} O_i & X_{ij} \\ 0 & O_j \end{pmatrix},$$

where $X_{ij} \neq 0$, which is impossible by lemma 2.4.11. Analogously there is no arrow which connects the vertex corresponding to the division ring $D$ with a vertex corresponding to a ring $H_{n_i}(O_i)$, by the same lemma. Therefore all vertices in the prime quiver $PQ(A)$ corresponding to the rings $H_{n_i}(O_i)$ are sources.

All the remaining statements of the corollary can be obtained in a similar way from lemma 2.4.11 and theorem 2.4.16. □

**Theorem 2.4.20.** [126, Theorem 3.8]. *The prime quiver $PQ(A)$ of a right hereditary SPSD-ring $A$ is an acyclic simply laced quiver such that any two vertices of it are connected by only one path. Moreover, $PQ(A)$ coincides with the Hasse diagram of some finite poset $S$ with weights $H_{n_i}(O_i)$ or D, where the $O_i$ are discrete valuation rings with a common classical division ring of fractions D, and all points with weights $H_{n_i}(O_i)$ are sources in $PQ(A)$ and correspond to the minimal elements of S ($i = 1, \ldots, k$). Conversely, if a quiver $\Gamma$ is an acyclic simply laced quiver such that any two vertices of it are connected by only one path, then there is a right hereditary SPSD-ring $A$ such that $PQ(A) = \Gamma$.*

*Proof.* By [103, proposition 8.5.25], the prime quiver $PQ(A)$ is an acyclic simply laced quiver. As has been shown in the proof of corollary 2.4.18 all vertices corresponding to the rings $H_{n_i}(O_i)$ are sources.

We proceed to show now that $PQ(A)$ has no extra arrows. Assume that $PQ(A)$ contains an extra arrow. Then the two-sided Peirce decomposition of $A$ has the minor which is of one the following forms:

1. $B = \begin{pmatrix} O & B_{12} & B_{13} \\ 0 & D & B_{23} \\ 0 & 0 & D \end{pmatrix}$,

where $O$ is a discrete valuation ring with classical division ring of fractions $D$, and $B_{12}, B_{13}, B_{23}$ are one-dimensional left and right $D$-vector spaces, and the prime quiver of $B$ is of the form:

$\qquad$ 1 $\qquad$ 2 $\qquad$ 3

$$2.\ C = \begin{pmatrix} D & C_{12} & C_{13} \\ 0 & D & C_{23} \\ 0 & 0 & D \end{pmatrix},$$

where $D$ is a division ring, and $C_{12}, C_{13}, C_{23}$ are one-dimensional left and right $D$-vector spaces, and the prime quiver of $C$ is of the form:

In case 1 the prime radical of $B$ has the form:

$$I = \begin{pmatrix} 0 & B_{12} & B_{13} \\ 0 & 0 & B_{23} \\ 0 & 0 & 0 \end{pmatrix},$$

and therefore

$$I^2 = \begin{pmatrix} 0 & 0 & B_{12}B_{23} \\ 0 & 0 & 0 \\ 0 & 0 & 0 \end{pmatrix}.$$

Since $B$ is a piecewise domain, $B_{12}B_{23} \neq 0$. Since $B_{13}$ is a one-dimensional vector space and $B_{12}B_{23} \subseteq B_{13}$, it follows that $B_{12}B_{23} = B_{13}$. This shows that in the prime quiver $PQ(B)$ there is no arrow from 1 to 3.

Case 2 can be handled analogously.

By [100, proposition 11.3.11], $PQ(A)$ is the Hasse diagram of a finite poset.

Let's show that in the prime quiver $PQ(A)$ any two vertices are connected only by one path. If there are two vertices which are connected by two different pathes, then the two-sided Peirce decomposition of $A$ contains one of the following minors

$$B = \begin{pmatrix} D & B_{12} & B_{13} & B_{14} \\ 0 & D & 0 & B_{24} \\ 0 & 0 & D & B_{34} \\ 0 & 0 & 0 & D \end{pmatrix} \quad \text{or} \quad C = \begin{pmatrix} O & C_{12} & C_{13} & C_{14} \\ 0 & D & 0 & C_{24} \\ 0 & 0 & D & C_{34} \\ 0 & 0 & 0 & D \end{pmatrix},$$

where the $B_{ij}$ and $C_{ij}$ are one-dimensional right and left $D$-vector spaces for $i, j = 1, 2, 3, 4$.

In both cases these minors are not right hereditary. So $A$ is not right hereditary as well. A contradiction.

Thus, $PQ(A)$ is an acyclic simply laced quiver without extra arrows which does not contain rhombuses of the form (1.1.5) or subquivers of the form (1.3.15). Therefore $PQ(S)$ is the Hasse diagram of some finite poset without rhombuses and subdiagrams of the form (1.3.15).

The last statement of the theorem follows from theorem 1.3.16 because the quiver $\Gamma$ has no multiple arrows. □

**Corollary 2.4.21.** *Any indecomposable right hereditary basic SPSD-ring is exactly a ring of the form $A(O, S)$ for some finite poset $S$ whose diagram does not contain rhombuses or subdiagrams of the form* (1.3.15).

## 2.5 Semihereditary SPSD-Rings

Note that by [103, proposition 4.6.11] a semiperfect ring is right semihereditary if and only if it is left semihereditary. Therefore in the case of SPSD-rings we can talk simply about semihereditary rings.

**Lemma 2.5.1.** *A right uniserial semiprime ring $O$ is prime.*

*Proof.* Let $L$ and $N$ be non-zero two-sided ideals of $O$ and $LN = 0$. One can assume that $N \subseteq L$. Therefore $N^2 = 0$, which implies that $N = 0$, since $O$ is semiprime. Consequently, $O$ is a prime ring. $\square$

**Proposition 2.5.2.** (V.V. Kirichenko [127, Corollary 3.9]). *A semiprime SPSD-ring $A$ decomposes into a finite direct product of prime rings.*

*Proof.* Let $A$ be a semiprime SPSD-ring with a decomposition of $1 = e_1 + \ldots + e_n \in A$ into a sum of pairwise orthogonal local idempotents. Then, by [103, theorem 1.10.9] and [100, proposition 9.2.13], all the rings $e_i A e_i$ are uniserial semiprime, $i = 1, 2, \ldots, n$. So, by lemma 2.5.1, they are prime and our assertion is true by [103, theorem 6.4.1]. $\square$

**Theorem 2.5.3.** (V.V. Kirichenko [127, Theorem 3.11]). *A prime semihereditary SPSD-ring $A$ is serial.*

*Proof.* Without loss of generality, one can assume that $A$ is a basic ring. Suppose that $A$ is not right serial. So assume that a principal right $A$-module $P = eA$ is not uniserial. Then there exist two cyclic submodules $I$, $J$ of $P$ such that both $I \not\subseteq J$ and $J \not\subseteq I$. Since $A$ is a prime semidistributive ring, every principal right $A$-module is uniform, by lemma 2.1.13(2). Therefore $I \cap J \neq 0$. Thus, $I + J$ is a two-generated indecomposable right $A$-module. Since $A$ is right semihereditary $I + J$ is a projective right $A$-module. Since $A$ is a semiperfect ring, any indecomposable projective module is isomorphic to a principal module, i.e. is cyclic. A contradiction.

In a similar way one can show that $A$ is a left serial ring. $\square$

**Corollary 2.5.4.** *A weakly prime right semihereditary SPSD-ring is a serial ring.*

*Proof.* Since every weakly prime right semihereditary semiperfect ring is prime, by [103, corollary 8.5.10], the proposition follows immediately from theorem 2.5.3. $\square$

**Proposition 2.5.5.** (V.V. Kirichenko [127, Proposition 3.14]). *Every indecomposable semihereditary SPSD-ring $A$ whose identity decomposes into a sum of two pairwise orthogonal local idempotents is a serial ring.*

*Proof.* Let $A$ be an indecomposable semihereditary SPSD-ring with $1 = e_1 + e_2$, where $e_1, e_2$ are orthogonal local idempotents. Denote $A_i = e_i A e_i$ $(i = 1, 2)$, $X = e_1 A e_2$, $Y = e_2 A e_1$.

If $X$ and $Y$ are simultaneously non-zero, then the ring is weakly prime, by [101, theorem 6.9.2]. Therefore it is a serial ring, by corollary 2.5.4.

Suppose that $X$ or $Y$ is zero. Since $A$ is an indecomposable ring, either $X$ or $Y$ must be non-zero. Without loss of generality one can assume that $X \neq 0$ and $Y = 0$. Let us prove that $\alpha X = X$ for every $\alpha \in A_1$, $\alpha \neq 0$. Suppose that there exists $\alpha \in A_1$ such that $\alpha X \neq X$. Since $X$ is a left $A_1$-module, $\alpha X \subset X$. Let $x_0 \in X \setminus \alpha X$, then $x_0 A_2 \subset X$. Since $X$ is a uniserial right $A_2$-module by [103, theorem 1.10.9], there must be a strict inclusion $\alpha X \subset x_0 A_2$. Then the right ideal

$$M = \begin{pmatrix} \alpha A_1 & \alpha X \end{pmatrix} + \begin{pmatrix} 0 & x_0 A_2 \end{pmatrix} = \begin{pmatrix} \alpha A_1 & x_0 A_2 \end{pmatrix}$$

is generated by two generators $a = \begin{pmatrix} \alpha & 0 \end{pmatrix}$ and $b = \begin{pmatrix} 0 & x_0 \end{pmatrix}$. On the other hand, $M$ is indecomposable since $aA \cap bA = \begin{pmatrix} 0 & x_0 A_2 \end{pmatrix} \neq 0$. Since $A$ is a right semihereditary ring, $M$ is a projective right $A$-module. Since $A$ is a semiperfect ring, any indecomposable projective module is isomorphic to a principal module, i.e. is cyclic. This gives a contradiction. This contradiction shows that $\alpha X = X$ for every $\alpha \in A_1$. Analogously one can show that $X = X\beta$ for every $\beta \in A_2$, since $A$ is also a left semihereditary ring.

We now show that the ring $A$ is serial, i.e. all the modules $P_1 = e_1 A = \begin{pmatrix} A_1 & X \end{pmatrix}$, $P_2 = \begin{pmatrix} 0 & A_2 \end{pmatrix}$, $Q_1 = \begin{pmatrix} A_1 \\ 0 \end{pmatrix}$, $Q_2 = \begin{pmatrix} X \\ A_2 \end{pmatrix}$ are uniserial. Since $A$ is a SPSD-ring, the modules $Q_1$ and $P_2$ are uniserial, by [103, theorem 1.10.9].

Consider the module $P_1$. Let $K_1$ and $K_2$ be submodules of the module $P_1$. Since $A_1$ is a uniserial ring, either $K_2 e_1 \subset K_1 e_1$ or $K_1 e_1 \subset K_2 e_1$. Suppose $K_2 e_1 \subset K_1 e_1$. If $K_1 e_1 \neq 0$, then, as has been shown above, $K_1 = \begin{pmatrix} K_1 e_1 & X \end{pmatrix}$ and $K_2 = \begin{pmatrix} K_2 e_1 & X \end{pmatrix}$. Thus $K_2 \subset K_1$. If $K_1 e_1 = 0$ then $K_1$ and $K_2$ are submodules of the uniserial $A_2$-module $X$. So $P_1$ is a uniserial module. Analogously one can show that $Q_2$ is a uniserial module. $\square$

Since any serial semihereditary ring has a classical ring of fractions of the form (9.2.16) [103], the following corollary follows from proposition 2.5.5.

**Corollary 2.5.6.** *Let $A$ be as in proposition 2.5.5. Then $A$ has a classical ring of fractions $\widetilde{A}$, and either $\widetilde{A} \simeq M_2(D)$ or $\widetilde{A} \simeq T_2(D)$. Moreover, in the last case $e_{11} A e_{22} = e_{11} T_2(D) e_{22}$.*

**Remark 2.5.7.** Obviously, a division ring of fractions $D$ of a uniserial semihereditary non-zero ring $O$ is an infinitely generated left and right module over $O$.

**Proposition 2.5.8.** *A semihereditary SPSD-ring is finite dimensional. In particular, u.dim $A_A < \infty$.*

*Proof.* Assume the contrary, i.e. that there is a semihereditary SPSD-ring $A$ which is infinite dimensional. Then $A$ contains an infinite direct sum of non-zero ideals, and so it contains an infinite direct sum of principal right ideals. Since $A$ is a semihereditary ring, all these ideals are projective. Since $A$ is a semiperfect ring, any projective right $A$-module is a direct sum of principal right $A$-modules. So we obtain that $A$ contains an ideal which is isomorphic to an infinite sum of principal right $A$-modules. Since $A$ is semiperfect, the number of distinct principal $A$-modules is finite. So in this infinite sum there exist at least two identical summands, which contradicts that $A$ is semidistributive, by the Camillo theorem [100, theorem 14.1.5]. Thus, $A$ is finite dimensional. In particular, u.dim $A_A < \infty$. $\square$

Let $A$ be a semihereditary SPSD-ring. By [103, corollary 8.5.7], $A$ is a primely triangular ring, i.e. there is a decomposition $1 = f_1 + \ldots + f_n$ of the identity of $A$ into a sum of pairwise orthogonal idempotents such that the two-sided Peirce decomposition of $A$ has the upper triangular matrix form:

$$A = \begin{pmatrix} A_{11} & A_{12} & \ldots & A_{1n} \\ 0 & A_{22} & \ldots & A_{2n} \\ \vdots & \vdots & \ddots & \vdots \\ 0 & 0 & \ldots & A_{nn} \end{pmatrix}, \tag{2.5.9}$$

where the $A_{ii}$ are prime rings and $A_{ij} = 0$ for all $i > j$, $i, j = 1, \ldots, n$. Then the prime radical $N = Pr(A)$ has the following form:

$$N = \begin{pmatrix} 0 & A_{12} & \ldots & A_{1n} \\ 0 & 0 & \ldots & A_{2n} \\ \vdots & \vdots & \ddots & \vdots \\ 0 & 0 & \ldots & 0 \end{pmatrix}. \tag{2.5.10}$$

Therefore

$$A = A_0 \oplus N \tag{2.5.11}$$

as a direct sum of two Abelian subgroups, where $A_0 \cong A/N$ is the diagonal of $A$. By theorem 2.5.3, $A_0 \simeq A/N$ is a serial semihereditary ring.

**Theorem 2.5.12.** (V.V. Kirichenko [127, Theorem 3.21]). *A semihereditary SPSD-ring $A$ has a classical ring of fractions $\widetilde{A}$ which is of the form $\widetilde{A} = \widetilde{A}_0 \oplus N$, where $\widetilde{A}_0$ is the classical ring of fractions of the serial semihereditary ring $A_0$.*

*Proof.* Since $A_0 \simeq A/I$ is a finite direct sum of prime semihereditary SPSD-rings, $A_0$ is a finite direct product of serial prime semihereditary semidistributive rings by theorem 2.5.3. Therefore, by [103, lemma 9.2.14], $A_0$ has a classical ring of fractions $\widetilde{A}_0$ which is semisimple.

Since u.dim $A < \infty$, by proposition 2.5.8, $A$ satisfies all conditions of [103, theorem 8.5.31]. Therefore $A$ has an Artinian classical ring of fractions $\widetilde{A}$.

The decomposition $\widetilde{A} = \widetilde{A}_0 \oplus N$ follows from corollary 2.5.6. This proves the result. $\square$

From this theorem it is easy to obtain the following corollaries.

**Corollary 2.5.13.** (V.V. Kirichenko [127, Corollary 3.22]). *The classical ring of fractions $\widetilde{A}$ of a semihereditary SPSD-ring A is an Artinian hereditary SPSD-ring and the prime radical N of A coincides with the Jacobson radical of the ring $\widetilde{A}$.*

**Corollary 2.5.14.** (V.V. Kirichenko [127, Corollary 3.23]). *The prime quiver PQ(A) of a semihereditary SPSD-ring A coincides with the quiver $Q(\widetilde{A})$ of the Artinian hereditary SPSD-ring $\widetilde{A}$.*

**Corollary 2.5.15.** (V.V. Kirichenko [127, Corollary 3.24]). *The prime quiver PQ(A) of a semihereditary SPSD-ring A coincides with the diagram of a finite poset without rhombuses.*

**Example 2.5.16.** Let $O$ be a discrete valuation ring with unique maximal ideal $M = \pi O = O\pi$. Consider the ring

$$
A = \begin{pmatrix} O & O & O \\ 0 & O & \pi O \\ 0 & 0 & O \end{pmatrix}.
$$

Clearly, the ring $A$ is a piecewise domain which is an SPSD-ring, but it is not semihereditary. Let $N$ be the prime radical of $A$. Then

$$
N = \begin{pmatrix} 0 & O & O \\ 0 & 0 & \pi O \\ 0 & 0 & 0 \end{pmatrix}, \quad N^2 = \begin{pmatrix} 0 & 0 & \pi O \\ 0 & 0 & 0 \\ 0 & 0 & 0 \end{pmatrix} \text{ and } N/N^2 = \begin{pmatrix} 0 & O & O/\pi O \\ 0 & 0 & \pi O \\ 0 & 0 & 0 \end{pmatrix}.
$$

Therefore

$$
PQ(A) = \underset{1}{\bullet} \longrightarrow \underset{2}{\bullet} \longrightarrow \underset{3}{\bullet}
$$

Thus, the condition that the prime quiver of a piecewise SPSD-domain is the diagram of a finite poset seems to be necessary for it to be a semihereditary SPSD-ring.

**Definition 2.5.17.** A semiperfect ring $A$ is called a **prime block** if its prime quiver $PQ(A)$ is connected.

Obviously, a prime block is an indecomposable ring. So from [100, theorem 2.4.11] and [100, corollary 11.4.4] one immediately obtains the following result:

**Theorem 2.5.18.** [89, Theorem 3.11]. *A semihereditary SPSD-ring uniquely decomposes into a finite direct product of prime blocks.*

Recall that the quotient ring $\overline{A} = A/N$, where $N$ is a prime radical of a ring $A$, is called the **diagonal** of $A$.

Let $A$ be a semihereditary SPSD-ring. Then the two-sided Peirce decomposition of $N$ has the form (2.5.9). If $A$ has a Noetherian diagonal, then in the decomposition (2.5.9) all the $A_{ii}$ are prime semihereditary Noetherian SPSD-rings, and so each minor of the first order is either a division ring or a discrete valuation ring, by [100, corollary 14.4.11].

**Lemma 2.5.19.** *Let A be a semihereditary SPSD-ring with Noetherian diagonal, and e an idempotent of A. Then eAe is also a semihereditary SPSD-ring with Noetherian diagonal.*

*Proof.* The proof follows immediately from what was said above, [100, proposition 14.4.7] and [103, theorem 8.2.16]. □

## 2.6 Notes and References

Commutative distributive rings are also called arithmetical rings. The properties of these rings were studied by C.U. Jensen in [111], [112], [113]. The first results which are devoted to noncommutative distributive rings were obtained in the fifties-sixties of the 20-th century by R.I. Blair [29], E.A. Behrens [26], [27], and W. Menzel [155], [156].

The systematic study of distributive modules and noncommutative distributive rings began with the papers of W. Stephenson [203], H. Achkar [2], V. Camillo [39], and H.H. Brungs [33], [34], [35], [36]. W. Stephenson obtained some basic properties of distributive modules in terms of homomorphisms of factor modules and submodules (in particular proposition 2.1.11 and corollary 2.1.12). He also characterized the classes of perfect and semiperfect distributive rings (e.g. lemma 2.1.30, corollary 2.1.31, theorem 2.1.32) and their connections with Artinian rings and valuation rings. He also studied invariance properties of distributive modules. In particular, he proved that every submodule of an Artinian or Noetherian distributive module is fully invariant [6]. From this result it follows that any right ideal of a right Noetherian distributive ring is a two-sided ideal, i.e. a right Noetherian distributive ring is a right duo ring [203, Corollary 3]. For right Noetherian right valuation rings this result was obtained by H.H.Brungs.

V. Camillo in [39] studied the properties of semiprime Noetherian distributive rings. The first basic result in his paper [39, Theorem 1] gives a characterization of distributive modules in terms of the socles of their quotients (see also [100, theorem 14.1.5]). In this paper he also showed that a ring which is both a semiprime right distributive right Goldie ring and a duo ring is a product of a finite number of domains. The main result of his paper [39, theorem 6] describes the structure of semiprime Noetherian distributive rings. This result for the case of one-sided Noetherian and one-sided distributed rings was generalized by A.A. Tuganbaev in [206], [207], [208].

Various properties and the structure of semidistributive rings and their properties are studied by C.E. Caballero [38], R. Mazurek and E.R. Puczylowski [150], and others. Two books of A. Tuganbaev [213] and [214] are devoted to the study of distributive and semidistributive rings. These books contain a large amount of information and an extensive bibliography on this theme. The results and last word

---

[6] A submodule $N$ of a module $M$ is called **fully invariant** if $f(N) \subseteq N$ for all $f \in \mathrm{End}(M)$.

on distributive and semidistributive modules and ring are also most fully present in the papers of A.A. Tuganbaev [215], [216].

Properties and the structure of semiperfect semidistributive rings were studied by V.V. Kirichenko and others in [125], [126], [127], [128]. Semihereditary semidistributive rings were considered in [89].

# CHAPTER 3

# The Group of Extensions

The more fundamental notions and results of homological algebra were discussed in various sections of [100], [101]. This chapter deals with some additional matters from homological algebra.

The notions of direct limits and inverse limits for a 'poset of modules' were considered in [103, section 1.4] (see also [100, section 4.7]). In the particular case, when this set has only three modules arranged in a particular way these constructions have their own names: pullback and pushout. They are very important and useful in studying rings and modules. Therefore, for the convenience of the reader these module constructions and their main properties are considered in more detail in section 3.1.

In the theory of homological algebra and its applications there is a very important result which is often known as the "snake lemma" (it is also called "zig-zag lemma", or "serpent lemma"). It holds in every Abelian category and is an important tool in the construction of long exact sequences. This lemma consists of two parts: 1) the construction of an exact sequence, which is often called "kernel-cokernel sequence", for any commutative diagram of a special type; and 2) the construction of long exact sequences of homology groups for any given short exact sequence of complexes. This second part of the lemma was proved in [100, theorem 6.1.1]. The proof of the first part of the lemma is given in section 3.2.

The functors $\text{Ext}^n$ as right derived functors for the contravariant left exact functor Hom using projective resolutions were introduced in [100, section 6.4]. These functors are closely related to module extensions.

Let $A$ be a ring, and let $X, Y \in \text{Mod}_r A$. In this chapter an interpretation of the group $\text{Ext}^1_A(Y, X)$ is given in terms of short exact sequences. Section 3.3 is devoted to studying extensions of modules in terms of short exact sequences.

Following R. Baer the addition of extensions of modules, which makes the set of equivalence classes of all extensions an Abelian group, is introduced in section 3.4. Some main properties of the group $\text{Ext}^1_A(Y, X)$ are considered in section 3.5. The results of section 3.6 show that there is an isomorphism between equivalence classes of the group of extensions of $X$ by $Y$ and elements of $\text{Ext}^1_A(Y, X)$ as seen in [100].

## 3.1 Module Constructions Pushout and Pullback

This section contains two useful constructions for a pair of homomorphisms in $\text{Mod}_r A$.

**Definition 3.1.1.** Let $X, Y, Z \in \text{Mod}_r A$. A **pullback** (also called a **fibered product** or a **cartesian square**) for a pair of $A$-homomorphisms $\alpha : X \to Z$ and $\beta : Y \to Z$ is a triple $(T, \gamma, \delta)$, where $T \in \text{Mod}_r A$, $\gamma : T \to Y$, $\delta : T \to X$ such that $\beta\gamma = \alpha\delta$, i.e. the diagram

$$
\begin{array}{ccc}
T & \xrightarrow{\delta} & X \\
\gamma \downarrow & & \downarrow \alpha \\
Y & \xrightarrow{\beta} & Z
\end{array}
$$

**Diag. 3.1.2.**

is commutative. Moreover, it is required that the pullback $(T, \gamma, \delta)$ is universal with respect to such diagrams. That is, if there is any other module $U$ with homomorphisms $f : U \to X$ and $g : U \to Y$ such that $\alpha f = \beta g$ then there exists a unique homomorphism $\varphi : U \to T$ such that the following diagram

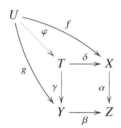

**Diag. 3.1.3.**

is commutative.

**Lemma 3.1.4.** *For any pair of homomorphisms $\alpha$, $\beta$ the pullback $(T, \gamma, \delta)$ as specified in definition 3.1.1 is unique up to isomorphism.*

*Proof.* Suppose there is another pullback $(T_1, \gamma_1, \delta_1)$ for the pair of homomorphisms $(\alpha, \beta)$. Then, by definition 3.1.1, there exists a homomorphism $\varphi : T_1 \to T$ such that $\delta\varphi = \delta_1$, $\gamma\varphi = \gamma_1$, and there exists a homomorphism $\varphi_1 : T \to T_1$ such that $\delta_1\varphi_1 = \delta$, $\gamma_1\varphi_1 = \gamma$. Therefore $\delta = \delta_1\varphi_1 = \delta\varphi\varphi_1$ and $\gamma = \gamma_1\varphi_1 = \gamma\varphi\varphi_1$. From the uniqueness property of the pullback $(T, \gamma, \delta)$ it follows that $\varphi\varphi_1 = 1_T$. Analogously one sees that $\gamma_1 = \gamma_1\varphi_1\varphi$ and $\delta_1 = \delta_1\varphi_1\varphi$. And therefore $\varphi_1\varphi = 1_{T_1}$. This shows that $\varphi$ and $\varphi_1$ are mutually inverse isomorphisms. $\square$

**Lemma 3.1.5.** *For any pair of $A$-homomorphisms $\alpha : X \to Z$ and $\beta : Y \to Z$ there exists a pullback $(T, \gamma, \delta)$.*

*Proof.* We set $T = \{(x, y) \in X \oplus Y \mid \alpha x = \beta y\}$ and define $\gamma$ and $\delta$ to be the restrictions of the canonical projections $X \oplus Y \longrightarrow Y$, $X \oplus Y \longrightarrow X$ to $T$: $\gamma(x, y) = y$ and $\delta(x, y) = x$. Suppose there is another triple $(U, f, g)$ with $\alpha f = \beta g$. Then one can define a morphism $\varphi : U \to T$ by setting $\varphi(u) = (f(u), g(u)) \in T$, since $\alpha f u = \beta g u$. We now show that the morphism $\varphi$ is unique. Assume there is another morphism $\xi : U \to T$ such that $\delta\xi = f$ and $\gamma\xi = g$. Suppose $\xi(u) = (x, y) \in T$. Then $\delta\xi(u) = x$, $\gamma\xi(u) = y$. Since $\delta\xi = f$ and $\gamma\xi = g$, it follows that $x = f(u)$ and $y = g(u)$, i.e. $\xi(u) = (f(u), g(u))$, and so $\xi = \varphi$, as required. $\square$

**Remarks 3.1.6.**

1. The pullback for a pair of $A$-homomorphisms $\alpha : X \to Z$ and $\beta : Y \to Z$ is the inverse limit of the diagram

2. (Terminological comment). Consider a pullback diagram 3.1.2. Then $\gamma$ is said to be the pullback of $\alpha$ along $\beta$ and $\delta$ is the pullback of $\beta$ along $\alpha$. This also explains the terminology "pullback".

Some main properties of pullbacks are given in the following statement.

**Lemma 3.1.7.** *Let $(T, \gamma, \delta)$ be a pullback for a pair of $A$-homomorphisms $\alpha : X \to Z$ and $\beta : Y \to Z$. Then:*

1. *If $\alpha$ is an epimorphism, then $\gamma$ is an epimorphism.*
2. *If $\alpha$ is a monomorphism, then $\gamma$ is a monomorphism.*

*Proof.*

1. Suppose $\alpha$ is an epimorphism. Let $y \in Y$ and $\beta(y) = z \in Z$. Since $\alpha$ is an epimorphism, there exists an element $x \in X$ such that $z = \beta(y) = \alpha(x)$. Then $\gamma(x, y) = y$, by construction, so $\gamma$ is an epimorphism.
2. Let $U = \mathrm{Ker}\gamma \subseteq T$ and let $h : \mathrm{Ker}\gamma \to T$ be the natural injection. Then $g = \gamma h = 0$. Set $f = \delta h$. As the diagram

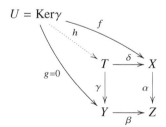

is commutative, $0 = \beta g = \alpha f$. Since $\alpha$ is a monomorphism, $f = 0$. Since $h = 0$ also makes the diagram commute, by the property of uniqueness of the pullback, it follows that $\mathrm{Ker}\gamma = 0$, that is, $\gamma$ is a monomorphism. $\square$

**Corollary 3.1.8.** *A pullback of an isomorphism is an isomorphism.*

In a dual way one can introduce the following notion.

**Definition 3.1.9.** Let $X, Y, Z \in \mathrm{Mod}_r A$. A **pushout** (also called a **fibered coproduct** or **fibered sum**, or **co-Cartesian square**) for a pair of $A$-homomorphisms $\alpha : Z \rightarrow X$ and $\beta : Z \rightarrow Y$ is a triple $(T, \gamma, \delta)$, where $T \in \mathrm{Mod}_r A$, $\gamma : Y \rightarrow T$, $\delta : X \rightarrow T$ such that $\gamma\beta = \delta\alpha$, i.e. the diagram

$$
\begin{array}{ccc}
Z & \xrightarrow{\;\beta\;} & Y \\
\alpha\downarrow & & \downarrow\gamma \\
X & \xrightarrow[\delta]{} & T
\end{array}
$$

Diag. 3.1.10.

is commutative. Moreover, it is required that the a pushout $(T, \gamma, \delta)$ is universal with respect to such diagrams. That is, if there is any other module $U$ with homomorphisms $f : Y \rightarrow U$ and $g : X \rightarrow U$ such that $g\alpha = f\beta$ then there exists a unique homomorphism $\varphi : T \rightarrow U$ such that the following diagram

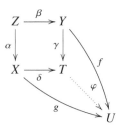

Diag. 3.1.11.

is commutative.

**Lemma 3.1.12.** *For any pair of homomorphisms* $\alpha$, $\beta$, *a pushout* $(T, \gamma, \delta)$, *as specified by definition 3.1.9, is unique up to isomorphism.*

*Proof.* Suppose there is another pushout $(T_1, \gamma_1, \delta_1)$ for a pair of homomorphisms $(\alpha, \beta)$. Then, by definition 3.1.9, there exists a homomorphism $\varphi : T \rightarrow T_1$ such that $\varphi\delta = \delta_1$, $\varphi\gamma = \gamma_1$, and there exists a homomorphism $\varphi_1 : T_1 \rightarrow T$ such that $\varphi_1\delta_1 = \delta$, $\varphi_1\gamma_1 = \gamma$. Therefore $\delta = \varphi_1\delta_1 = \varphi_1\varphi\delta_1$ and $\gamma = \varphi_1\gamma_1 = \varphi_1\varphi\gamma$. From the uniqueness property of a pushout $(T, \gamma, \delta)$ we obtain that $\varphi_1\varphi = 1_T$. Analogously, one sees that

$\gamma_1 = \varphi \varphi_1 \gamma_1$ and $\delta_1 = \varphi \varphi_1 \delta_1$. Therefore $\varphi \varphi_1 = 1_{T_1}$. This shows that $\varphi$ and $\varphi_1$ are mutually inverse isomorphisms. $\square$

**Lemma 3.1.13.** *For any pair of A-homomorphisms $\alpha : Z \to X$ and $\beta : Z \to Y$ there exists a pushout $(T, \gamma, \delta)$.*

*Proof.* Set $T = X \oplus Y / S$, where $S = \{(\alpha z, -\beta z) \mid z \in Z\}$. Write elements from $T$ as $\overline{(x, y)} = (x, y) + S$. Define $\gamma$ and $\delta$ in the following way: $\gamma(y) = \overline{(0, y)}$ and $\delta(x) = \overline{(x, 0)}$. Then $\delta \alpha(z) = \overline{(\alpha z, 0)}$ and $\gamma \beta(z) = \overline{(0, \beta z)}$. Therefore $\delta \alpha(z) = \gamma \beta(z)$, since $(\alpha z, -\beta z) \in S$.

Suppose there is a module $U$ and two homomorphisms $f, g$ such that $f \beta = g \alpha$. Then one defines a homomorphism $\varphi : T \to U$ by setting $\varphi(t) = \varphi(\overline{(x, y)}) = gx + fy \in U$. The homomorphism $\varphi$ is well defined because $\varphi(\overline{(\alpha z, -\beta z)}) = g \alpha z + f(-\beta z) = (g \alpha - f \beta) z = 0$ for $(\alpha z, -\beta z) \in S$.

It remains to show that the homomorphism $\varphi : T \to U$ is unique. Suppose there is another homomorphism $\varphi_1 : T \to U$ making the diagram 3.1.11 commute with $\varphi_1 \gamma = f$ and $\varphi_1 \delta = g$. Then

$$\varphi_1 \gamma(y) = \varphi_1(\overline{(0, y)}) = f(y) = \varphi \gamma(y)$$

$$\varphi_1 \delta(x) = \varphi_1(\overline{(x, 0)}) = g(x) = \varphi \delta(x).$$

Therefore

$$\varphi_1(\overline{(x, y)}) = \varphi_1(\gamma y + \delta x) = \varphi_1 \gamma(y) + \varphi_1 \delta(x) = f(y) + g(x) = \varphi \gamma(y) + \varphi \delta(x) =$$

$$= \varphi(\gamma y + \delta x) = \varphi(\overline{(x, y)})$$

for all $\overline{(x, y)} \in T$, which means that $\varphi$ is unique. $\square$

**Remarks 3.1.14.**

1. A pushout for a pair of A-homomorphisms $\alpha : Z \to X$ and $\beta : Z \to Y$ is a

   direct limit of the diagram
   $$\begin{array}{ccc} Z & \xrightarrow{\beta} & Y \\ {\scriptstyle \alpha} \downarrow & & \\ X & & \end{array}$$

2. (Terminological comment). In the pushout diagram 3.1.10 $\gamma$ is termed the pushout of $\alpha$ along $\beta$ and $\delta$ is the pushout of $\beta$ along $\alpha$.

**Remark 3.1.15.** The pullback and the pushout which were formed in lemma 3.1.5 and lemma 3.1.13 will be called **standard**. Because of their uniqueness by lemma 3.1.4 and 3.1.12, as a rule, all pullbacks and pushouts will be taken standard.

Some main properties of pushouts are given in the following statement.

**Lemma 3.1.16.** *Let $(T, \gamma, \delta)$ be a pushout for a pair of A-homomorphisms $\alpha : Z \to X$ and $\beta : Z \to Y$. Then:*

1. *If $\alpha$ is an epimorphism, then $\gamma$ is an epimorphism.*
2. *If $\alpha$ is a monomorphism, then $\gamma$ is a monomorphism.*

*Proof.* Without loss of generality suppose that the given pushout $(T, \gamma, \delta)$ is standard, that is, $T = X \oplus Y / S$, where $S = \{(\alpha z, -\beta z) \mid z \in Z\}$. Denote elements from $T$ by $\overline{(x, y)}$.

1. Suppose $\alpha$ is an epimorphism and $t = \overline{(x, y)} \in T$. Then there exists an element $z \in Z$ such that $x = \alpha z = -\beta z$. Therefore $\gamma(y + \beta z) = \overline{(0, y + \beta z)} = \overline{(0, y + \beta z)} + \overline{(\alpha z, -\beta z)} = \overline{(x, y)} = t$, that is, $\gamma$ is an epimorphism.
2. Suppose $\alpha$ is a monomorphism, and $\gamma(y) = 0$. Then by the construction of the standard pushout, $\gamma(y) = \overline{(0, y)}$ and there exists an element $z \in Z$ such that $y = \beta z = -\alpha z$. So $0 = \alpha(z)$. Since $\alpha$ is a monomorphism, there results that $z = 0$ and so $y = 0$, that is, $\gamma$ is a monomorphism. □

**Corollary 3.1.17.** *A pushout of an isomorphism is an isomorphism.*

In the final part of this section some applications of pullbacks and pushouts to short exact sequences will be considered.

**Lemma 3.1.18.** *Let $0 \to X \xrightarrow{f} Z \xrightarrow{g} Y \to 0$ be an exact sequence, and $\gamma : Y_1 \to Y$ a homomorphism. Then there exists a commutative diagram*

$$
\begin{array}{ccccccccc}
0 & \longrightarrow & X & \xrightarrow{f_1} & Z_1 & \xrightarrow{g_1} & Y_1 & \longrightarrow & 0 \\
& & {\scriptstyle 1_X}\Big\| & & {\scriptstyle \beta}\Big\downarrow & & {\scriptstyle \gamma}\Big\downarrow & & \\
0 & \longrightarrow & X & \xrightarrow{f} & Z & \xrightarrow{g} & Y & \longrightarrow & 0
\end{array}
$$

**Diag. 3.1.19.**

*with exact rows.*

**Remark.** Later it will be shown that the top short exact sequence of this diagram is unique up to a suitable notion of equivalence of short exact sequences.

*Proof.* For the pair of homomorphisms $(\gamma, g)$ one takes the standard pullback

$$
\begin{array}{ccc}
Z_1 & \xrightarrow{g_1} & Y_1 \\
{\scriptstyle \beta}\Big\downarrow & & {\scriptstyle \gamma}\Big\downarrow \\
Z & \xrightarrow{g} & Y
\end{array}
$$

where $Z_1 = \{(z, y_1) \in Z \oplus Y_1 \mid gz = \gamma y_1\}$, and $g_1(z, y_1) = y_1$, $\beta(z, y_1) = z$. Construct a homomorphism $f_1 : X \to Z_1$ by setting $f_1(x) = (f(x), 0)$. Then

$g_1 f_1(x) = g_1(f(x), 0) = 0$, hence $g_1 f_1 = 0$, and $\beta f_1(x) = \beta(f(x), 0) = f(x)$, hence $\beta f_1 = f$. Therefore the diagram

$$
\begin{array}{ccccc}
X & \xrightarrow{f_1} & Z_1 & \xrightarrow{g_1} & Y_1 \\
{\scriptstyle 1_X}\big\| & & {\scriptstyle \beta}\big\downarrow & & {\scriptstyle \gamma}\big\downarrow \\
0 \longrightarrow X & \xrightarrow{f} & Z & \xrightarrow{g} & Y \longrightarrow 0
\end{array}
$$

is commutative. It remains to show that the top row of this diagram is exact and that $f_1$ is a monomorphism and $g_1$ is an epimorphism. Since $g$ is an epimorphism, $g_1$ is also an epimorphism, by lemma 3.1.7. Suppose $f_1(x) = 0$. Since $\beta f_1 = f$, there results that $f(x) = \beta f_1(x) = 0$. Since $f$ is a monomorphism, $x = 0$, that is, $f_1$ is also a monomorphism. Since $g_1 f_1 = 0$, Im $f_1 \subseteq$ Ker $g_1$. So it remains to show that Ker $g_1 \subseteq$ Im $f_1$. Suppose $g_1(z_1) = 0$. Let $z_1 = (z, y_1)$, then $0 = g_1(z_1) = g_1(z, y_1) = y_1$. So $z_1 = (z, 0) \in Z_1$, therefore $g(z) = \gamma(0) = 0$, which means that $z \in$ Ker $g =$ Im $f$. So there exists $x \in X$ such that $z = f(x)$. Then $z_1 = (z, 0) = (f(x), 0) = f_1(x)$, that is, Ker $g_1 \subseteq$ Im $f_1$. Therefore one obtains the commutative diagram 3.1.19 with exact rows, as required. $\square$

**Lemma 3.1.20.** *Let* $0 \to X \xrightarrow{f} Z \xrightarrow{g} Y \to 0$ *be an exact sequence, and* $\alpha : X \to X_1$ *a homomorphism. Then there exists a commutative diagram*

$$
\begin{array}{ccccccccc}
0 \longrightarrow & X & \xrightarrow{f} & Z & \xrightarrow{g} & Y & \longrightarrow 0 \\
& {\scriptstyle \alpha}\big\downarrow & & {\scriptstyle \beta}\big\downarrow & & {\scriptstyle 1_Y}\big\| & \\
0 \longrightarrow & X_1 & \xrightarrow{f_1} & Z_1 & \xrightarrow{g_1} & Y & \longrightarrow 0
\end{array}
$$

**Diag. 3.1.21.**

*with exact rows.*

A similar uniqueness remark applies as the one above in connection with lemma 3.1.18.

*Proof.* For the pair of homomorphisms $(\alpha, f)$ construct the standard pushout

$$
\begin{array}{ccc}
X & \xrightarrow{f} & Z \\
{\scriptstyle \alpha}\big\downarrow & & {\scriptstyle \beta}\big\downarrow \\
X_1 & \xrightarrow{f_1} & Z_1
\end{array}
$$

where $Z_1 = (X_1 \oplus Z)/S$, and $S = \{(\alpha x, -fx) \mid x \in X\}$, $\beta(z) = \overline{(0, z)}$, $f_1(x_1) = \overline{(x_1, 0)}$. Define a homomorphism $g_1 : Z_1 \to Y$ by setting $g_1(\overline{(x_1, z)}) = g(z)$. This is well defined because $g_1(\overline{0}) = g_1(\overline{(\alpha x, -fx)}) = g(-fx) = -gf(x) = 0$.

Since $g_1\beta(z) = g_1(\overline{(0,z)}) = g(z)$, it follows that $g_1\beta = g$. So there results a commutative diagram

$$
\begin{array}{ccccccccc}
0 & \longrightarrow & X & \xrightarrow{\;f\;} & Z & \xrightarrow{\;g\;} & Y & \longrightarrow & 0 \\
& & \downarrow{\alpha} & & \downarrow{\beta} & & \downarrow{1_Y} & & \\
& & X_1 & \xrightarrow{\;f_1\;} & Z_1 & \xrightarrow{\;g_1\;} & Y & &
\end{array}
$$

It remains to show that the lower row of this diagram is exact, that $f_1$ is a monomorphism and that $g_1$ is an epimorphism.

Since $f$ is a monomorphism, $f_1$ is also a monomorphism, by lemma 3.1.16. Let $y \in Y$. Since $g$ is an epimorphism, there is an element $z \in Z$ such that $g(z) = y$. Since $g = g_1\beta$, it follows that $y = g(z) = g_1\beta(z) = g_1(\overline{(0,z)})$, where $\overline{(0,z)} \in Z_1$, that is, $g_1$ is an epimorphism.

Since $g_1 f_1(x_1) = g_1(\overline{(x_1, 0)}) = g(0) = 0$, one has $g_1 f_1 = 0$, i.e. $\operatorname{Im} f_1 \subseteq \operatorname{Ker} g_1$. So it remains to show that $\operatorname{Ker} g_1 \subseteq \operatorname{Im} f_1$. Suppose $g_1(z_1) = g_1(\overline{(x_1, z)}) = 0$, then $g(z) = 0$, i.e. $z \in \operatorname{Ker} g$. Since $\operatorname{Ker} g = \operatorname{Im} f$, $z = f(x)$ for some $x$. Therefore $z_1 = \overline{(x_1, f(x))} = \overline{(x_1, f(x))} + \overline{(\alpha x, -f(x))} = \overline{(x_1 + \alpha x, 0)} = f_1(x_1 + \alpha x) \in \operatorname{Im} f_1$, which means that $\operatorname{Ker} g_1 \subseteq \operatorname{Im} f_1$. Therefore one obtains the commutative diagram 3.1.21 with exact rows, as required. $\square$

Recall that for a homomorphism $f : X \to Y$ we write $\operatorname{Coker}(f) = Y/\operatorname{Im}(f)$.

**Lemma 3.1.22.** *Given a commutative diagram*

$$
\begin{array}{ccc}
Z & \xrightarrow{\;\beta\;} & Y \\
\downarrow{\alpha} & & \downarrow{\gamma} \\
X & \xrightarrow{\;\delta\;} & T
\end{array}
$$

*there is an induced commutative diagram with exact rows*

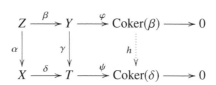

Diag. 3.1.23.

*Further, if $(T, \gamma, \delta)$ is a pushout of $(Z, \alpha, \beta)$, $h$ is an isomorphism.*

*Proof.* The morphism $h$ is defined as follows. Let $x \in \operatorname{Coker}(\beta)$ and $y \in Y$ such that $\varphi(y) = x$. Then $h(x)$ is taken to be $\psi\gamma(y)$. This is well-defined. Indeed if $y' \in Y$ is another element such that $\varphi(y') = x$, then $\varphi(y - y') = x$ and by exactness of the top row of the diagram 3.1.23, there is a $z \in Z$ such that $\beta(z) = y - y'$. Then $\psi\gamma(y - y') = \psi\gamma\beta(z) = \psi\delta\alpha(z) = 0$ because $\psi\delta = 0$.

Now let $(T, \gamma, \delta)$ be the pushout of $(Z, \alpha, \beta)$. It can be assumed that $(T, \gamma, \delta)$ is standard, i.e. $T = X \oplus Y/S$, where $S = \{(\alpha z, -\beta z) \mid z \in Z\}$. As before the elements from $T$ are denoted $\overline{(x, y)} = (x, y) + S$. The morphisms $\gamma$ and $\delta$ are defined by $\gamma(y) = \overline{(0, y)}$ and $\delta(x) = \overline{(x, 0)}$.

As before let $h\varphi(y) = 0$. Then $\psi\gamma(y) = 0$. So $\gamma(y) \in \text{Ker}(\psi)$ and there is an $x \in X$ with $\gamma(y) = \delta(x)$. But $\delta(x) = \overline{(x, 0)}$. Also $\gamma(y) = \overline{(0, y)}$. So $\overline{(x, 0)} = \overline{(0, y)}$ or $(x, -y) \in S$. So there is a $z \in Z$ with $(x, -y) = (\alpha(z), \beta(z))$. So $y \in \text{Im}(\beta)$ and $\psi\gamma(y) = 0$ proving that $\varphi(y) = 0$. This shows that $h$ is a monomorphism.

Further, an element of $\text{Coker}(\delta)$ is of the form $\psi(\overline{(x, y)}) = \psi(\overline{(x, 0)}) + \psi(\overline{(0, y)}) = \psi(\overline{(0, y)}) = \psi(\gamma(y)) = h(\varphi(y))$ (as $\psi(\overline{(x, 0)}) = \psi(\delta(x)) = 0$). So $h$ is surjective and in all $h$ is an isomorphism, as required. $\square$

**Lemma 3.1.24.** *Given a commutative diagram*

$$
\begin{array}{ccc}
T & \xrightarrow{\delta} & X \\
\gamma \downarrow & & \downarrow \alpha \\
Y & \xrightarrow{\beta} & Z
\end{array}
$$

*there is always an induced commutative diagram with exact rows*

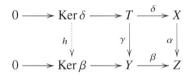

**Diag. 3.1.25.**

*Moreover if $(T, \gamma, \delta)$ is the pullback of $(Z, \alpha, \beta)$ the morphism $h : \text{Ker}\delta \to \text{Ker}\beta$ is an isomorphism.*

*Proof.* The morphism $h$ is defined as follows. Let $t \in \text{Ker }\delta$. Then $\beta\gamma(t) = \alpha\delta(t) = 0$. So $\gamma(t) \in \text{Ker }\beta$. Put $h(t) = \gamma(t)$.

Now let $(T, \gamma, \delta)$ be the (standard) pullback of $(Z, \alpha, \beta)$. Then $T = \{(x, y) \in X \oplus Y : \alpha x = \beta y\}$, $\delta((x, y)) = x$, $\gamma((x, y)) = y$. So $\text{Ker }\delta = \{(0, y) : \beta(y) = 0\}$. This identifies $\text{Ker }\delta$ with $\text{Ker }\beta$ (via $\gamma$) which takes care of things. $\square$

**Proposition 3.1.26.** *Let $M$ be an $A$-module, and let*

$$0 \to X \to P \xrightarrow{\varphi} M \to 0$$

$$0 \to Y \to Q \xrightarrow{\psi} M \to 0$$

*be two exact sequences with $P$ and $Q$ projective. Then*

$$X \oplus Q \simeq Y \oplus P.$$

*Proof.* Taking a (standard) pullback on $(\varphi, \psi)$ (which is also a pullback of $(\psi, \varphi)$) one obtains (as in lemma 3.1.18) a commutative diagram

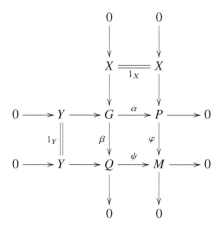

with exact rows and columns. Since $P$ and $Q$ are projective, $G \simeq P \oplus Y$ and $G \simeq Q \oplus X$, hence $P \oplus Y \simeq Q \oplus X$, as required. $\square$

**Corollary 3.1.27.** *Let $M$ be a module and let*

$$\ldots \longrightarrow P_i \xrightarrow{d_i} P_{i-1} \xrightarrow{d_{i-1}} \ldots \xrightarrow{d_2} P_1 \xrightarrow{d_1} P_0 \xrightarrow{d_0} M \longrightarrow 0$$

$$\ldots \longrightarrow Q_i \xrightarrow{d'_i} Q_{i-1} \xrightarrow{d'_{i-1}} \ldots \xrightarrow{d'_2} Q_1 \xrightarrow{d'_1} Q_0 \xrightarrow{d'_0} M \longrightarrow 0$$

*be two projective resolutions of $M$. Then*

$$X_i \oplus Q_i \oplus P_{i-1} \oplus \ldots \oplus Q_1 \oplus P_0 \simeq Y_i \oplus P_i \oplus Q_{i-1} \oplus \ldots \oplus P_1 \oplus Q_0,$$

*for $i = 0, 1, 2, \ldots$, where $X_i = \mathrm{Ker}(d_i)$, and $Y_i = \mathrm{Ker}(d'_i)$.*

*Proof.* This is proved by induction on $i$. For $i = 0$ the statement follows from proposition 3.1.26. Suppose the statement is true for $i = k$, i.e.

$$M_k = X_k \oplus Q_k \oplus P_{k-1} \oplus \ldots \oplus Q_1 \oplus P_0 \simeq Y_k \oplus P_k \oplus Q_{k-1} \oplus \ldots \oplus P_1 \oplus Q_0 = N_k.$$

Consider the two exact sequences:

$$0 \longrightarrow X_{k+1} \longrightarrow P_{k+1} \oplus M_k \longrightarrow M_k \longrightarrow 0$$

$$0 \longrightarrow Y_{k+1} \longrightarrow Q_{k+1} \oplus N_k \longrightarrow N_k \longrightarrow 0.$$

Then as in the proof of proposition 3.1.26, one obtains that $X_{k+1} \oplus Q_{k+1} \oplus N_k \simeq Y_{k+1} \oplus P_{k+1} \oplus M_k$, i.e. the statement is also true for $i = k + 1$. So it is true for all $i = 0, 1, \ldots$. $\square$

**Lemma 3.1.28.** *Let A be an Artinian ring, and X, Y, Z finitely generated A-modules. Let*

$$0 \longrightarrow X \xrightarrow{f} Z \xrightarrow{g} Y \longrightarrow 0$$

$$\downarrow{\alpha} \qquad \downarrow{\beta} \qquad \downarrow{\gamma}$$

$$0 \longrightarrow X \xrightarrow{f} Z \xrightarrow{g} Y \longrightarrow 0$$

*be a commutative diagram with exact rows, which are not split. Then:*

1. *If X is indecomposable and $\gamma = 1_Y$, then $\alpha$ and $\beta$ are automorphisms.*
2. *If Y is indecomposable and $\alpha = 1_X$, then $\gamma$ and $\beta$ are automorphisms.*

*Proof.*

1. Since $A$ is also a Noetherian ring, all finitely generated $A$-modules are Noetherian. Therefore, by [103, proposition 1.1.9], any endomorphism of $X$ is either an automorphism or nilpotent. If $\alpha$ is an automorphism, then by [103, corollary 1.2.7] $\beta$ is also an automorphism. Suppose $\alpha$ is not automorphism, then $\alpha$ is nilpotent, i.e. there is $n$ such that $\alpha^n = 0$. Then $\beta^n f = f\alpha^n = 0$, therefore there is a homomorphism $\varphi : Y \to Z$ such that $\beta^n = \varphi g$. Since $g\beta^n = g$, one obtains that $g\varphi g = g$, and hence $g\varphi = 1_Y$ as $g$ is an epimorphism of Artinian modules. So $g$ is a split epimorphism, and the sequence is split. A contradiction.

Statement 2 is proved analogously. $\square$

## 3.2 The Snake Lemma

This section contains the proof of the first part of the statement which is well known as the **snake lemma**. The second part of this lemma was proved in [100, section 6.1]. Though this lemma is valid in any Abelian category, here it will only be proved for the category of modules over a ring $A$.

**Lemma 3.2.1.** *Suppose $X_i$, $Y_i$ are A-modules for $i = 1, 2$, and let*

$$X_1 \xrightarrow{\alpha} Y_1$$

$$\downarrow{f} \qquad \downarrow{g}$$

$$X_2 \xrightarrow{\beta} Y_2$$

*be a commutative diagram. Then there exist unique well-defined homomorphisms $\varphi : \mathrm{Ker}\,\alpha \to \mathrm{Ker}\,\beta$ and $\psi : \mathrm{Coker}\,\alpha \to \mathrm{Coker}\,\beta$ which make the following diagram*

$$\mathrm{Ker}\,\alpha \longrightarrow X_1 \xrightarrow{\alpha} Y_1 \longrightarrow \mathrm{Coker}\,\alpha$$

$$\downarrow{\varphi} \qquad \downarrow{f} \qquad \downarrow{g} \qquad \downarrow{\psi}$$

$$\mathrm{Ker}\,\beta \longrightarrow X_2 \xrightarrow{\beta} Y_2 \longrightarrow \mathrm{Coker}\,\beta$$

*commutative.*

*Proof.* It is easy to show that the homomorphism $\varphi$ can be defined by setting $\varphi(x) = f(x)$ for all $x \in \operatorname{Ker}\alpha$ and the homomorphism $\psi$ can be defined by setting $\psi(y + \operatorname{Im}\alpha) = g(y) + \operatorname{Im}\beta$. See also lemmas 3.1.22 and 3.1.24. $\square$

**Lemma 3.2.2. (Kerner-Cokerner Lemma).** *Let*

$$
\begin{array}{ccc}
X & \xrightarrow{\ f\ } Z \xrightarrow{\ g\ } Y \\
\downarrow{\scriptstyle\alpha} & \downarrow{\scriptstyle\beta} & \downarrow{\scriptstyle\gamma} \\
X_1 & \xrightarrow{\ f_1\ } Z_1 \xrightarrow{\ g_1\ } Y_1
\end{array}
$$

*be a commutative diagram with exact rows.*

1. *If $f_1$ is a monomorphism then the induced sequence*

$$
\operatorname{Ker}\alpha \xrightarrow{\ \varphi\ } \operatorname{Ker}\beta \xrightarrow{\ \psi\ } \operatorname{Ker}\gamma \tag{3.2.3}
$$

   *is exact.*
2. *If $f$ is a monomorphism then $\varphi : \operatorname{Ker}\alpha \longrightarrow \operatorname{Ker}\beta$ is a monomorphism.*
3. *If $g$ is an epimorphism then the induced sequence*

$$
\operatorname{Coker}\alpha \xrightarrow{\ \sigma\ } \operatorname{Coker}\beta \xrightarrow{\ \tau\ } \operatorname{Coker}\gamma \tag{3.2.4}
$$

   *is exact.*
4. *If $g_1$ is an epimorphism then $\tau : \operatorname{Coker}\beta \longrightarrow \operatorname{Coker}\gamma$ is an epimorphism.*

*Proof.*

1. Suppose $f_1$ is a monomorphism. In a natural way, by lemma 3.2.1, one obtains the commutative diagram

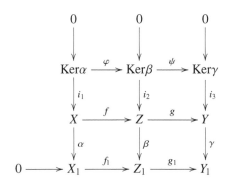

We must only show that the first row is exact, i.e. $\text{Im}\varphi = \text{Ker}\psi$. Let $a \in \text{Ker}\alpha$. Then $i_3\psi\varphi(a) = gfi_1(a) = 0$. So as $i_3$ is injective, $\psi\varphi(a) = 0$ and $\varphi(a) \in \text{Ker}\psi$. This argument gives $\text{Im}\varphi \subseteq \text{Ker}\psi$. Let $z \in \text{Ker}\beta$ and $\psi(z) = 0$. Then $gi_2(z) = i_3\psi(z) = 0$, which means that $i_2(z) = f(x)$, for some $x \in X$. Therefore $f_1\alpha(x) = \beta f(x) = \beta i_2(y) = 0$, hence $\alpha(x) = 0$, since $f_1$ is a monomorphism. So $x \in \text{Ker}\alpha$, $x = i_1(a)$, where $a \in \text{Ker}\alpha$ and $y = \varphi(a) \in \text{Im}\varphi$ as $i_2\varphi(a) = fi_1(a) = i_2(z)$ and $i_2$ is a monomorphism.

2. If $f$ is a monomorphism then $fi_1 = i_2\varphi$ is a monomorphism, and so $\varphi$ is a monomorphism.

3. Suppose that $g$ is an epimorphism. By lemma 3.2.1, one has the induced commutative diagram

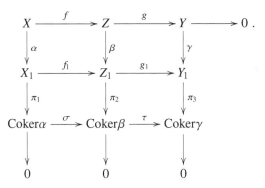

It will be shown that the bottom row is exact. Let $x \in \text{Coker}\alpha$, then there is $x_1 \in X_1$ such that $x = \pi_1(x_1)$. Then $\tau\sigma(x) = \tau\sigma\pi_1(x_1) = \pi_3g_1f_1(x_1) = 0$, so $\text{Im}\sigma \subseteq \text{Ker}(\tau)$. To show the opposite inclusion let $a \in \text{Ker}(\tau)$. There is a $z_1 \in Z_1$ such that $a = \pi_2(z_1)$. Let $g_1(z_1) = y_1$. Then $\pi_3(y_1) = \pi_3g_1(z_1) = \tau\pi_2(z_1) = \tau(a) = 0$. Therefore $y_1 \in \text{Im}\gamma$, i.e. $y_1 = \gamma(y)$ for some $y \in Y$. Since $g$ is an epimorphism, $y = g(z)$ for some $z \in Z$. By commutativity of the diagram it follows that $g_1(z_1) = y_1 = \gamma(y) = \gamma g(z) = g_1\beta(z)$. Therefore $z_2 = z_1 - \beta(z) \in \text{Ker}g_1 = \text{Im}f_1$, i.e. $z_2 = f_1(x_1)$ for some $x_1 \in X_1$. So $a = \pi_2(z_1) = \pi_2(z_2 + \beta z) = \pi_2 f_1(x_1) + \pi_2\beta z = \sigma\pi_1(x_1) \in \text{Im}(\sigma)$, which shows that $\text{Ker}(\tau) \subseteq \text{Ker}(\sigma)$ and completes the argument.

4. If $g_1$ is an epimorphism then $\pi_3g_1 = \tau\pi_2$ is an epimorphism, and so is $\tau$. $\square$

**Theorem 3.2.5 (Snake lemma).** *Let*

$$X \xrightarrow{f} Z \xrightarrow{g} Y \longrightarrow 0$$
$$\downarrow{\alpha} \qquad \downarrow{\beta} \qquad \downarrow{\gamma}$$
$$0 \longrightarrow X_1 \xrightarrow{f_1} Z_1 \xrightarrow{g_1} Y_1$$

**Diag. 3.2.6.**

*be a commutative diagram with exact rows. Then this diagram is a part of a commutative diagram*

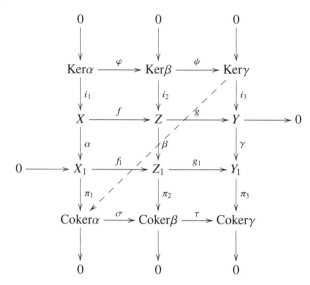

In addition, there is a homomorphism $\Delta : \text{Ker}(\gamma) \to \text{Coker}(\alpha)$ such that the long sequence

$$\text{Ker}\,\alpha \xrightarrow{\varphi} \text{Ker}\,\beta \xrightarrow{\psi} \text{Ker}\,\gamma \xrightarrow{\Delta}$$

$$\xrightarrow{\Delta} \text{Coker}(\alpha) \xrightarrow{\sigma} \text{Coker}(\beta) \xrightarrow{\tau} \text{Coker}(\gamma) \qquad (3.2.7)$$

is exact. Moreover, if $f$ is a monomorphism, then $0 \to \text{Ker}\,\alpha \xrightarrow{\varphi} \text{Ker}\,\beta$ is exact; and if $g_1$ is an epimorphism then $\text{Coker}(\beta) \xrightarrow{\tau} \text{Coker}(\gamma) \to 0$ is exact.

**Remark.** The picture above shows why this is sometimes called the **zig-zag lemma**. The zig-zag is the long exact sequence (3.2.7). Now draw $\Delta$ as follows. Starting at $\text{Ker}\,\gamma$ go right-down-left around $Y$ cross $\gamma$, $\beta$, $\alpha$ and go left-down-right around $X_1$ to reach $\text{Ker}\,\alpha$. The resulting picture explains the most used terminology: snake lemma.

*Proof.* Let's first define the connecting homomorphism $\Delta : \text{Ker}\,\gamma \to \text{Coker}\,\alpha$. Let $y \in \text{Ker}\gamma$. Since $g$ is an epimorphism, $i_3(y) = g(z)$ for some $z \in Z$. Since $g_1\beta(z) = \gamma g(z) = \gamma i_3(y) = 0$, $\beta(z) \in \text{Ker}(g_1) = \text{Im}f_1$ and so there is $x_1 \in X_1$ such that $\beta(z) = f_1(x_1)$. Define $\Delta(y) = \pi_1(x_1)$. It remains to check that $\Delta$ is well defined, i.e. $\Delta(x)$ depends only on $x$ and not on the choices of $z$ and $x_1$. Suppose $g(z') = g(z) = i_3(y)$. Then $z' - z \in \text{Ker}\,g = \text{Im}(f)$, and so $z' = z + f(x)$ for some $x \in X$. If $\beta z' = f_1(x_1')$, then, by definition, $\Delta(y) = \pi_1(x_1')$. But since the square

$$
\begin{array}{ccc}
x & \xrightarrow{\;\;f\;\;} & z - z' \\
\Big\uparrow{\alpha} & & \Big\uparrow{\beta} \\
x_1 - x_1' & \xrightarrow{\;\;f_1\;\;} & \beta(z - z')
\end{array}
$$

is commutative, it follows that $\pi_1(x - x_1') = \pi_1\alpha x = 0$, that is, $\pi_1(x_1) = \pi_1(x_1')$. Thus $\Delta$ is well defined.

So, there is the long sequence (3.2.7), and, by lemma 3.2.2, it remains to show that this sequence is exact at $\Delta$, i.e. that the sequences

$$\mathrm{Ker}\,\beta \xrightarrow{\psi} \mathrm{Ker}\,\gamma \xrightarrow{\Delta} \mathrm{Coker}(\alpha) \tag{3.2.8}$$

$$\mathrm{Ker}\,\gamma \xrightarrow{\Delta} \mathrm{Coker}(\alpha) \xrightarrow{\sigma} \mathrm{Coker}(\beta) \tag{3.2.9}$$

are both exact.

We now proceed to prove the exactness of sequence (3.2.8). Suppose $z \in \mathrm{Ker}\,\beta$ and consider $\Delta\psi(z)$. Then there is a commutative diagram

$$
\begin{array}{ccc}
z & \xrightarrow{\ \psi\ } & \psi(z) \\
\downarrow{\scriptstyle i_3} & & \downarrow{\scriptstyle i_3} \\
i_2(z) & \xrightarrow{\ g\ } & i_3\psi(z) = g i_2(z) \\
\downarrow{\scriptstyle \beta} & & \\
0 & &
\end{array}
$$

Hence, by definition, $\Delta\psi(z) = 0$, i.e. $\mathrm{Im}\,\psi \subseteq \mathrm{Ker}\,\Delta$. Suppose that $y \in \mathrm{Ker}\,\gamma$ and $\Delta(y) = 0$. Let $z \in Z$ and $x_1 \in X_1$ be defined as in the construction of $\Delta$, i.e. $\Delta(y) = \pi_1(x_1) = 0$. Therefore there is an $x \in X$ such that $\alpha(x) = x_1$. Then $\beta z = f_1(x_1) = f_1\alpha(x) = \beta f(x) = 0$, hence $z - f(x) \in \mathrm{Ker}(\beta)$, and so $y = \psi(z - f(x)) \in \mathrm{Im}(\psi)$, i.e. $\mathrm{Ker}\,\Delta \subseteq \mathrm{Im}(\psi)$. This proves that (3.2.8) is an exact sequence.

The exactness of sequence (3.2.9) can be proved in a similar way. The remaining statements are obvious. $\square$

The proofs above are known in the track as "diagram chasing".

As immediately corollaries of lemma 3.2.2 and the snake lemma we obtain the next two lemmas.

**Lemma 3.2.10. (Diagram lemma).** *Let $L_1, L_2, L_3, M_1, M_2, M_3 \in \mathrm{Mod}_r(A)$. Consider a commutative diagram*

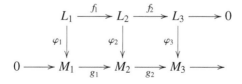

**Diag. 3.2.11.**

1. *If $\varphi_1$ and $\varphi_3$ are monomorphisms (epimorphisms) then $\varphi_3$ is also a monomorphism (epimorphism).*

2. *If $\varphi_1$ is an epimorphism and $\varphi_2$ is a monomorphism then $\varphi_3$ is a monomorphism.*

3. *If $\varphi_2$ is an epimorphism and $\varphi_3$ is a monomorphism then $\varphi_1$ is a monomorphism.*

**Lemma 3.2.12.** *Let $L_1, L_2, L_3, M_1, M_2, M_3 \in \mathrm{Mod}_r(A)$. Consider a commutative diagram*

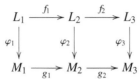

Diag. 3.2.13.

1. *If the lower row is exact, $\varphi_1$ is an epimorphism, $\varphi_2$ and $\varphi_3$ are monomorphisms, then the top row is also exact.*

2. *If the top row is exact, $\varphi_3$ is a monomorphism, $\varphi_1$ and $\varphi_2$ are epimorphisms, then the lower row is also exact.*

3. *If $\varphi_1$ is an epimorphism, $\varphi_3$ is a monomorphism, $\varphi_2$ is an isomorphism, the the top row is exact if and only if the lower row is exact.*

**Lemma 3.2.14.   (Homotopy lemma).**   *Let $L_1, L_2, L_3, M_1, M_2, M_3 \in \mathrm{Mod}_r(A)$. Consider a commutative diagram*

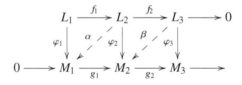

Diag. 3.2.15.

*Then the following statements are equivalent:*

1. *There exists $\alpha : L_2 \longrightarrow M_1$ such that $\varphi_1 = \alpha f_1$.*
2. *There exists $\beta : L_3 \longrightarrow M_2$ such that $\varphi_3 = g_2\beta$.*

*Proof.*
$1 \Longrightarrow 2$. Suppose there exists a homomorphism $\alpha : L_2 \longrightarrow M_1$ such that $\varphi_1 = \alpha f_1$. Then $g_1\varphi_1 = g_1\alpha f_1 = \varphi_2 f_1$ which implies that $(g_1\alpha - \varphi_2)f_1 = 0$. Since $f_2 = \mathrm{Coker} f_1$, there exists $\beta : L_3 \longrightarrow M_2$ such that $\beta f_2 = g_1\alpha - \varphi_2$. Therefore $g_2\beta f_2 = g_2 g_1\alpha - g_2\varphi_2 = \varphi_3 f_2$. So we obtain that $g_2\beta = \varphi_3$, since $f_2$ is an epimorphism.
    $2 \Longrightarrow 1$. This is proved in a similar way. $\square$

## 3.3 Extensions of Modules

This section is devoted to studying the $\text{Ext}^1$ in terms of short exact sequences, that is extensions of modules.

**Definition 3.3.1.** A right $A$-module $Z$ is said to be an **extension** of $X$ by (cokernel) $Y$ (or an **extension** of $Y$ with kernel $X$) if $X \subset Z$ and $Z/X \simeq Y$, which is equivalent to the existence of the following short exact sequence

$$0 \to X \xrightarrow{f} Z \xrightarrow{g} Y \to 0 \qquad (3.3.2)$$

Therefore in what follows an extension of $X$ by $Y$ will simply be a short exact sequence (3.3.2).

In the set of all short exact sequences there is a natural notion of morphism. Let $\mathcal{E} : \ 0 \to X \to Z \to Y \to 0$ and $\mathcal{E}_1 : \ 0 \to X_1 \to Z_1 \to Y_1 \to 0$ be two short exact sequences.

**Definition 3.3.3.** A **morphism** $\Gamma : \mathcal{E} \to \mathcal{E}_1$ is a triple $\Gamma = (\alpha, \beta, \gamma)$ of $A$-morphisms such that the diagram

**Diag. 3.3.4.**

is commutative.

The composition of morphisms is defined by $(\alpha_1, \ \beta_1, \ \gamma_1)(\alpha, \ \beta, \ \gamma) = (\alpha_1\alpha, \ \beta_1\beta, \ \gamma_1\gamma)$.

Consider the particular case when $X_1 = X$ and $Y = Y_1$. The intention is to construct a category whose objects are short exact sequences of the form (3.3.2), and whose morphisms are defined as above, when $\alpha = 1_X, \gamma = 1_Y$. Let $E_A(Y, X)$ be a set of all extensions of $X$ by $Y$. There is the following notion of equivalence of such extensions.

**Definition 3.3.5.** Two exact sequences $\mathcal{E} : \ 0 \to X \to Z \to Y \to 0$ and $\mathcal{E}_1 : \ 0 \to X \to Z_1 \to Y \to 0$ are said to be **equivalent** (and this is written by $\mathcal{E} \sim \mathcal{E}_1$) if there is a commutative diagram

**Diag. 3.3.6.**

Since this diagram is commutative, $\varphi$ is an isomorphism by [103, corollary 1.2.7]. Therefore it is easy to see that $\sim$ is in fact an equivalence relation on the set of all extensions of $X$ by $Y$, because it is reflexive ($\mathcal{E} \sim \mathcal{E}$), symmetric ($\mathcal{E} \sim \mathcal{E}_1$ implies $\mathcal{E}_1 \sim \mathcal{E}$) and transitive ($\mathcal{E} \sim \mathcal{E}_1$ and $\mathcal{E}_1 \sim \mathcal{E}_2$ implies $\mathcal{E} \sim \mathcal{E}_2$). Denote the set of all equivalence classes of $E_A(Y, X)$ by $\mathcal{E}xt_A(Y, X)$. The equivalence class of a short exact sequence $\mathcal{E}$ is denoted by $[\mathcal{E}]$. The next step is to define a group structure on $\mathcal{E}xt_A(Y, X)$ and to show that there is a group isomorphism between $\mathcal{E}xt_A(Y, X)$ and $\mathrm{Ext}_A^1(Y, X)$.

Recall that a short exact sequence (3.3.2) is said to be **split** if there exist homomorphisms $\overline{f} : Z \to X$ and $\overline{g} : Y \to X$ such that $\overline{f}f = 1_X$ and $g\overline{g} = 1_Y$.

The set of extensions $E_A(Y, X)$ always contains the (canonical) split sequence

$$0 \to X \to X \oplus Y \to Y \to 0 \tag{3.3.7}$$

From [103, proposition 1.2.3] it follows that any short exact sequence which is equivalent to this one is also split. The equivalence class which contains all split extensions of $X$ by $Y$ is called the **zero** of the set $\mathcal{E}xt_A(Y, X)$ and denoted by $[0]$.

We first study the set of extensions $E_A(Y, X)$ with $X$ fixed and varying $Y$.

**Lemma 3.3.8.** *Let $\mathcal{E} \in E_A(Y, X)$, then for any A-homomorphism $\gamma : Y_1 \to Y$ there exists an extension $\mathcal{E}\gamma \in E_A(Y_1, X)$ and a morphism $\gamma^* = (1_X, \beta, \gamma) : \mathcal{E}\gamma \to \mathcal{E}$. Moreover, if $\mathcal{E}$ is a split sequence, then $\mathcal{E}\gamma$ is split, as well. If $\mathcal{E} \sim \mathcal{E}'$ then $\mathcal{E}\gamma \sim \mathcal{E}'\gamma$.*

*Proof.* The existence of an extension $\mathcal{E}\gamma \in E_A(Y_1, X)$ and a morphism $\gamma^* = (1_X, \beta, \gamma) : \mathcal{E}\gamma \to \mathcal{E}$ follow from lemma 3.1.18, by constructing the commutative diagram:

$$
\begin{array}{ccccccccc}
\mathcal{E}\gamma : & 0 & \longrightarrow & X & \xrightarrow{f_1} & Z_1 & \xrightarrow{g_1} & Y_1 & \longrightarrow 0 \\
 & & & {\scriptstyle 1_X}\big\| & & {\scriptstyle \beta}\big\downarrow & & {\scriptstyle \gamma}\big\downarrow & \\
\mathcal{E} : & 0 & \longrightarrow & X & \xrightarrow{f} & Z & \xrightarrow{g} & Y & \longrightarrow 0
\end{array}
$$

where $(Z_1, g_1, \beta)$ is a pullback of $(g, \gamma)$.

Suppose that the short exact sequence $\mathcal{E}$ is split. Then, by [103, proposition 1.2.3], there is a homomorphism $h : Y \to Z$ such that $gh = 1_Y$. Hence $gh\gamma = \gamma$, and so there is a commutative diagram:

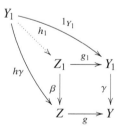

where $(Z_1, g_1, \beta)$ is the previous used pullback of $g, \gamma$. Therefore by the main property of pullbacks there is a homomorphism $h_1 : Y_1 \to Z_1$ making this diagram commute.

In particular, there results that $g_1 h_1 = 1_{Y_1}$. Hence, the extension $\mathcal{E}\gamma$ is split, by [103, proposition 1.2.3].

It remains to show that equivalent extensions by means of the morphism $\gamma^*$ correspond to equivalent ones. Suppose $\mathcal{E} \sim \mathcal{E}'$. Then one can write down the following commutative diagram with exact rows

$$
\begin{array}{ccccccccc}
\mathcal{E}\gamma: & 0 & \longrightarrow & X & \xrightarrow{f_1} & Z_1 & \xrightarrow{g_1} & Y_1 & \longrightarrow 0 \\
& & & {\scriptstyle 1_X}\big\| & & {\scriptstyle \beta}\big\downarrow & & {\scriptstyle \gamma}\big\downarrow & \\
\mathcal{E}: & 0 & \longrightarrow & X & \xrightarrow{f} & Z & \xrightarrow{g} & Y & \longrightarrow 0 \\
& & & {\scriptstyle 1_X}\big\| & & {\scriptstyle \varphi^{-1}}\big\uparrow\big\downarrow{\scriptstyle \varphi} & & {\scriptstyle 1_Y}\big\| & \\
\mathcal{E}': & 0 & \longrightarrow & X & \xrightarrow{f'} & Z' & \xrightarrow{g'} & Y & \longrightarrow 0 \\
& & & {\scriptstyle 1_X}\big\| & & {\scriptstyle \beta'}\big\uparrow & & {\scriptstyle \gamma}\big\uparrow & \\
\mathcal{E}'\gamma: & 0 & \longrightarrow & X & \xrightarrow{f_1'} & Z_1' & \xrightarrow{g_1'} & Y_1 & \longrightarrow 0 \\
\end{array}
$$

**Diag. 3.3.9.**

where $\varphi$ and $\varphi^{-1}$ are mutually inverse isomorphisms. So there is the following commutative diagram

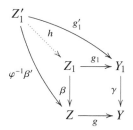

as $(Z_1, g_1, \beta)$ is a pullback of $(g, \gamma)$. Therefore by the universal property of a pullback there is a homomorphism $h : Z_1' \to Z_1$ yielding commutative the diagram above. In particular, $g_1 h = g_1'$, and $\beta h = \varphi^{-1}\beta'$.

It remains to prove that the diagram

$$
\begin{array}{ccccccccc}
\mathcal{E}\gamma: & 0 & \longrightarrow & X & \xrightarrow{f_1} & Z_1 & \xrightarrow{g_1} & Y_1 & \longrightarrow 0 \\
& & & {\scriptstyle 1_X}\big\| & & {\scriptstyle h}\big\uparrow & & {\scriptstyle 1_{Y_1}}\big\| & \\
\mathcal{E}'\gamma: & 0 & \longrightarrow & X & \xrightarrow{f_1'} & Z_1' & \xrightarrow{g_1'} & Y_1 & \longrightarrow 0 \\
\end{array}
$$

is commutative. Since $g_1 h = g_1'$, by the construction of $h$, it remains to show that

$hf_1' = f_1$. From the commutativity of the diagram 3.3.9 with exact rows it follows that $\beta f_1 = f$, $g_1 f_1 = 0$, and $\beta h f_1' = \varphi^{-1}\beta' f_1 = f$, $g_1 h f_1' = g_1' f_1'$, therefore there are the two commutative diagrams:

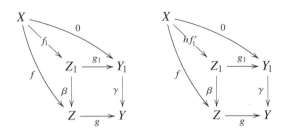

Since $(Z_1, g_1, \beta)$ is a pullback, the uniqueness of homomorphism from $X$ to $Z_1$ implies that $hf_1' = f_1$, as required. $\square$

From this lemma it follows that the morphism $\gamma^*$ is really a morphism of the set of equivalent classes $\gamma^* : \mathcal{E}xt_A(Y, X) \to \mathcal{E}xt_A(Y_1, X)$ which split sequences maps to split ones, i.e. the zero element of the set $\mathcal{E}xt_A(Y, X)$ maps to the zero element of the set $\mathcal{E}xt_A(Y_1, X)$.

Suppose there are two homomorphisms $\gamma : Y_1 \to Y$ and $\gamma_1 : Y_2 \to Y_1$. Then there is a homomorphism $\gamma\gamma_1 : Y_2 \to Y$, and one can construct a morphism $(\gamma\gamma_1)^* : \mathcal{E}xt_A(Y, X) \to \mathcal{E}xt_A(Y_2, X)$ such that $\mathcal{E}(\gamma\gamma_1) = (\mathcal{E}\gamma)\gamma_1$.

Let $\gamma = 1_Y$, then $\gamma^* = (1_X, \beta, 1_Y)$. Since in this case $\mathcal{E}\gamma = \mathcal{E}1_Y \sim \mathcal{E}$, lemma 3.3.8 implies that $\gamma^* = (1_X, \beta, 1_Y)$ is the identity map on the set $\mathcal{E}xt_A(Y, X)$, i.e. $(1_Y)^*[\mathcal{E}] = [\mathcal{E}]$.

**Corollary 3.3.10.** *For any A-homomorphism $\gamma : Y_1 \to Y$ there is a morphism $\gamma^* : \mathcal{E}xt_A(Y, X) \to \mathcal{E}xt_A(Y_1, X)$ such that $\gamma^*[\mathcal{E}] = [\mathcal{E}\gamma]$ and $\gamma^*[0] = [0]$, $(1_Y)^*[\mathcal{E}] = [\mathcal{E}]$.*

Consider now the set $E_A(Y, X)$ of all short exact sequences with a fixed module $Y$. In the dual, using pushouts instead of pullbacks, one can prove the following statement.

**Lemma 3.3.11.** *Let $\mathcal{E} \in E_A(Y, X)$, then for any A-homomorphism $\alpha : X \to X_1$ there exists an extension $\alpha\mathcal{E} \in E_A(Y, X_1)$ and a morphism $\alpha_* = (\alpha, \beta, 1_Y) : \mathcal{E} \to \alpha\mathcal{E}$. Moreover, if $\mathcal{E}$ is a split sequence, then $\alpha\mathcal{E}$ is split, as well. If $\mathcal{E} \sim \mathcal{E}'$ then $\alpha\mathcal{E} \sim \alpha\mathcal{E}'$.*

**Corollary 3.3.12.** *For any A-homomorphism $\gamma : X \to X_1$ there is a morphism $\alpha_* : \mathcal{E}xt_A(Y, X) \to \mathcal{E}xt_A(Y, X_1)$ such that $\alpha_*[\mathcal{E}] = [\alpha\mathcal{E}]$ and $\alpha_*[0] = [0]$, $(1_X)_*[\mathcal{E}] = [\mathcal{E}]$.*

If there are two $A$-homomorphisms $\gamma : Y_1 \to Y$ and $\alpha : X \to X_1$, then one can construct the morphisms $\alpha_*$ and $\gamma^*$, and so there is the following diagram

$$
\begin{array}{ccc}
\mathcal{E}xt_A(Y, X) & \xrightarrow{\;\gamma^*\;} & \mathcal{E}xt_A(Y_1, X) \\
\downarrow{\scriptstyle \alpha_*} & & \downarrow{\scriptstyle \alpha_*} \\
\mathcal{E}xt_A(Y, X_1) & \xrightarrow[\;\gamma^*\;]{} & \mathcal{E}xt_A(Y_1, X_1)
\end{array}
$$

This diagram is commutative, i.e.

$$\alpha_* \gamma^* = \gamma^* \alpha_*$$

Indeed let $[\mathcal{E}] \in \mathcal{E}xt_A(Y, X)$, then $\gamma^*[\mathcal{E}] = [\mathcal{E}\gamma] \in \mathcal{E}xt_A(Y_1, X)$, $\alpha^*[\mathcal{E}] = [\alpha\mathcal{E}] \in \mathcal{E}xt_A(Y, X_1)$ and so one has the following commutative diagram:

$$
\begin{array}{clcccccccc}
\alpha(\mathcal{E}\gamma): & 0 & \longrightarrow & X_1 & \xrightarrow{\bar{f_1}} & \bar{Z}_1 & \xrightarrow{\bar{g_1}} & Y_1 & \longrightarrow & 0 \\
& & & \uparrow{\scriptstyle \alpha} & & \uparrow{\scriptstyle \beta_1} & & \| {\scriptstyle 1_{Y_1}} \\
\mathcal{E}\gamma: & 0 & \longrightarrow & X & \xrightarrow{f_1} & Z_1 & \xrightarrow{g_1} & Y_1 & \longrightarrow & 0 \\
& & & \| {\scriptstyle 1_X} & & \downarrow{\scriptstyle \beta} & & \downarrow{\scriptstyle \gamma} \\
\mathcal{E}: & 0 & \longrightarrow & X & \xrightarrow{f} & Z & \xrightarrow{g} & Y & \longrightarrow & 0 \\
& & & \downarrow{\scriptstyle \alpha} & & \downarrow{\scriptstyle \beta'} & & \| {\scriptstyle 1_Y} \\
\alpha\mathcal{E}: & 0 & \longrightarrow & X_1 & \xrightarrow{f'} & Z' & \xrightarrow{g'} & Y & \longrightarrow & 0 \\
& & & \| {\scriptstyle 1_{X_1}} & & \uparrow{\scriptstyle \beta'_1} & & \uparrow{\scriptstyle \gamma} \\
(\alpha\mathcal{E})\gamma: & 0 & \longrightarrow & X_1 & \xrightarrow{f'_1} & Z'_1 & \xrightarrow{g'_1} & Y_1 & \longrightarrow & 0
\end{array}
$$

From the commutativity of this diagram it follows in particular that $g'\beta'\beta = \gamma g_1$.

Since $(Z'_1, g'_1, \beta'_1)$ is a pullback of $(g', \gamma)$, one has the following commutative diagram:

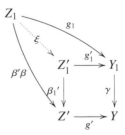

And so, by the universal property of a pullback, there is a unique homomorphism $\xi : Z_1 \to Z'_1$ such that $g'_1\xi = g_1$, $\beta_1'\xi = \beta'\beta$.

Since $(\bar{Z}_1, \bar{f}_1, \beta_1)$ is a pushout of $(f_1, \alpha)$, one has a unique $\eta : \bar{Z}_1 \longrightarrow Z'_1$ such that the following diagram commutes:

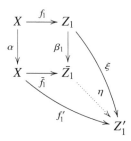

This proves the important associativity property $(\alpha\mathcal{E})\gamma \sim \alpha(\mathcal{E}\gamma)$. Therefore $\alpha_*\gamma^* = \gamma^*\alpha_*$, as required.

## 3.4 Baer Sum of Extensions

This section deals with an addition on $\mathcal{E}xt_A(Y, X)$ following R.Baer. This sum is called the **Baer sum**, and as it will be shown below this operation makes $\mathcal{E}xt_A(Y, X)$ an Abelian group.

Let

$$\mathcal{E}_1 : \quad 0 \to X \to Z_1 \to Y \to 0 \tag{3.4.1}$$

and

$$\mathcal{E}_2 : \quad 0 \to X \to Z_2 \to Y \to 0 \tag{3.4.2}$$

be two short exact sequences, then one can construct a new short exact sequence by taking direct sums

$$\mathcal{E}_1 \oplus \mathcal{E}_2 : \quad 0 \to X \oplus X \to Z_1 \oplus Z_2 \to Y \oplus Y \to 0 \tag{3.4.3}$$

Consider the **diagonal map** $\nabla : X \oplus X \to X$ which is the morphism given by $\nabla(x_1, x_2) = x_1 + x_2$, and the **codiagonal map** $\Delta : Y \to Y \oplus Y$ which is the morphism given by $\Delta(y) = (y, y)$. Then by the constructions given in the previous section one has the maps

$$\Delta^* : \mathcal{E}xt_A(Y \oplus Y, X) \longrightarrow \mathcal{E}xt_A(Y, X)$$

and

$$\nabla_* : \mathcal{E}xt_A(Y \oplus Y, X \oplus X) \longrightarrow \mathcal{E}xt_A(Y \oplus Y, X).$$

Then

$$\Delta^*\nabla_* : \mathcal{E}xt_A(Y \oplus Y, X \oplus X) \longrightarrow \mathcal{E}xt_A(Y, X)$$

(defining) $\Delta^*\nabla_*(\mathcal{E}_1 \oplus \mathcal{E}_2) \in \mathcal{E}xt_A(Y, X)$.

Therefore one can define an addition in $E_A(X, Y)$ as

$$\mathcal{E}_1 + \mathcal{E}_2 = \Delta^*\nabla_*(\mathcal{E}_1 \oplus \mathcal{E}_2), \tag{3.4.4}$$

and an addition in $\mathcal{E}xt_A(Y, X)$ as

$$[\mathcal{E}_1] + [\mathcal{E}_2] = [\mathcal{E}_1 + \mathcal{E}_2]. \tag{3.4.5}$$

Let $\gamma : Y_1 \rightarrow Y$ and $\alpha : X \rightarrow X_1$ be $A$-homomorphisms. Then, naturally, one expects that the corresponding morphisms $\gamma^*$ and $\alpha_*$ preserve the operation of addition (3.4.5).

To that end consider the following commutative diagram:

$$
\begin{array}{ccc}
Y_1 & \xrightarrow{\gamma} & Y \\
\downarrow{\scriptstyle \Delta} & & \downarrow{\scriptstyle \Delta} \\
Y_1 \oplus Y_1 & \xrightarrow{\gamma \oplus \gamma} & Y \oplus Y
\end{array}
$$

This diagram being commutative implies that

$$\gamma^* \Delta^* = \Delta^* (\gamma \oplus \gamma)^*. \tag{3.4.6}$$

One now has the following commutative diagram:

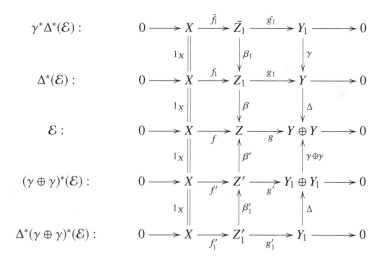

**Diag. 3.4.7.**

From the commutativity of this diagram it follows in particular that $g\beta\beta_1 = \Delta\gamma\bar{g}_1$.

Taking into account that $(Z', g', \beta')$ is a pullback of $(g, \gamma \oplus \gamma)$, there is a unique $\xi : \bar{Z}_1 \longrightarrow Z'$ such that the following diagram commutes:

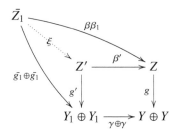

**Diag. 3.4.8.**

Analogously, taking into account that $(Z'_1, g'_1, \beta'_1)$ is a pullback of $(g', \Delta)$, there is a unique $\eta : \bar{Z}_1 \longrightarrow Z'_1$ making the following diagram commutative:

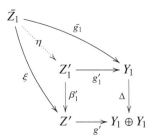

**Diag. 3.4.9.**

So there is the following diagram

$$
\begin{array}{ccccccccc}
\gamma^*\Delta^*(\mathcal{E}) : & & 0 & \longrightarrow & X & \xrightarrow{\bar{f}_1} & \bar{Z}_1 & \xrightarrow{\bar{g}_1} & Y_1 & \longrightarrow & 0 \\
 & & & & \big\Vert 1_X & & \Big\downarrow \eta & & \big\Vert 1_{Y_1} & & \\
\Delta^*(\gamma \oplus \gamma)^*(\mathcal{E}) : & & 0 & \longrightarrow & X & \xrightarrow{f'_1} & Z'_1 & \xrightarrow{g'_1} & Y_1 & \longrightarrow & 0
\end{array}
$$

**Diag. 3.4.10.**

It remains to show that this diagram is commutative, i.e. $g'_1\eta = \bar{g}_1$ and $\eta\bar{f}_1 = f'_1$. The first equality follows from the commutativity of diagram 3.4.9. So it remains to prove the second equality.

From the commutativity of diagrams 3.4.7 and 3.4.8 one has

$$g'f' = 0, \quad \beta'f' = f,$$

and

$$g'\eta\bar{f}_1 = (\bar{g}_1 \oplus \bar{g}_1)\bar{f}_1 = 0 \oplus 0 = 0$$

$$\beta' \eta \bar{f}_1 = \beta \beta_1 \bar{f}_1 = \beta f_1.$$

Therefore there are two commutative diagrams:

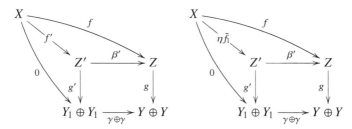

Since $(Z', g', \beta')$ is a pullback, from the universal property of a pullback it follows that $\eta \bar{f}_1 = f'$. Hence $\beta'_1 \eta \bar{f}_1 = \xi \bar{f}_1 = f'$. From the commutativity of diagrams 3.4.7 and 3.4.9 one has $g'_1 \eta \bar{f}_1 = \bar{g}_1 \bar{f}_1 = 0$. Therefore there are again two commutative diagrams:

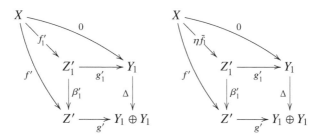

Since $(Z'_1, g'_1, \beta'_1)$ is a pullback, from the universal property of a pullback one has $\eta \bar{f}_1 = f'_1$, as required.

Thus, diagram 3.4.10 is commutative, and therefore, by definition, $\gamma^* \Delta^*(\mathcal{E}) \sim \Delta^*(\gamma \oplus \gamma)^*(\mathcal{E})$, which proves (3.4.6).

Now one can prove that the morphism $\gamma^*$ respects the addition (3.4.5) in $\mathcal{E}xt_A(Y, X)$. Let $\mathcal{E}_1, \mathcal{E}_2 \in E_A(Y, X)$. Then one sees

$$\gamma^*(\mathcal{E}_1 + \mathcal{E}_2) \sim \gamma^* \Delta^* \nabla_*(\mathcal{E}_1 \oplus \mathcal{E}_2) \sim (\Delta\gamma)^* \nabla_*(\mathcal{E}_1 \oplus \mathcal{E}_2) \sim \Delta^*(\gamma \oplus \gamma)^* \nabla_*(\mathcal{E}_1 \oplus \mathcal{E}_2) \sim$$

$$\sim \Delta^* \nabla_*(\gamma \oplus \gamma)^*(\mathcal{E}_1 \oplus \mathcal{E}_2) \sim \Delta^* \nabla_*(\gamma^* \mathcal{E}_1 \oplus \gamma^* \mathcal{E}_2) \sim \gamma^* \mathcal{E}_1 + \gamma^* \mathcal{E}_2,$$

hence

$$\gamma^*[\mathcal{E}_1 + \mathcal{E}_2] = \gamma^*[\mathcal{E}_1] + \gamma^*[\mathcal{E}_2].$$

Analogously one can show that if $\alpha : X \to X_1$ is an $A$-homomorphism then

$$\alpha_*[\mathcal{E}_1 + \mathcal{E}_2] = \alpha_*[\mathcal{E}_1] + \alpha_*[\mathcal{E}_2]$$

One now further shows that the addition defined by (3.4.5) makes $\mathcal{E}xt_A(Y, X)$ an Abelian group, whose zero element is the equivalence class of all split sequences, and for which the inverse of the class $[\mathcal{E}]$, where

$$\mathcal{E}: \quad 0 \longrightarrow X \overset{f}{\longrightarrow} Z \overset{g}{\longrightarrow} Y \longrightarrow 0, \tag{3.4.11}$$

is the class $[\mathcal{E}_1]$, where

$$\mathcal{E}_1 : 0 \longrightarrow X \xrightarrow{f} Z \xrightarrow{-g} Y \longrightarrow 0 \qquad (3.4.12)$$

Let [0] be the equivalence class of all split exact sequences, and let $\mathcal{E} \in \mathcal{E}xt_A(Y, X)$ be of form (3.4.11). To see that [0] is the zero element with respect to addition, that is,

$$[\mathcal{E}] + [0] = [\mathcal{E}]. \qquad (3.4.13)$$

Let the element 0 be the following (canonical) split sequence

$$0 \longrightarrow X \longrightarrow X \oplus Y \longrightarrow Y \longrightarrow 0.$$

Then one can have the following commutative diagram:

$$
\begin{array}{ccccccccc}
\mathcal{E}: & & 0 \longrightarrow & X & \xrightarrow{f} & Z & \xrightarrow{g} & Y & \longrightarrow 0 \\
& & & \downarrow 0 & & \downarrow \beta & & \| 1_Y & \\
0: & & 0 \longrightarrow & X & \longrightarrow & X \oplus Y & \longrightarrow & Y & \longrightarrow 0
\end{array}
$$

where $\beta : Z \to X \oplus Y$ is defined as $\beta(z) = (0, g(z))$. Thus the split sequence 0 can be written as $0_X \mathcal{E}$, where $0_X : X \to X$ is the zero homomorphism. So $\mathcal{E} + 0 \sim \mathcal{E} + 0_X \mathcal{E} \sim 1_X \mathcal{E} + 0_X \mathcal{E} \sim (1_X + 0_X)\mathcal{E} \sim 1_X \mathcal{E} \sim \mathcal{E}$, that is, $[\mathcal{E}] + [0] = [\mathcal{E}]$. Therefore the following main theorem has been proved.

**Theorem 3.4.14.** *The set $\mathcal{E}xt_A(Y, X)$ is an Abelian group with its operation of addition defined by (3.4.5) and (3.4.4), and with as zero element, the equivalence class of all split exact sequences in $E_A(Y, X)$, and the inverse element of an equivalence class $[\mathcal{E}]$ given by a sequence (3.4.11) is the equivalence class $[\mathcal{E}_1]$ given by the sequence (3.4.12).*

## 3.5 Properties of $\mathrm{Ext}^1$

In the previous section the group of extensions $\mathcal{E}xt_A(Y, X)$ was considered and some of its properties were studied. In the next section it will be shown that this group is closely connected with $\mathrm{Ext}^1_A(Y, X)$ as e.g. was considered in [100]. In this section some main definitions and some further properties of $\mathrm{Ext}^1_A(Y, X)$ will be considered, and it will be shown that $\mathrm{Ext}^1_A(Y, X)$ is also an Abelian group. It is a (co)homology group.

**Proposition 3.5.1.** *If $\mathrm{Ext}^1_A(Y, X) = 0$, then every extension of $X$ by $Y$ is split.*

*Proof.* Let $\mathcal{E} : \quad 0 \longrightarrow X \xrightarrow{f} Z \xrightarrow{g} Y \longrightarrow 0$ be an exact sequence. Apply $\mathrm{Hom}_A(*, X)$ to this sequence. Then, by [103, proposition 1.7.2], one has an exact sequence

$$\text{Hom}_A(X, Z) \xrightarrow{h(f)} \text{Hom}_A(X, X) \xrightarrow{\partial} \text{Ext}^1_A(Y, X) = 0.$$

So $h(f)$ is an epimorphism. In particular, there is a homomorphism $\bar{f} \in \text{Hom}_A(X, Z)$ such that $h(f)(\bar{f}) = \bar{f}f = 1_X$. This shows, by [103, proposition 1.2.3], that $\mathcal{E}$ is a split sequence. $\square$

The following statements show that $\text{Ext}^1$ behaves as Hom does on direct sums and direct products.

**Proposition 3.5.2.** *Let $A$ be a ring and $X$, $Y_i$ ($i \in I$) be $A$-modules. Then there exists a natural isomorphism*

$$\text{Ext}^1_A(\underset{i \in I}{\oplus} Y_i, X) \simeq \prod_{i \in I} \text{Ext}^1_A(Y_i, X). \tag{3.5.3}$$

*Proof.* For each $i$ consider an exact sequence

$$0 \longrightarrow Z_i \longrightarrow P_i \longrightarrow Y_i \longrightarrow 0$$

with $P_i$ projective. Then there is an induced exact sequence

$$0 \longrightarrow \underset{i \in I}{\oplus} Z_i \longrightarrow \underset{i \in I}{\oplus} P_i \longrightarrow \underset{i \in I}{\oplus} Y_i \longrightarrow 0,$$

with $\underset{i \in I}{\oplus} P_i$ projective, by [103, proposition 1.5.2]. So, taking into account [103, proposition 1.7.3], one obtains a commutative diagram with exact rows:

$$
\begin{array}{ccccccc}
\text{Hom}_A(\underset{i \in I}{\oplus} P_i, X) & \longrightarrow & \text{Hom}_A(\underset{i \in I}{\oplus} Y_i, X) & \longrightarrow & \text{Ext}^1_A(\underset{i \in I}{\oplus} Y_i, X) & \longrightarrow & 0 \\
\downarrow{\scriptstyle i_1} & & \downarrow{\scriptstyle i_2} & & & & \\
\prod_{i \in I} \text{Hom}_A(P_i, X) & \longrightarrow & \prod_{i \in I} \text{Hom}_A(Y_i, X) & \longrightarrow & \prod_{i \in I} \text{Ext}^1_A(Y_i, X) & \longrightarrow & 0
\end{array}
$$

where $i_1$ and $i_2$ are isomorphisms, by [103, proposition 1.2.16(1)]. Then, (see also lemma 3.1.22), there is a homomorphism $\varphi : \text{Ext}^1_A(\underset{i \in I}{\oplus} Y_i, X) \to \prod_{i \in I} \text{Ext}^1_A(Y_i, X)$, making this diagram commutative. By [103, corollary 1.2.9], $\varphi$ is an isomorphism. $\square$

In a dual way, using [103, proposition 1.2.16(2)], and [103, proposition 1.7.4], one can prove the following statement. (Here lemma 3.1.24 applies.)

**Proposition 3.5.4.** *Let $A$ be a ring and $Y$, $X_i$, ($i \in I$) be $A$-modules. Then there exists a natural isomorphism*

$$\text{Ext}^1_A(Y, \prod_{i \in I} X_i) \simeq \prod_{i \in I} \text{Ext}^1_A(Y, X_i). \tag{3.5.5}$$

From [100, theorem 6.2.7] we get the following result:

**Proposition 3.5.6.** *Let*

$$
\begin{array}{ccccccccc}
\mathcal{E}: & 0 & \longrightarrow & X & \overset{f}{\longrightarrow} & Z & \overset{g}{\longrightarrow} & Y & \longrightarrow & 0 \\
 & & & {\scriptstyle\alpha}\downarrow & & {\scriptstyle\beta}\downarrow & & {\scriptstyle\gamma}\downarrow & & \\
\mathcal{E}_1 & 0 & \longrightarrow & X_1 & \overset{f_1}{\longrightarrow} & Z_1 & \overset{g_1}{\longrightarrow} & Y_1 & \longrightarrow & 0
\end{array}
$$

**Diag. 3.5.7.**

*be a commutative diagram with exact rows. Then the following induced diagram*

$$
\begin{array}{ccc}
\operatorname{Hom}_A(M,Y) & \overset{\partial_{\mathcal{E}}}{\longrightarrow} & \operatorname{Ext}^1_A(M,X) \\
{\scriptstyle h(\gamma)}\downarrow & & \downarrow{\scriptstyle H_1(\alpha)} \\
\operatorname{Hom}_A(M,Y_1) & \overset{\partial_{\mathcal{E}_1}}{\longrightarrow} & \operatorname{Ext}^1_A(M,X_1)
\end{array}
$$

**Diag. 3.5.8.**

*is commutative, where $\partial_{\mathcal{E}}$, (resp. $\partial_{\mathcal{E}_1}$) is a connecting homomorphism of the sequence $\mathcal{E}$ (resp. $\mathcal{E}_1$).*

## 3.6 Ext$^1$ and Extensions

The results of this section will show that the group of extensions $\mathcal{E}xt_A(Y,X)$ and the homology group $\operatorname{Ext}^1_A(Y,X)$ considered in [100], are really the same group. The main goal of this section is to prove the following theorem:

**Theorem 3.6.1.** *For any pair of A-modules $X$ and $Y$ there is the functorial isomorphism*

$$
\chi : \mathcal{E}xt_A(Y,X) \to \operatorname{Ext}^1_A(Y,X) \tag{3.6.2}
$$

*Proof.* Let $\mathcal{E}$ : $0 \to X \overset{f}{\to} Z \overset{g}{\to} Y \to 0$ be an exact sequence. Then, by [103, proposition 1.7.2], we get a connecting homomorphism $\partial_{\mathcal{E}} : \operatorname{Hom}_A(Y,Y) \to \operatorname{Ext}^1_A(Y,X)$. The element $\chi[\mathcal{E}] = \partial_{\mathcal{E}}(1_Y)$ is called the **characteristic class** of the extension $\mathcal{E}$ [1]. If $\mathcal{E} \sim \mathcal{E}_1$, then from proposition 3.5.6 it follows that the diagram

---

[1] It can also be viewed as an obstruction (to the splitting of the exact sequence $\mathcal{E}$); see [223, p.77].

$$\begin{array}{ccc}
\mathrm{Hom}_A(Y,Y) & \xrightarrow{\ \partial_{\mathcal{E}}\ } & \mathrm{Ext}^1_A(Y,X) \\
id_1 \Big\| & & \Big\| id_2 \\
\mathrm{Hom}_A(Y,Y) & \xrightarrow{\ \partial_{\mathcal{E}_1}\ } & \mathrm{Ext}^1_A(Y,X)
\end{array}$$

**Diag. 3.6.3.**

is commutative, where $id_1$ and $id_2$ are identity morphisms. Therefore

$$\chi[\mathcal{E}] = \partial_{\mathcal{E}}(1_Y) = \partial_{\mathcal{E}_1}(1_Y) = \chi[\mathcal{E}_1],$$

so one obtains a well defined map

$$\chi : \mathcal{E}xt_A(Y,X) \to \mathrm{Ext}^1_A(Y,X),$$

where $\chi[\mathcal{E}] = \partial_{\mathcal{E}}(1_Y)$.

We will now show that the map $\chi$ defines an isomorphism of Abelian groups and that it is a functorial isomorphism.

To this end we first construct an inverse map

$$\omega : \mathrm{Ext}^1_A(Y,X) \to \mathcal{E}xt_A(Y,X)$$

such that $\chi\omega$ and $\omega\chi$ are identity morphisms of the corresponding sets.

Since any module can be embedded into an injective module, one can write down a short exact sequence:

$$\mathcal{F}: \ 0 \to X \xrightarrow{\ \varphi\ } Q \xrightarrow{\ \psi\ } M \to 0, \tag{3.6.4}$$

with an injective module $Q$. Then from [103, proposition 1.7.1], it follows that there is an exact sequence:

$$\mathrm{Hom}_A(Y,Q) \xrightarrow{\ h(\psi)\ } \mathrm{Hom}_A(Y,M) \xrightarrow{\ \partial\ } \mathrm{Ext}^1_A(Y,X) \to 0,$$

because $\mathrm{Ext}^1_A(Y,Q) = 0$ for an injective module $Q$. Therefore $\partial$ is an epimorphism, and so for any $\xi \in \mathrm{Ext}^1_A(Y,X)$ there is a homomorphism $\gamma \in \mathrm{Hom}_A(Y,M)$ such that $\xi = \partial(\gamma)$. Then, by corollary 3.3.10, one has a morphism $\gamma^* : \mathcal{E}xt_A(M,X) \to \mathcal{E}xt_A(Y,X)$, such that $\gamma^*[\mathcal{F}] = [\mathcal{F}\gamma]$. So, by setting $\omega(\xi) = [\mathcal{F}\gamma]$, one defines a map

$$\omega : \mathrm{Ext}^1_A(Y,X) \to \mathcal{E}xt_A(Y,X). \tag{3.6.5}$$

The commutative diagram

$$\begin{array}{ccccccccc}
\mathcal{F}\gamma: & & 0 & \longrightarrow & X & \xrightarrow{\ f\ } & Z & \xrightarrow{\ g\ } & Y & \longrightarrow & 0 \\
& & & & 1_X \Big\| & & \beta \Big\downarrow & & \gamma \Big\downarrow & & \\
\mathcal{F}: & & 0 & \longrightarrow & X & \xrightarrow{\ \varphi\ } & Q & \xrightarrow{\ \psi\ } & M & \longrightarrow & 0
\end{array}$$

where $(Z, g, \beta)$ is a pullback of $\psi, \gamma$, yields, in view of proposition 3.5.6, a commutative diagram:

$$\begin{array}{ccc}
\operatorname{Hom}_A(Y, Y) & \xrightarrow{\partial_{\mathcal{F}\gamma}} & \operatorname{Ext}^1_A(Y, X) \\
{\scriptstyle h(\gamma)} \downarrow & & \parallel {\scriptstyle id_2} \\
\operatorname{Hom}_A(Y, M) & \xrightarrow{\partial} & \operatorname{Ext}^1_A(Y, X)
\end{array}$$

Then one obtains:

$$\chi\omega(\xi) = \chi[\mathcal{F}\gamma] = \partial_{\mathcal{F}\gamma}(1_Y) = \partial(\gamma) = \xi.$$

So it remains to show that $\omega\chi[\mathcal{E}] = [\mathcal{E}]$ for an arbitrary extension

$$\mathcal{E}: \quad 0 \longrightarrow X \xrightarrow{f} Z \xrightarrow{g} Y \longrightarrow 0.$$

Suppose $\chi[\mathcal{E}] = \xi = \partial_{\mathcal{E}}(1_Y) \in \operatorname{Ext}^1_A(Y, X)$. Any module can be embedded into an injective module $Q$. Take such an inclusion $\varphi : X \to Q$. Then there is a homomorphism $\beta : Z \to Q$ such that $\beta f = \varphi$. Thus there results a commutative diagram

$$\begin{array}{ccccccccc}
\mathcal{E}: & 0 & \longrightarrow & X & \xrightarrow{f} & Z & \xrightarrow{g} & Y & \longrightarrow 0 \\
& & & {\scriptstyle 1_X}\parallel & & {\scriptstyle \beta}\downarrow & & {\scriptstyle \gamma}\downarrow & \\
\mathcal{E}_1: & 0 & \longrightarrow & X & \xrightarrow{\varphi} & Q & \xrightarrow{\psi} & M & \longrightarrow 0
\end{array}$$

which induces the following commutative diagram:

$$\begin{array}{ccc}
\operatorname{Hom}_A(Y, Y) & \xrightarrow{\partial_{\mathcal{E}}} & \operatorname{Ext}^1_A(Y, X) \\
{\scriptstyle h(\gamma)} \downarrow & & \parallel {\scriptstyle id_2} \\
\operatorname{Hom}_A(Y, M) & \xrightarrow{\partial} & \operatorname{Ext}^1_A(Y, X)
\end{array}$$

so that $\xi = \partial(\gamma)$. Now using the homomorphism $\gamma$ one can form the exact sequence $\mathcal{E}_1\gamma$, and by definition $\omega(\xi) = [\mathcal{E}_1\gamma]$. So it remains to show that $[\mathcal{E}_1\gamma] = [\mathcal{E}]$. To that end consider the following commutative diagram with exact rows:

$$\begin{array}{ccccccccc}
\mathcal{E}: & 0 & \longrightarrow & X & \xrightarrow{f} & Z & \xrightarrow{g} & Y & \longrightarrow 0 \\
& & & {\scriptstyle 1_X}\parallel & & {\scriptstyle \beta}\downarrow & & {\scriptstyle \gamma}\downarrow & \\
\mathcal{E}_1: & 0 & \longrightarrow & X & \xrightarrow{\varphi} & Q & \xrightarrow{\psi} & Y & \longrightarrow 0 \\
& & & {\scriptstyle 1_X}\parallel & & {\scriptstyle \beta_1}\uparrow & & {\scriptstyle \gamma}\uparrow & \\
\mathcal{E}_1\gamma: & 0 & \longrightarrow & X & \xrightarrow{f_1} & T & \xrightarrow{g_1} & Y & \longrightarrow 0
\end{array}$$

**Diag. 3.6.6.**

where $(T, g_1, \beta_1)$ is a pullback of $(\gamma, \psi)$. From the commutativity of this diagram it follows in particular, that $\gamma g = \psi \beta$.

Since $(T, g_1, \beta_1)$ is a pullback of $(\gamma, \psi)$, there is a unique homomorphism $\alpha : Z \to T$ with commutative diagram:

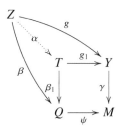

Diag. 3.6.7.

And thus

$$\beta_1 \alpha = \beta, \quad g_1 \alpha = g. \tag{3.6.8}$$

Therefore there results the following diagram with exact rows:

$$
\begin{array}{ccccccccc}
\mathcal{E}: & 0 & \longrightarrow & X & \overset{f}{\longrightarrow} & Z & \overset{g}{\longrightarrow} & Y & \longrightarrow & 0 \\
 & & & \big\| 1_X & & \big\downarrow \alpha & & \big\| 1_Y & & \\
\mathcal{E}_1 \gamma: & 0 & \longrightarrow & X & \overset{f_1}{\longrightarrow} & T & \overset{g_1}{\longrightarrow} & Y & \longrightarrow & 0
\end{array}
$$

Diag. 3.6.9.

It remains to show that this diagram is commutative, i.e. $g_1 \alpha = g$ and $\alpha f = f_1$. The first equality is follows from (3.6.8). To prove the second equality construct the following diagrams:

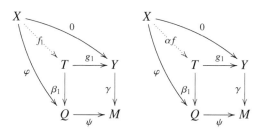

Diag. 3.6.10.

Since $g_1f_1 = 0$, $\beta_1f_1 = \varphi$ and $g_1\alpha f = gf = 0$, $\beta_1\alpha f = \beta f = \varphi$ using the commutativity of diagrams 3.6.6 and 3.6.7, it follows that the diagrams 3.6.10 are commutative. Since $(T, g_1, \beta_1)$ is a pullback of $(\gamma, \psi)$, by the universal property of pullbacks it follows that $f_1 = \alpha f$, i.e. $\mathcal{E} \sim \mathcal{E}_1\gamma$. So $\omega(\xi) = [\mathcal{E}]$, as required. $\square$

From this theorem it follows that the elements of $\mathrm{Ext}^1_A(Y, X)$ can be identified with equivalence classes of extensions in $\mathcal{E}xt_A(Y, X)$.


## 3.7 Additive and Abelian Categories

Additive and Abelian categories are the most interesting, indeed fundamental, in homological algebra. The categories of Abelian groups and modules over rings are examples of an additive category and an Abelian category.

**Definition 3.7.1.** A category $\mathfrak{C}$ is said to be **additive** if $\mathfrak{C}$ satisfies the following extra conditions:

1. There is a null (or zero) object in $C$, usually denoted by 0, such that for any object $X \in \mathrm{Ob}\,\mathfrak{C}$ the sets $\mathrm{Hom}(X, 0)$ and $\mathrm{Hom}(0, X)$ are singletons.
2. Every set of morphisms $\mathrm{Hom}(X, Y)$ has an addition endowing it with the structure of an Abelian group.
3. For any two objects $X, Y \in \mathrm{Ob}\,\mathfrak{C}$ there is a direct product $X \sqcap Y \in \mathrm{Ob}\,\mathfrak{C}$ and a coproduct $X \sqcup Y \in \mathrm{Ob}\,\mathfrak{C}$.
4. For any objects $X, Y, Z \in \mathrm{Ob}\,\mathfrak{C}$ the composition of morphisms

$$\mathrm{Hom}_{\mathfrak{C}}(X, Y) \times \mathrm{Hom}_{\mathfrak{C}}(Y, Z) \to \mathrm{Hom}_{\mathfrak{C}}(X, Z)$$

is a bilinear mapping.

Note that in an additive category the direct product and coproduct of a finite number of objects $X_1, X_2, \ldots, X_n \in \mathrm{Ob}\,\mathfrak{C}$ are isomorphic. Their common value is denoted by $X_1 \oplus X_2 \oplus \cdots \oplus X_n$ and is called their **direct sum**. So that in an additive category for two direct sums of objects $X = X_1 \oplus X_2 \oplus \cdots \oplus X_n$ and $Y = Y_1 \oplus Y_2 \oplus \cdots \oplus Y_n$ there holds:

$$\coprod_{i=1}^{n} X_i \simeq \bigoplus_{i=1}^{n} X_i \simeq \prod_{i=1}^{n} X_i$$

and

$$\mathrm{Hom}_{\mathfrak{C}}(X, Y) = \bigoplus_{i,j} \mathrm{Hom}_{\mathfrak{C}}(X_i, Y_j).$$

A non-zero object $X \in \mathrm{Ob}\,\mathfrak{C}$ is called **indecomposable** if $X = X_1 \oplus X_2$ implies that $X_1 = 0$ or $X_2 = 0$.

The category of Abelian groups is an additive category with trivial group as the zero element. The category of modules over a ring is another example of an additive category with zero module as the zero element.

The motivating prototype of Abelian categories was the category of Abelian groups **Ab**. In an Abelian category objects and morphisms can be added, there exist finite direct products and direct sums, there exist kernels and cokernels satisfying some special properties. More precisely one has the following definition.

**Definition 3.7.2.** A category $\mathfrak{C}$ is said to be **Abelian** if it is an additive category such that:

1. Every morphism has a kernel and a cokernel.
2. Every monomorphism is the kernel of some morphism and every epimorphism is the cokernel of some morphism.

Note that this definition is equivalent to the following:

A additive category $\mathfrak{C}$ is said to be **Abelian** if it satisfies the following extra conditions:

1. It has all pullbacks and pushouts.
2. Every monomorphism is the kernel of some morphism and every epimorphism is the cokernel of some morphism.

**Examples 3.7.3.** The following categories are Abelian.

1. The category of all Abelian groups.
2. The category of all finitely generated Abelian groups.
3. The category of all finite Abelian groups.
4. The category of all left (or right) modules over an associative ring with identity.
5. The category of finitely generated right modules over a right Noetherian ring.
6. The category of vector spaces over a fixed field $k$.
7. The category of finite-dimensional vector spaces over a field $k$.

Note that there are examples of additive categories which are not Abelian. One of such examples is given by M. Auslander in [14].

**Example 3.7.4.** Let $A$ be a Noetherian ring and let $\mathcal{M}$ be the category of finitely generated two-sided $A$-modules. Let $M \in \mathcal{M}$ with dual module $M^* = \mathrm{Hom}_A(M, A) \in \mathcal{M}$. Consider the canonical homomorphism $\delta_M : M \longrightarrow M^{**}$ defined by $\delta_M(m)(f) = f(m)$ for $m \in M$ and $f \in M^*$. A module $M$ is semi-reflexive if $\delta_M$ is a monomorphism. The category of all semi-reflexive modules $\mathcal{F}$ is a sub-category in $\mathcal{M}$. This category is additive, but it is not Abelian in the general case. The category $\mathcal{F}$ is Abelian iff $\mathcal{F} = \mathcal{M}$ iff any module $M \in \mathcal{M}$ is self-reflexive iff any finitely generated module is a submodule of a free module, by [103, proposition

4.5.5]. For a Noetherian ring $A$ this condition is equivalent that $A$ is a self-injective ring[2], by [101, Lemma 4.12.13].

Since there are Noetherian rings which are not self-injective, $\mathcal{F}$ is additive but not Abelian category for such rings. An example of such a ring is $A = k[x, y]/(x, y)^2$ where $k$ is a field.

In any additive category one can introduce the important notions of a two-sided ideal and a radical.

Let $\mathfrak{C}$ be a category. We say that $\mathfrak{R}$ is an **equivalence relation** on $\mathfrak{C}$ if it is an equivalence relation on the morphisms of $\mathfrak{C}$, i.e. if it satisfies the following conditions:

(1) if $f_1, f_2 \in \mathrm{Mor}\,\mathfrak{C}$, then $f_1 \mathfrak{R} f_2$ can occur if and only if there are $X, Y \in \mathrm{Ob}\,\mathfrak{C}$ such that $f_1, f_2 \in \mathrm{Hom}_{\mathfrak{C}}(X, Y)$;

(2) if $f_1 \mathfrak{R} f_2$ and $g \in \mathrm{Hom}_{\mathfrak{C}}(Y, Z)$, $h \in \mathrm{Hom}_{\mathfrak{C}}(W, X)$, then $(g f_1 h)\mathfrak{R}(g f_2 h)$, where $W, X, Y, Z \in \mathrm{Ob}\,\mathfrak{C}$.

An ideal in an additive category $\mathfrak{C}$ is a special case of an equivalence relation on that category.

**Definition 3.7.5.** An **ideal** $\mathfrak{I}$ in an additive category $\mathfrak{C}$ is a collection of subgroups $\mathfrak{I}(A, B)$ of $\mathrm{Hom}_{\mathfrak{C}}(A, B)$ that is stable under composition whenever defined. That is if $\varphi \in \mathfrak{I}(A, B)$ and $f : A' \longrightarrow A$ and $g : B \longrightarrow B'$ are morphisms in $\mathfrak{C}$ then $\varphi f \in \mathfrak{I}(A', B)$ and $g\varphi \in \mathfrak{I}(A, B')$.

The notion seems to have originated with [123]. The notion of an ideal $\mathfrak{I}$ in a category can be succinctly expressed by saying that it is a sub-bi-functor of the Hom functor.

Note that for any additive category $\mathfrak{C}$ and any object $X \in \mathrm{Ob}\,\mathfrak{C}$ in the natural way we can introduce on the sub-category $\mathrm{End}_{\mathfrak{C}}(X)$ the structure of a ring by setting $(\varphi + \psi)(X) = \varphi(X) + \psi(X)$ and $(\varphi\psi)(X) = \varphi(\psi(X))$, for all $\varphi, \psi \in \mathrm{End}_{\mathfrak{C}}(X)$. We write $J(\mathrm{End}_{\mathfrak{C}}(X))$ the Jacobson radical of $\mathrm{End}_{\mathfrak{C}}(X)$.

Let $\mathfrak{C}$ be an additive category. Given $X, Y \in \mathrm{Ob}\,\mathfrak{C}$ we define the **radical**

$$\mathrm{Rad}_{\mathfrak{C}}(X, Y) = \{\varphi \in \mathrm{Hom}_{\mathfrak{C}}(X, Y) \mid \varphi\psi \in J(\mathrm{End}_{\mathfrak{C}}(Y))\ \text{for all}\ \psi \in \mathrm{Hom}_{\mathfrak{C}}(Y, X)\}.$$

Taking into account that the Jacobson radical of a ring $A$ coincides with the set of all elements $r \in A$ such that the element $1 - ra$ is right invertible for all $a \in A$ (see [100, Proposition 3.4.5]), we obtain that $\varphi \in \mathrm{Rad}_{\mathfrak{C}}(X, Y)$ if and only if $1_Y - \varphi\psi$ has a right inverse for each $\psi \in \mathrm{Hom}_{\mathfrak{C}}(Y, X)$.

Another important class of additive categories are Krull-Schmidt categories, which can be considered as a generalization of categories in which Krull-Schmidt theorem holds.

---

[2]Recall that a ring $A$ is **self-injective** if its right and left regular modules are injective.

**Definition 3.7.6.** An additive category $\mathfrak{C}$ is said to be a **Krull-Schmidt category** if every its object decomposes into a finite direct sum of objects having local endomorphism ring.

The uniqueness of decompositions of an object $X \in \mathrm{Ob}\,\mathfrak{C}$ into a direct sum of objects with local endomorphism rings is stated by the following theorem which is analog to the Krull-Remark-Schmidt theorem for semiperfect rings [103, theorem 1.9.5]:

**Theorem 3.7.7.** ([135, Theorem 4.2]). *Let $X$ be an object of an additive category and suppose $X$ has two different decompositions*

$$X = X_1 \oplus \cdots \oplus X_n = Y_1 \oplus \cdots \oplus Y_m$$

*as a finite direct sum of objects whose endomorphism rings are local. Then $m = n$ and there is a permutation $\tau$ of the numbers $i = 1, 2, \ldots, n$ such that $X_i \simeq Y_{\tau(i)}$ for $1 \le i \le n$.*

We say that an additive category $\mathfrak{C}$ has **split idempotents** if for any object $X \in \mathrm{Ob}\,\mathfrak{C}$ every idempotent endomorphism $\varphi = \varphi^2 \in \mathrm{End}_{\mathfrak{C}}(X)$ splits, i.e. there exist morphisms $\pi : X \longrightarrow Y$ and $\iota : Y \longrightarrow X$ such that $\varphi = \iota\pi$ and $\pi\iota = 1_Y$.

From theorem 3.7.7 one obtains the following corollary which gives the equivalent definition of Krull-Schmidt category.

**Corollary 3.7.8.** *An additive category is a Krull-Schmidt category if and only if it has split idempotents and the endomorphism ring of every object is semiperfect.*

The important example of Krull-Schmidt categories is the category of modules having finite composition length. The particular case of this category is the category of finite dimensional modules over an algebra. Another example of Krull-Schmidt categories is the category of finitely generated projective modules over a semiperfect ring.

## 3.8 Notes and References

Most of this chapter consists of a very detailed and very explicit treatment of the Baer group $\mathcal{E}xt_A(Y, X)$ of equivalence classes. This is a (co)homology group and this work dates from 1934, quite a bit of time before (co)homology became important in algebra. As noted it is functorially isomorphic to its derived functor $\mathrm{Ext}^1_A(Y, X)$. There are more sophisticated and shorter (but less detailed) treatments. See e.g. [223, Section 3.4].

A natural question is whether something similar can be done for the higher $\mathrm{Ext}^n_A(Y, X)$. The answer is 'yes': $\mathrm{Ext}^n_A(Y, X)$ can be interpreted as being an Abelian group of equivalence classes of exact sequences:

$$0 \longrightarrow X \longrightarrow Z_1 \longrightarrow Z_2 \longrightarrow \cdots \longrightarrow Z_n \longrightarrow Y \longrightarrow 0 \qquad (3.8.1)$$

See [223, p.79], [144, pp.82-87], [31, §7.5].

This interpretation gives rise to a product structure: given an exact sequence (3.8.1) and a second one

$$0 \longrightarrow Y \longrightarrow Z'_1 \longrightarrow \cdots \longrightarrow Z'_m \longrightarrow Y' \longrightarrow 0 \qquad (3.8.2)$$

they can be spliced together to give an exact sequence

$$0 \longrightarrow X \longrightarrow Z_1 \longrightarrow \cdots \longrightarrow Z_n \longrightarrow Z'_1 \longrightarrow \cdots \longrightarrow Z'_m \longrightarrow Y' \longrightarrow 0$$

This so called **Yoneda splice** respects equivalences of exact sequences and gives rise to Yoneda products, an associative and unital system of pairings of Ext groups. See [148].

Another advantage of the approach to $\mathrm{Ext}^n_A$ via exact sequences is that it does not rely on the presence of enough projectives and/or injectives.

Pullbacks and pushouts (or fibered products and sums) are very important in many parts of mathematics. They exist in many contexts. E.g. there are pullbacks for sets, topological spaces and rings and pushouts for sets, topological spaces and commutative rings (where pushout is the tensor product).

Let

$$\begin{array}{ccc} T & \xrightarrow{\ \delta\ } & X \\ {\scriptstyle\gamma}\downarrow & & \downarrow{\scriptstyle\alpha} \\ Y & \xrightarrow{\ \beta\ } & Z \end{array}$$

be a topological pullback diagram. Then the fiber in $T$ over $y \in Y$, i.e. $\gamma^{-1}(y)$, is the same as the fiber in $X$ over $\beta(y)$. This is often a good thing to have in mind when working with pullbacks.

Abelian categories were introduced by Buchsbaum (1955) (under the name of "exact category") and Grothendieck (1957) in order to unify various cohomology theories. At the time, there was a cohomology theory for sheaves, and a cohomology theory for groups. These two theories were defined differently, but they had similar properties. In fact, much of category theory was developed as a language to study these similarities. Grothendieck unified the two theories: they both arise as derived functors on Abelian categories; the Abelian category of sheaves of Abelian groups on a topological space, and the Abelian category of $G$-modules for a given group $G$.

Krull-Schmidt categories with applications to sheaves were considered by M. Atiyah [11]. The connection of Krull-Schmidt categories with projective covers can be found in the paper of H. Krause [135].

Note that some authors define Krull-Schmidt category as the category in which every its object can be written as a finite direct sum of indecomposable objects in a unique way up to isomorphism and up to a permutation of the direct summands. Such concept of a Krull-Schmidt category can be found e.g. in [14].

# CHAPTER 4

# Modules Over Semiperfect Rings

In this chapter a number of the key results on modules over semiperfect rings are given.

Semiperfect rings are very nice rings to work over as should already be clear from [100, chapter 10]. They are also varied enough (rich enough in phenomena) to yield a large amount of theory, constructions and results, as well, hopefully, become to the realms of various dualities and stableness phenomena.

By definition a ring is semiperfect if it has the idempotent lifting property modulo $R$ the Jacobson radical of $A$ and $A/R$ is Artinian. If $A$ is semiperfect both categories of finitely generated modules, $\text{mod}_r A$ and $\text{mod}_l A$, are Krull-Schmidt, and projective covers do exist. Indeed by theorem 1.9.3, vol. I (Hyman Bass) the existence of projective covers in the category of finitely generated modules is equivalent to semiperfectness.

In [100, section 10.4] the structure of finitely generated projective modules over semiperfect rings was discussed. It was proved that any such right module can be uniquely decomposed into a direct sum of principal right modules. A generalization of this statement, the important theorem [103, theorem 7.2.16] states that any projective module over a semiperfect ring is a direct sum of principal modules. In section 4.1 this result is used to study the structure of finitely generated modules over semiperfect rings. The main result of this section, which was proved by R.B. Warfield, Jr. concerns the decompositions of any finitely generated module over a semiperfect ring. This theorem raises the possibility of introducing the notion of stably isomorphic modules.

In section 4.2 we prove that all modules over a semiperfect ring can be divided into the equivalence classes of stably isomorphic modules. Moreover, each stable isomorphism class of finitely generated modules over a semiperfect ring contains a unique (up to isomorphism) minimal element. Most of the results of this section were obtained by R.B. Warfield, Jr. (see [220], [222]).

In [101, section 4.10], we looked at the property of duality in Noetherian rings, which is given by the covariant functor $^* = \text{Hom}_A(-, A)$. For an arbitrary ring $A$ this functor induces a duality between the full subcategories of finitely generated projective right $A$-modules and left $A$-modules. In section 4.3 the main properties of this functor and torsionless modules are studied for the case of modules over semiperfect rings. This section presents an introduction to the duality theory of Auslander and Bridger [13], and yields a relationship between finitely presented right modules and finitely presented left modules over semiperfect rings. Some main properties of the Auslander-Bridger transpose, which is closely related to almost split sequences, are discussed. These sequences were first introduced and studied by M. Auslander and I. Reiten in [15] and [20], and they play an important role in the representation theory of rings and finite dimensional algebras. Section 4.4 is devoted to the study of some main properties of these sequences. In section 4.5 we give some useful identities for finitely presented modules which connect the tensor functor $\otimes$ with the functor Hom. The important concepts of pure submodules and pure sequences and some their properties are considered in section 4.6. In section 4.7 we consider almost split sequences over semiperfect rings and prove the existence of these sequences for strongly indecomposable modules. Section 4.8 presents an introduction to the theory of linkage of modules over semiperfect Noetherian rings using two types of functors: syzygy and transpose.

## 4.1 Finitely Generated Modules over Semiperfect Rings

Let $A$ be a ring with Jacobson radical $R$. Recall that $A$ is semiperfect if $A/R$ is Artinian and idempotents can be lifted modulo $R$. Another equivalent definition of a semiperfect ring is given by [103, theorem 1.9.3]. A principal right module $P$ over a semiperfect ring $A$ is an indecomposable projective right $A$-module. Any principal right $A$-module $P$ has the form $eA$, where $e$ is a local idempotent, i.e. $eAe$ is a local ring. Moreover, $P = eA$ has exactly one maximal submodule which is of the form $PR = eR$, where $R$ is the Jacobson radical of the ring $A$ (see [103, theorem 1.9.4]). Since $e$ is a local idempotent, $\text{End}_A(P) = eAe$ is a local ring. Therefore any principal module over a semiperfect ring is strongly indecomposable. Using this and [103, proposition 7.1.30], we immediately obtain the following result.

**Proposition 4.1.1.** *Every principal module over a semiperfect ring has the exchange property.*

See [103, section 7.1] for a discussion of the exchange property.

A Krull-Schmidt theory holds for finitely generated modules over a semiperfect ring. Namely we have the Krull-Remak-Schmidt theorem [103, theorem 1.9.5] and its corollary in the form of [103, proposition 1.9.6], which states that a finitely generated projective module over a semiperfect ring can be uniquely decomposed up to isomorphism into a finite direct sum of principal modules. This result was generalized by L.H. Rowen to the case of all f.g. modules over a semiperfect ring. Proposition [103, proposition 9.2.3] states that any f.g. module over a semiperfect ring can be decomposed into a finite direct sum of indecomposable submodules.

Then from [103, proposition 1.9.6, corollary 7.1.13] and proposition 4.1.1 we obtain the following result.

**Proposition 4.1.2.** *Every finitely generated projective module over a semiperfect ring has the exchange property.*

Recall that an $A$-module $M$ is said to have the **cancellation property** when $M \oplus X = M_1 \oplus Y$ with $M \simeq M_1$ implies $X \simeq Y$ for any pair of modules $X$ and $Y$.

**Proposition 4.1.3.** *Every finitely generated projective module over a semiperfect ring has the cancellation property.*

*Proof.* Since, by [103, proposition 1.9.6], any finitely generated projective module over a semiperfect ring uniquely decomposes into a finite direct sum of principal modules which have local endomorphism rings, the statement follows by induction from [103, proposition 7.3.5], applying it separately to each principal module. □

**Theorem 4.1.4.** (R.B. Warfield, Jr. [222, Theorem 1.4]). *Let $M$ be a finitely generated module over a semiperfect ring $A$. Then there is a decomposition $M = N \oplus P$, where $P$ is projective and $N$ has no projective summands. Further, if $M = N_1 \oplus P_1$ is another such decomposition, then $N \simeq N_1$ and $P \simeq P_1$.*

*Proof.* Since $A$ is semiperfect and $M$ is a finitely generated $A$-module, the indicated decomposition follows from [103, proposition 9.2.3]. Suppose that there are two such decompositions: $M = N \oplus P = N_1 \oplus P_1$. Since $M$ is a finitely generated module, any of its direct summands is finitely generated as well. So $P, P_1, N, N_1$ are finitely generated submodules. By proposition 4.1.2, $P$ and $P_1$ have the exchange property. Applying the exchange property for $P$ we obtain that there are submodules $\widetilde{N_1} \subseteq N_1$ and $\widetilde{P_1} \subseteq P_1$ such that $M = P \oplus \widetilde{N_1} \oplus \widetilde{P_1}$. Then from proposition 4.1.3 it follows that $N \simeq \widetilde{N_1} \oplus \widetilde{P_1}$. Since $N$ has no non-zero projective summands, this implies that $\widetilde{P_1} = 0$, i.e. $N \simeq \widetilde{N_1}$ and $M = P \oplus \widetilde{N_1}$. Decomposing $N_1$:

$$N_1 = N_1 \cap M = N_1 \cap (P \oplus \widetilde{N_1}) = (N_1 \cap P) \oplus \widetilde{N_1},$$

we obtain that

$$P \oplus \widetilde{N_1} = M = P_1 \oplus N_1 = P_1 \oplus (N_1 \cap P) \oplus \widetilde{N_1}.$$

Comparing the two complements to $\widetilde{N_1}$, one sees that $P_1 \oplus (N_1 \cap P) \simeq P$. Since, by hypothesis, $N_1$ has no non-zero projective summands, it follows that $N_1 \cap P = 0$ and $N_1 = \widetilde{N_1}$. The above formulas now imply that $N \simeq N_1$ and $P \simeq P_1$, as required. □

Recall also the important theorem [103, theorem 7.2.16], which is a generalization of [103, proposition 1.9.6]. It gives the full description of projective modules over a semiperfect rings and states that any projective module over a semiperfect ring is a direct sum of principal modules.

## 4.2 Stable Equivalence

In this section we show that using theorem 4.1.4 it is possible to divide all modules over a semiperfect ring into equivalence classes. The modules in such a class are called stably isomorphic, following to R.B. Warfield, Jr.

**Definition 4.2.1.** Let $A$ be a ring. Two $A$-modules $X$ and $Y$ are said to be **stably isomorphic** (or **projectively equivalent**) if there exist projective $A$-modules $P$ and $Q$ such that $X \oplus P \simeq Y \oplus Q$. A module $M$ is called **stable** if it has no projective summands.

If $X$ and $Y$ are finitely generated, one can assume that $P$ and $Q$ are also finitely generated.

It is easy to verify that stable isomorphism is an equivalence relation on the set of all $A$-modules. Using definition 4.2.1 there is the following corollary from theorem 4.1.4.

**Corollary 4.2.2.** *Let $A$ be a semiperfect ring. Then*

1. *Every finitely generated module is stably isomorphic to a stable module.*
2. *Two stable modules are stably isomorphic if and only if they are isomorphic.*

Let $M$, $N$ be right modules over a ring $A$.

**Definition 4.2.3.** A monomorphism $f : N \to M$ is called **essential** if for each sequence of $A$-modules

$$N \xrightarrow{f} M \xrightarrow{\varphi} X$$

such that $\varphi f$ is a monomorphism, $\varphi$ is a monomorphism.

An epimorphism $f : M \to N$ is called **essential** if for each sequence of $A$-modules

$$X \xrightarrow{\varphi} M \xrightarrow{f} N$$

such that $f \varphi$ is an epimorphism, $\varphi$ is an epimorphism.

Recall that a submodule $N$ of a module $M$ is called **essential** if $N \cap L \neq 0$ for any non-zero submodule $L \subseteq M$. A submodule $N$ of a module $M$ is called **small** if the equality $N + X = M$ implies that $X = M$ for any submodule $X$ of $M$.

The following proposition shows the close relationships between essential submodules and essential monomorphisms, and their dual concepts, small submodules and essential epimorphisms.

**Proposition 4.2.4.**

1. *Let $N, M$ be $A$-modules. Then a monomorphism $f : N \to M$ is an essential monomorphism if and only if $\operatorname{Im} f$ is an essential submodule in $M$.*
2. *Let $M, P$ be right $A$-modules. Then an epimorphism $f : P \to M$ is essential if and only if $\operatorname{Ker} f$ is a small submodule in $P$.*

*Proof.*

1. Let $K = \operatorname{Im} f$. Suppose that $\operatorname{Im} f \subseteq_e M$ and $\varphi f$ is a monomorphism, where $\varphi : M \to X$ is a homomorphism of $A$-modules. Let $x \in \operatorname{Ker}\varphi \cap \operatorname{Im} f$. Then $\varphi(x) = 0$ and there exists an $y \in N$ such that $x = f(y)$. Therefore $\varphi(x) = \varphi f(y) = 0$, which implies that $y = 0$ as $\varphi f$ is a monomorphism. So $x = f(y) = 0$, i.e. $\operatorname{Ker}\varphi \cap \operatorname{Im} f = 0$. Since $\operatorname{Im} f \subseteq_e M$, this means that $\operatorname{Ker}\varphi = 0$, i.e. $\varphi$ is a monomorphism and so $f$ is an essential monomorphism. Conversely, suppose that $f$ is an essential monomorphism. Suppose that $K$ is a submodule of $M$ and $\operatorname{Im} f \cap K = 0$. Consider a natural epimorphism $\varphi : M \to M/K$ and the composition $N \overset{f}{\to} M \overset{\varphi}{\to} M/K$. Then $\varphi f$ is a monomorphism as $\operatorname{Im} f \cap K = 0$. Therefore, by assumption, $\varphi$ is also a monomorphism, i.e. $K = 0$. This means that $\operatorname{Im} f \subseteq_e M$.

2. Suppose an epimorphism $f : P \to M$ is essential, and $X + \operatorname{Ker} f = P$ for some submodule $X$ of $P$. Let $\varphi : X \to P$ be the natural inclusion. Then there is a sequence $X \overset{\varphi}{\to} P \overset{f}{\to} M$, where $f\varphi$ is an epimorphism, since $f$ is an epimorphism and $X + \operatorname{Ker} f = P$. Then, by hypothesis, $\varphi$ is also an epimorphism, and so $\varphi$ is an isomorphism. Therefore $X = P$, i.e. $\operatorname{Ker} f$ is a small submodule in $P$.

   Conversely, suppose that there is an epimorphism $f : P \to M$ such that $\operatorname{Ker} f$ is a small submodule in $P$. Let $\varphi : N \to P$ be a homomorphism such that $f\varphi$ is an epimorphism. It follows that $\operatorname{Ker} f + \operatorname{Im} \varphi = P$. Since $\operatorname{Ker} f$ is small, $\operatorname{Im} \varphi = P$, i.e. $\varphi$ is an epimorphism, and so $f$ is essential. $\square$

Recall that a projective module $P(M)$ with an essential epimorphism

$$f : P(M) \longrightarrow M$$

is called a **projective cover** of a module $M$. Proposition 4.2.4 provides an equivalent definition of a projective cover.

Note that not every module has a projective cover. But if $A$ is a semiperfect ring, then any finitely generated $A$-module $M$ has a projective cover $P(M)$, by [103, theorem 1.9.3].

**Lemma 4.2.5.** *If $A$ is a semiperfect ring with Jacobson radical $R$, $P$ is a finitely generated projective $A$-module, and $N$ a submodule not contained in $PR$, then $N$ contains a non-zero summand of $P$.*

*Proof.* Since $P$ is a finitely generated $A$-module and $N$ is a submodule in $P$, $M = P/N$ is a finitely generated $A$-module. Therefore, by [103, theorem 1.9.3], $M$ has a projective cover $P(M)$. Then, by the Bass lemma [100, lemma 10.4.5], there is a decomposition $P = P(M) \oplus Q$, where $Q \subset N$ and $P(M) \cap N$ is a small submodule in $P(M)$. Suppose that $Q = 0$, then $P = P(M)$, and from [100, lemma 10.4.1] it follows that $N \subseteq \operatorname{rad}(P)$. Since $P$ is a non-zero projective $A$-module, $\operatorname{rad}(P) = PR$, by [100, proposition 5.1.8], and so we obtain $N \subseteq PR$. A contradiction. $\square$

**Definition 4.2.6.** An exact sequence

$$P_1 \xrightarrow{\pi_1} P_0 \xrightarrow{\pi_0} M \longrightarrow 0 \qquad (4.2.7)$$

is called a **minimal projective resolution** (or **minimal projective presentation**) of a module $M$ if $P_0 \xrightarrow{\pi_0} M$ and $P_1 \xrightarrow{\pi_1} \mathrm{Ker}(\pi_0)$ are projective covers.

Recall that a module $M$ is called **finitely presented** if it is finitely generated and there is an epimorphism $\varphi$ of a finitely generated projective module $P$ onto the module $M$ such that $\mathrm{Ker}\,\varphi$ is a finitely generated module.

As immediate consequence of [103, theorem 1.9.3], we have the following result.

**Proposition 4.2.8.** *For any finitely presented module $M$ over a semiperfect ring $A$ there is a minimal projective resolution*

$$Q \xrightarrow{\varphi} P \longrightarrow M \longrightarrow 0$$

*with finitely generated projective modules $P$ and $Q$. Moreover, the following statements hold:*

1. *The induced homomorphism $P/PR \to M/MR$ is an isomorphism.*
2. *If $K$ is the kernel of the homomorphism $P \to M$ then the induced homomorphism $Q/QR \to K/KR$ is an isomorphism.*
   *In this case the isomorphism types of $P/PR$ and $Q/QR$ are invariants of $M$.*

Here note that $\varphi$ maps $Q$ into $K$ because $Q \longrightarrow P \longrightarrow M \longrightarrow 0$ is exact.

The last sequence of the proposition means that if $M' \simeq M$ and

$$Q' \longrightarrow P' \longrightarrow M' \longrightarrow 0$$

is a minimal projective resolution of $M'$ then $P'/P'R \simeq P/PR$ and $Q'/Q'R \simeq Q/QR$.

Let $A$ be a semiperfect ring with Jacobson radical $R$. Let $M$ be a finitely generated $A$-module, and $S$ a simple $A$-module. Denote by $\mu(M)$ the number of direct summands in a decomposition of $M/MR$ as a direct sum of simple modules, and by $\mu(M, S)$ the number of such summands isomorphic to $S$. These numbers are well-defined by the Krull-Schmidt theorem [103, theorem 1.8.4].

Let $M$ be a finitely presented $A$-module, $f : P \to M$ a projective cover, and $K = \mathrm{Ker}\,f$. Write $\nu(M) = \mu(K)$, and $\nu(M, S) = \mu(K, S)$. These numbers are also well-defined by [100, lemma 13.1.1].

**Proposition 4.2.9.** *If $M = X \oplus Y$ is a finitely generated module over a semiperfect ring $A$, and $S$ is a simple $A$-module, then $\mu(M, S) = \mu(X, S) + \mu(Y, S)$. If, moreover, $M = X \oplus Y$ is a finitely presented module, then also $\nu(M, S) = \nu(X, S) + \nu(Y, S)$.*

The proof of this proposition immediately follows from the definition of projective cover.


## 4.3 Auslander-Bridger Duality

Denote by $\text{Mod}_r A$ ($\text{Mod}_l A$) the category of all right (left) $A$-modules, and denote by $\text{mod}_r A$ ($\text{mod}_l A$) the category of all right (left) finitely generated $A$-modules. From [103, proposition 4.5.9] it follows that the functor

$$* = \text{Hom}_A(-, A) : \text{Mod}_r A \to \text{Mod}_l A$$

induces a duality between the full subcategories of finitely generated projective modules in $\text{Mod}_r A$ and $\text{Mod}_l A$. Note that $M^*$ is projective (and finitely generated) if $M$ is projective (and finitely generated) because, being projective, $M$ is a direct summand of a (finitely generated) free module.

In this section we always assume that $A$ is a semiperfect ring. Some of the results hold without the Noetherianness assumption.

Let $A$ be a ring, and let $M$ be a right $A$-module with minimal projective resolution

$$P_1 \xrightarrow{\varphi} P_0 \longrightarrow M \longrightarrow 0 \tag{4.3.1}$$

where $P_0$ and $P_1$ are finitely generated projective modules. Such a resolution exists when, for example, $A$ is a semiperfect ring and $M$ is a finitely presented module, or $A$ is a semiperfect Noetherian ring and $M$ is a finitely generated module (since any f.g. module over a Noetherian ring is f.p.).

Applying the duality functor $* = \text{Hom}_A(-, A)$ to the exact sequence (4.3.1), there results an exact sequence:

$$0 \longrightarrow M^* \longrightarrow P_0^* \xrightarrow{\varphi^*} P_1^* \xrightarrow{\omega} \text{Tr}(M) \longrightarrow 0 \tag{4.3.2}$$

where

$$\text{Tr}(M) = \text{Coker}(\varphi^*) \tag{4.3.3}$$

is a left module over the opposite ring $A^{op}$, which is called the **transpose** of $M$. It is a simple matter to show that $\text{Tr}(M)$ is well-defined (up to isomorphism).

Given this, one naturally suspects that $\text{Tr}(-)$ should be a functor. However there arises trouble as soon as one tries to define $\text{Tr}(f)$ for a morphism $f : M \longrightarrow M'$. Given such a morphism the projectiveness of $P_1$ and $P_0$ immediately yields a commutative diagram

$$
\begin{array}{ccccccc}
P_1 & \longrightarrow & P_0 & \longrightarrow & M & \longrightarrow & 0 \\
\downarrow{\scriptstyle f_1} & & \downarrow{\scriptstyle f_0} & & \downarrow{\scriptstyle f} & & \\
P_1' & \longrightarrow & P_0' & \longrightarrow & M' & \longrightarrow & 0
\end{array}
$$

for suitable morphisms $f_0$ and $f_1$. However, there is nothing unique about these morphisms and that appears to prevent Tr from being a functor.

**Lemma 4.3.4.** *Let A be a semiperfect ring, and M a finitely presented module. Then M is projective if and only if* $\text{Tr}(M) = 0$.

*Proof.* If $M = P_0$ is projective, then we have a minimal projective resolution of the form

$$0 \longrightarrow 0 \longrightarrow P_0 \longrightarrow P_0 \longrightarrow 0,$$

and hence

$$\text{Tr}(M) = \text{Coker}(P_0^* \to 0) = 0.$$

Conversely, if $\text{Tr}(M) = 0$, then $\varphi^*$ in (4.3.2) is an epimorphism. Since $P_1^*$ is a projective module, $\varphi^*$ is a split epimorphism, and so $\varphi$ is a split monomorphism, by [103, lemma 4.5.14], i.e. $M$ is isomorphic to a direct summand of $P_0$, and so $M$ is projective. $\square$

If $A$ is a semiperfect ring, then from corollary 4.2.2 it follows that each stable isomorphism class of finitely generated modules has a canonical minimal element, which is a stable module, i.e. a module without non-zero projective summands.

**Proposition 4.3.5.** *Let A be a semiperfect ring with Jacobson radical R, M a finitely presented right A-module with minimal projective resolution*

$$P_1 \xrightarrow{\varphi} P_0 \longrightarrow M \longrightarrow 0$$

*and the corresponding induced presentation* (4.3.2) *for* Tr$M$. *Then*

1. *The map* $\omega : P_1^* \longrightarrow \text{Tr}(M)$ *is a projective cover.*
2. *M is a stable module if and only if the induced sequence*

$$P_0^* \xrightarrow{\varphi^*} P_1^* \xrightarrow{\omega} \text{Tr}(M) \longrightarrow 0 \qquad (4.3.6)$$

   *is a minimal projective resolution for* Tr$(M)$.
3. Tr$(M)$ *is a stable module.*
4. Tr$(\text{Tr}(M)) \simeq M$.

*Proof.*

1. Since $M$ is f.p. module over a semiperfect ring, $P_0$, $P_1$ are f.g. projective modules. Therefore $P_0^*$ and $P_1^*$ are f.g. projective $A^{op}$-modules and $P_0^{**} \cong P_0$, $P_1^{**} \cong P_1$, by [103, proposition 4.5.9]. To show that $\omega : P_1^* \longrightarrow \text{Tr}(M)$ is a projective cover it suffices to prove that $\text{Im}\varphi^* = \text{Ker}\,\omega \subseteq RP_1^*$, by [100, lemma 10.4.4]. If not, by lemma 4.2.5, we could decompose $P_0^* = X \oplus Y$, where $\varphi^*$ maps $Y$ isomorphically onto a non-zero summand of $P_1^*$. Dualizing again, and identifying $P_0^{**}$ with $P_0$, we would obtain a non-zero summand $Y^*$ of $P_0$ in the kernel of the homomorphism from $P_0$ to $M$, contradicting the fact that the projective resolution is minimal.

2. Suppose that $M$ is a stable module. By the previous statement $\omega$ is a projective cover. The homomorphism $\varphi^*$ is constructed as the superposition (Yoneda slice) of two exact sequences

$$0 \longrightarrow \mathrm{Ker}\omega \longrightarrow P_1^* \overset{\omega}{\longrightarrow} \mathrm{Tr}(M) \longrightarrow 0$$

and

$$0 \longrightarrow M^* \longrightarrow P_0^* \overset{\theta}{\longrightarrow} \mathrm{Ker}\omega \longrightarrow 0.$$

Suppose that $\theta : P_0^* \longrightarrow \mathrm{Ker}\omega$ is not a projective cover. By [100, lemma 10.4.4], $M^* \not\subseteq RP_0^*$. Then from lemma 4.2.5 it follows that $M^*$ and $P_0^*$ have a common non-zero projective summand, which is then also true for $P_0^{**}$ and $M^{**}$. Then from the commutative diagram

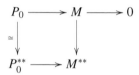

it follows that $M$ also has a projective summand, which contradicts the assumption that $M$ is stable.

Conversely, suppose that the induced sequence (4.3.6) is a minimal projective resolution for $\mathrm{Tr}\,M$. If $M$ is not stable, it has a non-zero projective summand. Then $M^*$ and $P_0^*$ have a common non-zero projective summand, which contradicts the assumption.

3. Let's now see that $\mathrm{Tr}(M)$ is a stable module. Suppose that $\mathrm{Tr}(M)$ has a non-zero projective summand, then as before $\varphi^*(P^*)$ would lie in a proper summand of $P_1^*$, and dualyzing, one would obtain a non-zero summand of $P_1$ in the kernel of $\varphi$, contradicting the minimality of the projective resolution (4.3.1).

4. Since (4.3.6) is a minimal projective resolution for $\mathrm{Tr}(M)$, and $\mathrm{Tr}(M)$ has no projective summands, we can form the induced sequence:

$$P_1^{**} \overset{\varphi^{**}}{\longrightarrow} P_0^{**} \longrightarrow \mathrm{Tr}(\mathrm{Tr}(M)) \longrightarrow 0$$

which is a minimal projective resolution for $\mathrm{Tr}(\mathrm{Tr}(M))$, by statement 2. Since $P_0^{**} \simeq P_0$ and $P_1^{**} \simeq P_1$, we have a commutative diagram

$$
\begin{array}{ccccccc}
P_1 & \longrightarrow & P_0 & \longrightarrow & M & \longrightarrow & 0 \\
{\scriptstyle g_1}\downarrow {\scriptstyle \simeq} & & {\scriptstyle g_0}\downarrow {\scriptstyle \simeq} & & \downarrow {\scriptstyle f'} & & \\
P_1^{**} & \longrightarrow & P_0^{**} & \longrightarrow & \mathrm{Tr}(\mathrm{Tr}(M)) & \longrightarrow & 0
\end{array}
$$

with exact rows. Since $g_0$ and $g_1$ are isomorphisms, from [103, corollary 1.2.9], it follows that $f'$ is also an isomorphism, i.e. $M \simeq \mathrm{Tr}(\mathrm{Tr}(M))$. $\square$

If a finitely presented module $M$ has a minimal projective resolution of the form

$$A^m \longrightarrow A^n \longrightarrow M \longrightarrow 0$$

(so that (in a minimal way) it is given by $n$ generators and $m$ relations), then $\text{Tr}(M)$ is given by $m$ generators and $n$ relations.

**Theorem 4.3.7.** *Let A be a semiperfect ring with Jacobson radical R, let $M$ be the class of finitely presented stable right A-modules, and $N$ the class of finitely generated stable left $A^{op}$-modules. Then for each $M \in M$ there exists an element $\text{Tr}(M) \in N$, uniquely determined up to isomorphism, such that*

1. $M \simeq N$ *if and only if* $\text{Tr}(M) \simeq \text{Tr}(N)$.
2. $\text{Tr}(M \oplus N) \simeq \text{Tr}(M) \oplus \text{Tr}(N)$.

   *Furthermore, if to each simple right A-module V we associate a corresponding simple left $A^{op}$-module U as in [103, lemma 4.5.11], then the following equalities hold:*

$$\mu(M, V) = \nu(\text{Tr}(M), U), \quad and \quad \nu(M, V) = \mu(\text{Tr}(M), U) \tag{4.3.8}.$$

*Proof.*

1. This follows immediately from [103, lemma 4.5.11].
2. This follows from the fact, that if

$$P_1 \xrightarrow{\varphi} P_0 \longrightarrow M \longrightarrow 0$$

and

$$Q_1 \xrightarrow{\varphi_1} Q_0 \longrightarrow N \longrightarrow 0$$

are minimal projective resolutions for $M$ and $N$, respectively, then

$$P_1 \oplus Q_1 \longrightarrow P_0 \oplus Q_0 \longrightarrow M \oplus N \longrightarrow 0$$

is a minimal projective resolution for $M \oplus N$. $\square$

Following M. Auslander [14] we now introduce the important notion of a category modulo some equivalence relation, which will play an important role in the further study.

Let $C$ be a category. We define the category $C/\mathcal{R}$, which is called the **category $C$ modulo $\mathcal{R}$**, as follows:

1. $\text{Ob}(C/\mathcal{R}) = \text{Ob}\,C$.
2. For any objects $X, Y \in \text{Ob}(C/\mathcal{R}) = \text{Ob}\,C$ we define the set of morphisms as $\text{Hom}_{C/\mathcal{R}}(X, Y) = \text{Hom}_C(X, Y)/\mathcal{R}$.
3. The composition of morphisms in $C/\mathcal{R}$:

$$\text{Hom}_{C/\mathcal{R}}(X, Y) \times \text{Hom}_{C/\mathcal{R}}(Y, Z) \longrightarrow \text{Hom}_{C/\mathcal{R}}(X, Z)$$

is defined as a unique map which makes the following diagram

$$\mathrm{Hom}_C(X, Y) \times \mathrm{Hom}_C(Y, Z) \longrightarrow \mathrm{Hom}_C(X, Z)$$

$$\mathrm{Hom}_{C/\mathcal{R}}(X, Y) \times \mathrm{Hom}_{C/\mathcal{R}}(Y, Z) \longrightarrow \mathrm{Hom}_{C/\mathcal{R}}(X, Z)$$

commutative, where each of the vertical arrows is the canonical map of a set $S$ to $S/\mathcal{R}$, where $S/\mathcal{R}$ is the set of equivalent classes of $S$ under the equivalence relation $\mathcal{R}$.

**Definition 4.3.9.** We say that a morphism $f : M \longrightarrow N$ in $\mathrm{mod}_r A$ **factors through a projective module** if it can be written as a composition $M \xrightarrow{g} P \xrightarrow{h} N$ with a projective $A$-module $P$, i.e. $f = hg$.

Consider the following construction. Let $\mathcal{P}_A(M, N)$ denote the subset of $\mathrm{Hom}_A(M, N)$ consisting of all homomorphisms which factor through a finitely generated projective $A$-modules. A first step is to show that this defines an ideal $\mathcal{P}$ in the category $\mathrm{mod}_r A$.

Let's first show that $\mathcal{P}_A(M, N)$ is a subgroup in $\mathrm{Hom}_A(M, N)$. Assume $f, f_1 \in \mathcal{P}(M, N)$, then there exist projective modules $P$, $P_1$ and homomorphisms $g, g_1, h, h_1$ such the following diagrams

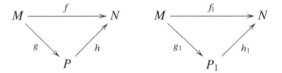

are commutative. Then the homomorphism

$$f + f_1 = hg + h_1 g_1 = \begin{bmatrix} h & h_1 \end{bmatrix} \begin{bmatrix} g \\ g_1 \end{bmatrix}$$

factors through the projective module $P \oplus P_1$ and so $f, f_1 \in \mathcal{P}(M, N)$.

Suppose $f \in \mathcal{P}(M, L)$ and $g \in \mathrm{Hom}_A(L, N)$, then there is a commutative diagram

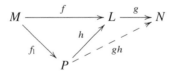

i.e. $gf = ghf_1 \in \mathcal{P}(M, N)$. Analogously if $g \in \mathrm{Hom}_A(M, L)$ and $f \in \mathcal{P}(L, N)$, then there is a commutative diagram

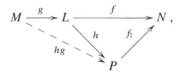

i.e. $fg = f_1gh \in \mathcal{P}(M, N)$. Thus, the relation factoring through a projective module is an equivalence relation $\mathcal{R}$ on the category $\text{mod}_r A$.

So one can consider the quotient category

$$\underline{\text{mod}}_r A = \text{mod}_r A/\mathcal{P}, \qquad (4.3.10)$$

which is called the **projective stable category** of $\text{mod}_r A$ modulo projectives. The objects of this category are the same as those in $\text{mod}_r A$, and the group of morphisms $\underline{\text{Hom}}_A(M, N)$ of morphisms from $M$ to $N$ in $\underline{\text{mod}}_r A$ is defined as the quotient group

$$\underline{\text{Hom}}_A(M, N) = \text{Hom}_A(M, N)/\mathcal{P}(M, N) \qquad (4.3.11)$$

of $\text{Hom}_A(M, N)$ with the composition of morphisms induced by a composition in $\text{mod}_r A$. For each $f \in \text{Hom}_A(M, N)$ denote by $\underline{f}$ the image of $f$ in $\underline{\text{Hom}}_A(M, N)$.

Let

$$P_1 \xrightarrow{\varphi} P_0 \xrightarrow{\psi} M \longrightarrow 0 \qquad (4.3.12)$$

$$Q_1 \xrightarrow{\varphi_1} Q_0 \xrightarrow{\psi_1} N \longrightarrow 0 \qquad (4.3.13)$$

be minimal projective resolutions for finitely generated right $A$-modules $M$ and $N$, respectively. Then, by [100, theorem 6.2.2], any homomorphism $f : M \longrightarrow N$ induces a commutative exact diagram:

$$
\begin{array}{ccccccc}
P_1 & \xrightarrow{\varphi} & P_0 & \xrightarrow{\psi} & M & \longrightarrow & 0 \\
\downarrow{g_1} & & \downarrow{g_0} & & \downarrow{f} & & \\
Q_1 & \xrightarrow{\varphi_1} & Q_0 & \xrightarrow{\psi_1} & N & \longrightarrow & 0
\end{array}
$$

**Diag. 4.3.14.**

**Lemma 4.3.15.** *Let $M$, $N$ be finitely presented $A$-modules with minimal projective resolutions (4.3.12) and (4.3.13) respectively. Then the commutative exact diagram (4.3.14) has the property that there is a homomorphism $\tau : M \longrightarrow Q_0$ such that $\psi_1\tau = f$ if and only if there is a homomorphism $\delta : P_0 \longrightarrow Q_1$ such that $\varphi_1\delta\varphi = g_0\varphi = \varphi_1 g_1$.*

*Proof.*

1. Suppose there exists a homomorphism $\delta : P_0 \longrightarrow Q_1$ such that $\varphi_1\delta\varphi = g_0\varphi = \varphi_1 g_1$.

$$
\begin{array}{ccccccc}
P_1 & \xrightarrow{\varphi} & P_0 & \xrightarrow{\psi} & M & \longrightarrow & 0 \\
\downarrow{g_1} & \swarrow{\delta} & \downarrow{g_0} & & \downarrow{f} & & \\
Q_1 & \xrightarrow{\varphi_1} & Q_0 & \xrightarrow{\psi_1} & N & \longrightarrow & 0
\end{array}
$$

Then $(\varphi_1\delta - g_0)\varphi = 0$, and so there exists a homomorphism $\tau : M \longrightarrow Q_0$ such that $\varphi_1\delta - g_0 = \tau\psi$.

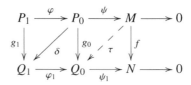

But then $\psi_1\tau\psi = \psi_1(\varphi_1\delta - g_0) = \psi_1 g_0 = f\psi$, hence $\psi_1\tau = f$, since $\psi$ is an epimorphism.

2. Suppose there is a homomorphism $\tau : M \longrightarrow Q_0$ such that $\psi_1\tau = f$. Then $\psi_1(g_0 - \tau\psi) = \psi_1 g_0 - \psi_1\tau\psi = \psi_1 g_0 - f\psi = 0$. Therefore there exists a homomorphism $\delta : P_0 \rightarrow Q_1$ such that $\varphi_1\delta = g_0 - \tau\psi$. Hence $\varphi_1\delta\varphi = g_0\varphi - \tau\psi\varphi = g_0\varphi$. $\square$

Applying to diagram (4.3.14) the duality functor $^*$, one gets a homomorphism $f' : \mathrm{Tr}(N) \rightarrow \mathrm{Tr}(M)$ such that the diagram

$$
\begin{array}{ccccccc}
Q_0^* & \xrightarrow{\varphi_1^*} & Q_1^* & \xrightarrow{\psi_1^*} & \mathrm{Tr}(N) & \longrightarrow & 0 \\
\downarrow{\scriptstyle g^*} & & \downarrow{\scriptstyle g_1^*} & & \downarrow{\scriptstyle f'} & & \\
P_0^* & \xrightarrow{\varphi^*} & P_1^* & \xrightarrow{\psi^*} & \mathrm{Tr}(M) & \longrightarrow & 0
\end{array}
$$

**Diag. 4.3.16.**

is commutative.

So if we choose a fixed minimal projective resolution of $M$ in $\mathrm{mod}_r A$ and define $\mathrm{Tr}(M) = \mathrm{Coker}(\varphi^*)$ we obtain a map from the objects of $\mathrm{mod}_r A$ (and so to the objects of $\underline{\mathrm{mod}}_r A$) to the objects of $\mathrm{mod}_l A$ (and so the objects of $\underline{\mathrm{mod}}_l A$). And there is a map

$$
\mathrm{Tr} : \mathrm{Hom}_A(M, N) \rightarrow \mathrm{Hom}_A(\mathrm{Tr}(N), \mathrm{Tr}(M))
$$

defined by $\mathrm{Tr}(f) = f'$.

Note that the constructions of diagrams (4.3.14) and (4.3.16) are not necessarily unique, since $f'$ depends on the choice of $g_0$ and $g_1$. Suppose we have another "factorization" of $f$, say by maps $h_0$ and $h_1$. Construct $h'$ like $f'$ from $g_0$ to $g_1$. Since any two extensions of $f$ are homotopic, by [100, theorem 6.2.2], there exists a homomorphism $\Delta_1 : P_0 \rightarrow Q_1$ such that $g_0 - h_0 = \varphi_1\Delta_0$. Then the difference $g_0 - h_0 \in \mathrm{Im}(\varphi_1) = \mathrm{Ker}(\psi_1)$ factors through $Q_1$, and so $f' - h'$ factors through the finitely generated projective module $Q_1^*$, and so there is the following commutative diagram:

$$
\begin{array}{ccccccc}
Q_0^* & \xrightarrow{\varphi_1^*} & Q_1^* & \xrightarrow{\psi_1^*} & \mathrm{Tr}(N) & \longrightarrow & 0 \\
\downarrow{\scriptstyle g_0^*-h_0^*} & {\scriptstyle \tau} & \downarrow{\scriptstyle g_1^*-h_1^*} & & \downarrow{\scriptstyle f'-h'} & & \\
P_0^* & \xrightarrow{\varphi^*} & P_1^* & \xrightarrow{\psi^*} & \mathrm{Tr}(M) & \longrightarrow & 0
\end{array}
$$

Then from lemma 4.3.15 it follows that there exists a homomorphism

$$\sigma : \mathrm{Tr}(N) \longrightarrow P_1^*$$

such that $\sigma \varphi^* = f' - h'$, i.e. the homomorphism $f' - g'$ factors through the projective module $P_1^*$. Thus we have a map

$$\mathrm{Tr} : \mathrm{Hom}_A(M, N) \longrightarrow \underline{\mathrm{Hom}}_A(\mathrm{Tr}(N), \mathrm{Tr}(M))$$

where $\mathrm{Tr}(f) = \underline{f'} = \underline{\mathrm{Tr}(f)}$ for each $f \in \mathrm{Hom}_A(M, N)$. Since by lemma 4.3.15 $\underline{\mathrm{Tr}(f)} = 0$ in $\underline{\mathrm{Hom}}_A(\mathrm{Tr}(N), \mathrm{Tr}(M))$ if and only if $\underline{f} = 0$ in $\underline{\mathrm{Hom}}_A(M, N)$, it follows that the functor

$$\mathrm{Tr} : \mathrm{mod}_r A \longrightarrow \underline{\mathrm{mod}}_l A$$

induces a contravariant functor

$$\mathrm{Tr} : \underline{\mathrm{mod}}_r A \longrightarrow \underline{\mathrm{mod}}_l A,$$

which we will also denote by Tr.

Thus from the considerations above we get the following theorem:

**Theorem 4.3.17.** (M. Auslander [17], see also [3, Proposition 2.2.1]). *Let A be a semiperfect ring, and let M and N be finitely presented A-modules, then*

1. *There is a group isomorphism*

$$\underline{\mathrm{Hom}}_A(M, N) \simeq \underline{\mathrm{Hom}}_A(\mathrm{Tr}N, \mathrm{Tr}M) \tag{4.3.18}$$

2. *The correspondence $M \mapsto \mathrm{Tr}(M)$ defines a duality functor between the projective stable categories of right and left modules:*

$$\mathrm{Tr} : \underline{\mathrm{mod}}_r A \longleftrightarrow \underline{\mathrm{mod}}_l A^{op}. \tag{4.3.19}$$

3. *$\mathrm{End}_A M$ is a local ring if and only if $\mathrm{End}_A(\mathrm{Tr}M)$ is a local ring.*

The duality functor Tr defined by (4.3.19) in theorem 4.3.17 is called the **Auslander-Bridger transpose**.

Similarly one can contemplate an ideal in $\mathrm{mod}_r A$ by considering for all pairs of A-modules M and N the subgroup $\mathcal{I}(M, N)$ consisting of all homomorphisms which factor through an injective A-module. The quotient category

$$\overline{\mathrm{mod}}_r A = \mathrm{mod}_r A / \mathcal{I}, \tag{4.3.20}$$

is called the **injective stable category** of $\mathrm{mod}_r A$ modulo injectives. The objects of this category are the same as those in $\mathrm{mod}_r A$, and the group of morphisms $\overline{\mathrm{Hom}}_A(M, N)$ of morphisms from M to N in $\overline{\mathrm{mod}}_r A$ is defined as the quotient group

$$\overline{\mathrm{Hom}}_A(M, N) = \mathrm{Hom}_A(M, N) / \mathcal{I}(M, N) \tag{4.3.21}$$

of $\mathrm{Hom}_A(M, N)$ with the composition of morphisms induced from the composition in $\mathrm{mod}_r A$. $\square$

The next step is to apply the theory of Auslander-Bridger duality considered above to semiperfect hereditary rings.

**Proposition 4.3.22.** *Let $A$ be a semiperfect Noetherian hereditary ring, and $M$ a finitely generated $A$-module. Then there exists a functorial isomorphism*

$$\mathrm{Tr}(M) \simeq \mathrm{Ext}^1_A(M, A).  \tag{4.3.23}$$

*Proof.* Since $A$ is a hereditary ring, $\mathrm{gl.dim}\, A \leq 1$. Therefore a minimal projective resolution of a module $M$ has the following form:

$$0 \longrightarrow Q \xrightarrow{\varphi} P \longrightarrow M \longrightarrow 0.  \tag{4.3.24}$$

Applying the functor $*$ to this exact sequence we obtain the following exact sequence

$$0 \longrightarrow M^* \longrightarrow P^* \xrightarrow{\varphi^*} Q^* \longrightarrow \mathrm{Ext}^1_A(M, A) \longrightarrow 0.$$

The statement $\mathrm{Tr}(M) \simeq \mathrm{Ext}^1_A(M, A)$ follows now at once from [103, corollary 1.2.9]. $\square$

## 4.4 Almost Split Sequences

In this section we study the main properties of almost split sequences (also called Auslander-Reiten sequences) which were introduced by M. Auslander and I. Reiten (see [15] and [20]). Roughly speaking, almost-split sequences are minimal non-split short exact sequences. They have become a central tool in the theory of representations of finite dimensional algebras and Artinian rings.

Let $A$ be a ring, and let, as before, $\mathcal{M} = \mathrm{mod}_r A$ be the category of all finitely generated right $A$-modules.

**Definition 4.4.1.** An $A$-module homomorphism $f : M \longrightarrow N$ is called **left minimal** if each $\varphi \in \mathrm{End}_A N$ with $\varphi f = f$ is an automorphism.

An $A$-module homomorphism $f : M \longrightarrow N$ is called **right minimal** if each $\varphi \in \mathrm{End}_A M$ with $f\varphi = f$ is an automorphism.

Obviously a surjective morphism is left minimal and an injective morphism is right minimal. But as proposition 4.4.2 below illustrates a morphism need not be surjective to be left minimal.

Recall that an **injective hull** of $M$ is an injective module $E(M)$ with an essential monomorphism $f : M \longrightarrow E(M)$. Note, that an injective hull exists for any $A$-module $M$, and that it is unique up to isomorphism extending the identity on $M$.

**Proposition 4.4.2.** *Let $E(M)$ be an injective hull of a module $M$, i.e. $f : M \rightarrow E(M)$ is an essential monomorphism. Then $f$ is a left minimal homomorphism.*

*Proof.* Assume that there is a sequence

$$M \xrightarrow{f} E(M) \xrightarrow{\varphi} E(M)$$

with $\varphi f = f$. Here $f$ is an essential monomorphism. This means, by definition, that $\varphi$ is a monomorphism. Consider the exact sequence

$$0 \longrightarrow E(M) \xrightarrow{\varphi} E(M)$$

which splits because $E(M)$ is injective. Therefore $E(M) \simeq \operatorname{Ker}\varphi \oplus \operatorname{Im}\varphi$. Since $\operatorname{Ker}\varphi = 0$, $\operatorname{Im}\varphi = E(M)$, i.e. $\varphi$ is an epimorphism, and so $\varphi$ is an isomorphism. Thus, $f$ is a left minimal homomorphism. $\square$

**Lemma 4.4.3.** *Let $M$ be a simple left $A$-module and let $I = E(M)$ be its injective hull. Then*

1. *Every non-zero submodule $X$ of $I$ contains $M$.*
2. *The endomorphism ring $\operatorname{End}_A I$ is a local ring.*

*Proof.*

1. Let $X$ be a non-zero submodule of $I$. Since $I$ is an essential extension of $M$, $M_1 = X \cap M \neq 0$. Since $M_1 \subseteq M$, $M_1 \subset X$ and $M$ is a simple module, $M_1 = M$, and so $M \subset X$.
2. From condition 1 it follows that $I$ is a uniform module. Then, by [103, proposition 5.2.1], $I = E(M)$ is indecomposable, and, by [103, theorem 5.2.2], $\operatorname{End}_A I$ is local. $\square$

Since for any module $M$ there is an injective hull, then one can always write down an exact sequence of the form

$$0 \longrightarrow M \xrightarrow{h_0} E_0 \xrightarrow{h_1} E_1, \tag{4.4.4}$$

where $E_0$ is an injective hull of $M$, and $E_1$ is an injective hull of $\operatorname{Im} h_0$. Such a sequence is called a **minimal injective resolution (presentation)** of $M$.

**Proposition 4.4.5.** *Let $P(M)$ be a projective cover of a module $M$, i.e. $f : P(M) \longrightarrow M$ is an essential epimorphism. Then $f$ is a right minimal homomorphism.*

*Proof.* Suppose $\varphi \in \operatorname{End}_A M$ with $f\phi = f$. Let $p \in P(M)$ and $\varphi(p) = p_1$, then $f\varphi(p) = f(p_1)$ and $f\varphi(p) = f(p)$, since $f\varphi = f$. Therefore $f(p) = f(p_1)$, and so $p - p_1 \in \operatorname{Ker} f$. Since $p = p - p_1 + p_1 = p - p_1 + \varphi(p)$, $P(M) = \operatorname{Ker} f + \operatorname{Im}\varphi$. By the definition of a projective cover, $\operatorname{Ker} f$ is a small submodule, therefore $P(M) = \operatorname{Im}\varphi$, i.e. $\varphi$ is an epimorphism. Since $P(M)$ is a projective module, the sequence

$$0 \longrightarrow \operatorname{Ker}\varphi \longrightarrow P(M) \xrightarrow{\varphi} P(M) \rightarrow 0$$

splits, and so $P(M) \simeq \operatorname{Ker} \varphi \oplus \operatorname{Im} \varphi$. Hence $\operatorname{Ker} \varphi = 0$, i.e. $\varphi$ is a monomorphism, and so $\varphi$ is an isomorphism. Thus, $f$ is a right minimal homomorphism. $\square$

**Definition 4.4.6.**

1. Let $M$ and $N$ be $A$-modules. A homomorphism $f : N \longrightarrow M$ is said to be **right almost split** if
    a. It is not a split epimorphism.
    b. For any homomorphism $g : X \longrightarrow M$ which is not a split epimorphism, there is a homomorphism $h : X \longrightarrow N$ such that $g = fh$.
2. A homomorphism $f : N \longrightarrow M$ is called **minimal right almost split** if it is both right minimal and right almost split.

The definition of a (minimal) left almost split homomorphism is dual:

**Definition 4.4.7.**

1. Let $M$ and $N$ be $A$-module. A homomorphism $g : M \longrightarrow N$ is said to be **left almost split** if
    a. It is not a split monomorphism.
    b.    For any homomorphism $h : M \longrightarrow Y$ which is not a split monomorphism, there is a homomorphism $h_1 : N \longrightarrow Y$ such that $h = gh_1$.
2. A homomorphism $g : M \rightarrow N$ is called **minimal left almost split** if it is both left minimal and left almost split.
3. A homomorphism $g : M \rightarrow N$ is called **almost split** if it is either right almost split or left almost split.

**Lemma 4.4.8.**

1. *Let $f : L \longrightarrow M$ be a left almost split homomorphism of $A$-modules. Then the module $L$ is indecomposable and has a local ring of endomorphisms.*
2. *Let $g : M \longrightarrow N$ be a right almost split homomorphism of $A$-modules. Then the module $N$ is indecomposable and has a local ring of endomorphisms.*

*Proof.*

1. Suppose $L$ decomposes into a direct sum of non-zero modules $L = L_1 \oplus L_2$. Let $\pi_i : L \longrightarrow L_i$ be the corresponding canonical projection for $i = 1, 2$. Since $\pi_i$ is not a split monomorphism and $f$ is left almost split, there is a homomorphism $h_i : M \longrightarrow L_i$ such that $h_i f = \pi_i$ for $i = 1, 2$. Then one can write down the homomorphism $h = \begin{bmatrix} h_1 \\ h_2 \end{bmatrix} : M \longrightarrow L$ and $hf = 1_L$. Thus, we obtain that $f$ is a split monomorphism. A contradiction.
   Let's now show that $\operatorname{End}_A(L)$ is a local ring. For a homomorphism $f : L \longrightarrow M$ consider the induced homomorphism $\bar{f} : \operatorname{Hom}_A(M, L) \longrightarrow \operatorname{Hom}_A(L, L)$ defined by $\bar{f}(h) = hf$ for any $h \in \operatorname{Hom}_A(M, L)$. We write $I = \operatorname{Im}(\bar{f}) \subseteq \operatorname{End}_A(L)$. Let $\varphi \in \operatorname{End}_A(L)$, then as $(\varphi \bar{f})(h) = \varphi(\bar{f}h) = \varphi hf = \bar{f}(\varphi h)$ we see

that $I$ is a left ideal in $\text{End}_A(L)$. Suppose $g \in I$ is left invertible, i.e. $ug = 1_L$, then $I = \text{End}_A(L)$. Therefore there exists an element $v \in \text{Hom}_A(M, L)$ such that $\bar{f}(v) = 1_L = vf$, which is not the case since $f$ is not a split monomorphism.

Let $g \in \text{End}_A(L)$ and $g$ not left invertible. Since $f$ is left almost split there is $f_1 : M \longrightarrow L$ such that $g = f_1 f$, hence $g = \bar{f}(f_1) \in I$. So $g \in I$ if and only if $g$ is not left invertible. From [100, Proposition 3.4.5] and [100, Theorem 10.1.1] it follows that $\text{End}_A(L)$ is a local ring with unique maximal ideal $I$.

2. This case is dual to the case 1, and so it can be proved similarly. $\square$

**Definition 4.4.9.** A short exact sequence

$$0 \longrightarrow L \xrightarrow{f} M \xrightarrow{g} N \longrightarrow 0 \qquad (4.4.10)$$

is called **almost split (Auslander-Reiten) sequence** provided that

    a. $f$ is a left minimal almost split homomorphism.

    b. $g$ is a right minimal almost split homomorphism.

Given such a sequence exists from lemma 4.4.8 it follows that $L$ and $N$ are indecomposable modules. From the definition of left minimal almost split and right minimal split homomorphisms it follows that this sequence is not split. Therefore $L$ is not injective and $N$ is not projective.

**Corollary 4.4.11.** *Let* $0 \longrightarrow L \xrightarrow{f} M \xrightarrow{g} N \longrightarrow 0$ *be an almost split sequence in the category of all right A-modules over a ring A. Then L and N are indecomposable rings with local endomorphism rings.*

*Proof.* This statement immediately follows from lemma 4.4.8. $\square$

**Lemma 4.4.12.** (M. Auslander, I. Reiten [21, Lemma 2.13]). *Let*

$$\mathcal{E}: \quad 0 \longrightarrow L \xrightarrow{f} M \xrightarrow{g} N \longrightarrow 0$$

*be a non-split exact sequence, where L, M, N are right A-modules, and let*

$$
\begin{array}{ccccccccc}
0 & \longrightarrow & L & \xrightarrow{f} & M & \xrightarrow{g} & N & \longrightarrow & 0 \\
& & \varphi \downarrow & & h \downarrow & & \psi \downarrow & & \\
0 & \longrightarrow & L & \xrightarrow{f} & M & \xrightarrow{g} & N & \longrightarrow & 0
\end{array}
$$

Diag. 4.4.13.

*be a commutative diagram.*

    1. *If* $\text{End}_A(L)$ *is a local ring and* $\psi = 1_N$, *then* $\varphi$ *and h are isomorphisms.*

    2. *If* $\text{End}_A(N)$ *is a local ring and* $\varphi = 1_L$, *then* $\psi$ *and h are isomorphisms.*

*Proof.*

1.  From results in section 3.6 it follows that the group of extensions $\mathcal{E}xt_A(N, L)$
    is isomorphic to the homology group $\text{Ext}^1_A(N, L)$, and it can be considered
    also as a module over a ring $\text{End}_A(L)$. Suppose that $\varphi$ is not an isomorphism,
    then $\varphi \in J = \text{rad}(\text{End}_A(L))$, since $U = \text{End}_A(L)$ is a local ring. Then, since
    diagram 4.4.13 is commutative, $[\mathcal{E}]J = [\mathcal{E}]U$. Hence by Nakayama's lemma
    $[\mathcal{E}] = 0$, i.e. the sequence $\mathcal{E}$ is split. A contradiction. So $\varphi$ is an isomorphism,
    and $h$ is also an isomorphism by [103, corollary 1.2.7].
2.  This is proved similarly. $\square$

**Theorem 4.4.14.** (M. Auslander, I. Reiten [21, Theorem 2.14]). *Let*

$$0 \longrightarrow L \overset{f}{\longrightarrow} M \overset{g}{\longrightarrow} N \longrightarrow 0$$

*be a non-split exact sequence, where L, M, N are right A-modules. Then the following
statements are equivalent:*

1.  *The given sequence is almost split.*
2.  *$\text{End}_A L$ is a local ring and g is right almost split.*
3.  *$\text{End}_A N$ is a local ring and f is left almost split.*

*Proof.*
$1 \Rightarrow 2$ and $1 \Rightarrow 3$ by lemma 4.4.8.

$3 \Rightarrow 2$. By lemma 4.4.8, it suffices to show that $g$ is right almost split. Since $f$ is
not a split monomorphism, $g$ is not a split epimorphism, by [103, proposition 1.2.3].
Suppose there is a homomorphism $\gamma : X \longrightarrow N$ which is not a split epimorphism.
Then, by lemma 3.1.18, there is a commutative diagram

$$
\begin{array}{ccccccccc}
0 & \longrightarrow & L & \overset{f_1}{\longrightarrow} & T & \overset{g_1}{\longrightarrow} & X & \longrightarrow & 0 \\
 & & \Vert 1_L & & \downarrow \beta & & \downarrow \gamma & & \\
0 & \longrightarrow & L & \overset{f}{\longrightarrow} & M & \overset{g}{\longrightarrow} & N & \longrightarrow & 0
\end{array}
$$

with exact rows. Here $(T, \beta, g_1)$ is the standard pullback of $(\gamma, g)$. If $f_1$ is a split
monomorphism, then $g_1$ is a split epimorphism, by [103, proposition 1.2.3], and so
there is a homomorphism $\alpha : X \longrightarrow T$ such that $g_1\alpha = 1_X$. In this case we have a
homomorphism

$$h = \beta\alpha : X \longrightarrow M.$$

Since $\gamma g_1 = g\beta$, $\gamma = \gamma 1_X = \gamma g_1 \alpha = g\beta\alpha = gh$. So in this case the desired morphism
$X \longrightarrow M$ exists. Suppose $f_1$ is not a split monomorphism. Since $f$ is left almost
split, there is a homomorphism $\delta : M \longrightarrow T$ such that $\delta f = f_1$. So there is a
homomorphism

$$h = g_1\delta : M \longrightarrow X,$$

such that $hf = g_1\delta f = g_1 f_1 = 0$. Hence $\mathrm{Im}(h) \subseteq \mathrm{Ker} f = \mathrm{Im} g$, so $h$ can be split as $h = \alpha g$, where $\alpha : N \longrightarrow X$. Since $g\beta\delta = \gamma g_1 \delta = \gamma h = \gamma \alpha g$, there results a commutative diagram:

$$
\begin{array}{ccccccccc}
0 & \longrightarrow & L & \xrightarrow{f} & M & \xrightarrow{g} & N & \longrightarrow & 0 \\
 & & \Big\| {\scriptstyle 1_L} & & \Big\downarrow {\scriptstyle \beta\delta} & & \Big\downarrow {\scriptstyle \gamma\alpha} & & \\
0 & \longrightarrow & L & \xrightarrow{f} & M & \xrightarrow{g} & N & \longrightarrow & 0
\end{array}
$$

with exact rows. Then, by lemma 4.4.12, $\gamma\alpha$ is an automorphism, i.e. $\gamma$ is a split epimorphism. A contradiction.

$2 \Rightarrow 3$. This is proved in a similar way as the previous implication.

2 and $3 \Rightarrow 1$. Suppose that $\mathrm{End}_A(L)$ and $\mathrm{End}_A(N)$ are local rings, $g$ is right almost split and $f$ is left almost split. Let $\varphi \in \mathrm{End}_A M$ and $\varphi f = f$, $g\varphi = g$. Then we have the commutative diagram

$$
\begin{array}{ccccccccc}
0 & \longrightarrow & L & \xrightarrow{f} & M & \xrightarrow{g} & N & \longrightarrow & 0 \\
 & & \Big\| {\scriptstyle 1_L} & & \Big\downarrow {\scriptstyle \varphi} & & \Big\| {\scriptstyle 1_N} & & \\
0 & \longrightarrow & L & \xrightarrow{f} & M & \xrightarrow{g} & N & \longrightarrow & 0
\end{array}
$$

with exact rows. Then, by lemma 4.4.12, $\varphi$ is an automorphism. Thus $f$ is left minimal and $g$ is right minimal. Therefore $f$ is left minimal almost split and $g$ is right minimal almost split, and so the given sequence is almost split. $\square$

**Definition 4.4.15.** Let $L, L_1, M, M_1, N, N_1 \in \mathrm{Mod}_r A$. Two short exact sequences

$$ 0 \longrightarrow L \xrightarrow{f} M \xrightarrow{g} N \longrightarrow 0 $$

and

$$ 0 \longrightarrow L_1 \xrightarrow{f_1} M_1 \xrightarrow{g_1} N_1 \longrightarrow 0 $$

are said to be **isomorphic** if there is a commutative diagram

$$
\begin{array}{ccccccccc}
0 & \longrightarrow & L & \xrightarrow{f} & M & \xrightarrow{g} & N & \longrightarrow & 0 \\
 & & \Big\downarrow {\scriptstyle \alpha} & & \Big\downarrow {\scriptstyle \beta} & & \Big\downarrow {\scriptstyle \gamma} & & \\
0 & \longrightarrow & L_1 & \xrightarrow{f_1} & M_1 & \xrightarrow{g_1} & N_1 & \longrightarrow & 0
\end{array}
$$

**Diag. 4.4.16.**

where $\alpha, \beta, \gamma$ are isomorphisms.

**Lemma 4.4.17.**

1. *A monomorphism* $f : L \longrightarrow M$ *is right split if an only if* $fu$ *is split for an isomorphism* $u : X \longrightarrow L$ *if and only if* $vf$ *is split for an isomorphism* $v : M \longrightarrow Y$.
2. *An epimorphism* $g : M \longrightarrow N$ *is left split if and only if* $gu$ *is split for an isomorphism* $u : X \longrightarrow M$ *if and only if* $vg$ *is split for an isomorphism* $v : N \longrightarrow Y$.

*Proof.*

1. If $f$ is a split monomorphism then there is a homomorphism $f_1 : M \longrightarrow L$ such that $f_1 f = 1_L$. If $u : X \longrightarrow M$ is an isomorphism then $(u^{-1}f_1)(fu) = 1_X$, i.e. $fu : X \longrightarrow M$ is a split monomorphism. Conversely, assume that $fu : X \longrightarrow M$ is a split monomorphism, where $u : X \longrightarrow M$ is an isomorphism. Then there is $f_1 : M \longrightarrow X$ such that $f_1(fu) = 1_X$. Then $f_1 f = u^{-1}$ and $(uf_1)f = uu^{-1} = 1_L$. Another statement is proved similarly.
2. This case is dual to the case 1, and so it can be proved similarly. $\square$

Almost split sequences are uniquely determined up to isomorphism by the isomorphism class of the first (last) term. More precisely, it holds the following statement.

**Proposition 4.4.18.** (M. Auslander, I. Reiten). *Let*

$$0 \longrightarrow L \xrightarrow{f} M \xrightarrow{g} N \longrightarrow 0 \qquad\qquad (4.4.19)$$

*and*

$$0 \longrightarrow L_1 \xrightarrow{f_1} M_1 \xrightarrow{g_1} N_1 \longrightarrow 0 \qquad\qquad (4.4.20)$$

*be almost split sequences. Then the following statement are equivalent.*

1. *Almost split sequences (4.4.19) and (4.4.20) are isomorphic.*
2. $L \simeq L_1$.
3. $N \simeq N_1$.

*Proof.*
$1 \Longrightarrow 2$ and $1 \Longrightarrow 3$ follows from definition 4.4.15.
$2 \Longrightarrow 1$. Let there exists an isomorphism $\alpha : L \longrightarrow L_1$. Since $f_1$ is not a split monomorphism, $f_1\alpha : L \longrightarrow M$ is also not a split monomorphism by lemma 4.4.17. Since $f_1$ is left almost split, there is a homomorphism $\beta : M \longrightarrow M_1$ such that $f_1\alpha = \beta f$. Analogously, since $f$ is not a split monomorphism, $f\alpha^{-1}$ is also not a split monomorphism. Since $f$ is left almost split, there is a homomorphism $\beta_1 : M_1 \longrightarrow M$ such that $\beta_1 f_1 = f\alpha$. So that $\beta_1\beta f = \beta_1 f_1\alpha = f$ and $\beta\beta_1 f_1 = \beta f\alpha^{-1} = f_1\alpha\alpha^{-1} = f_1$. Since $f$ and $f_1$ are left minimal, $\beta\beta_1$ and $\beta_1\beta$ are automorphism, so $\beta$ and $\beta_1$ are isomorphisms. Thus we obtain the commutative diagram:

$$0 \longrightarrow L \xrightarrow{\ f\ } M \xrightarrow{\ g\ } N \longrightarrow 0$$
$$\downarrow \alpha \qquad \downarrow \beta$$
$$0 \longrightarrow L_1 \xrightarrow{\ f_1\ } M_1 \xrightarrow{\ g_1\ } N_1 \longrightarrow 0$$

Since $N \simeq \operatorname{Coker} f$ and $N_1 \simeq \operatorname{Coker} f_1$, there exists a homomorphism $\gamma : N \longrightarrow N_1$ such that diagram 4.4.16 is commutative. Since $\alpha$ and $\beta$ are isomorphisms, $\gamma$ is also an isomorphism by [103, corollary 1.2.9].

$3 \Longrightarrow 1$. This statement proves similar as the previous one. $\square$

**Remark 4.4.21.** Note that almost split sequences do not always exist. Their existence is proved only for particular cases. The following statement about existence of almost split sequences, which is given without proof, was obtained by M. Auslander and I. Reiten in 1975 for the case of Artin algebras. The existence of almost split sequences for some classes of modules over semiperfect rings will be shown below in section 4.7.

**Proposition 4.4.22.** (M. Auslander, I. Reiten). *Let $A$ be an Artin algebra, and let* mod$_r A$ *be a category of finitely generated right $A$-modules.*

1. *If $N \in$ mod$_r A$ is indecomposable and non-projective, then there is an almost split sequence*

$$0 \longrightarrow L \xrightarrow{\ f\ } M \xrightarrow{\ g\ } N \longrightarrow 0.$$

2. *If $L \in$ mod$_r A$ is indecomposable and noninjective, then there is an almost split sequence*

$$0 \longrightarrow L \xrightarrow{\ f\ } M \xrightarrow{\ g\ } N \longrightarrow 0.$$

## 4.5 Cogenerators and Some Identities for Finitely Presented Modules

In this section we give some useful identities for finitely presented modules which connect the tensor functor $\otimes$ with the functor Hom. There are (of course) in addition to the tensor product defining adjunct functor equation

$$\operatorname{Hom}_{\mathbf{Z}}(M \otimes_A N, G) \simeq \operatorname{Hom}_A(M, \operatorname{Hom}_{\mathbf{Z}}(N, G)).$$

Here $M$ is a right $A$-module, $N$ is a left $A$-module and $G$ is an Abelian group.

**Lemma 4.5.1.** (H.Cartan, S.Eilenberg [40, Proposition 5.2]) *Given modules* $(X_T, {}_S M_T, {}_S Y)$, *where $S$, $T$ are rings. Then there is a natural homomorphism*

$$\tau : X \otimes_T \operatorname{Hom}_S(M, Y) \longrightarrow \operatorname{Hom}_S(\operatorname{Hom}_T(X, M), Y)$$

*defined by* $\tau(x \otimes f)(g) = f(g(x))$ *for each* $x \in X$, $f \in \mathrm{Hom}_S(M, Y)$ *and* $g \in \mathrm{Hom}_T(X, M)$.

*If* $X \cong T^n$ *is a finitely generated free* $T$-*module, then* $\tau$ *is an isomorphism.*

*Proof.* Since $M$ is an $(S, T)$-bimodule, both $\mathrm{Hom}_S(M, Y)$ and $\mathrm{Hom}_T(X, M)$ make sense. Define

$$\tau(x \otimes f)(g) = f(g(x))$$

for each $x \in X$, $f \in \mathrm{Hom}_S(M, Y)$ and $g \in \mathrm{Hom}_T(F, M)$. By standard calculations one can easily verify that $\tau$ is a natural homomorphism.

The second statement is obvious when $X = T$, since $T \otimes_T \mathrm{Hom}_S(M, Y) \simeq \mathrm{Hom}_S(M, Y)$ and $\mathrm{Hom}_T(T, M) \simeq M$. When $X \cong T^n$ the statement is also true since all functors considered are additive. $\square$

**Theorem 4.5.2.** (see [72, Theorem 3.2.1]). *Let $S$ and $T$ be rings.*

1. *If $X$ is a finitely presented right $T$-module, $M$ is an $(S, T)$-bimodule, and $Y$ is an injective left $S$-module, then the natural homomorphism*

$$\tau : X \otimes_T \mathrm{Hom}_S(M, Y) \longrightarrow \mathrm{Hom}_S(\mathrm{Hom}_T(X, M), Y)$$

*defined by* $\tau(x \otimes f)(g) = f(g(x))$ *for each* $x \in X$, $f \in \mathrm{Hom}_S(M, Y)$, *and* $g \in \mathrm{Hom}_T(X, M)$ *is an isomorphism.*

2. *If $X$ is a finitely presented left $S$-module, $M$ is an $(S, T)$-bimodule, and $Y$ is an injective right $T$-module, then the natural homomorphism*

$$\tau : \mathrm{Hom}_T(M, Y) \otimes_S X \longrightarrow \mathrm{Hom}_T(\mathrm{Hom}_S(X, M), Y)$$

*defined by* $\tau(f \otimes x)(g) = f(g(x))$ *for each* $x \in X$, $f \in \mathrm{Hom}_S(M, Y)$, *and* $g \in \mathrm{Hom}_T(X, M)$ *is an isomorphism.*

*Proof.*

1. Since $X$ is a finitely presented right $T$-module, there exists an exact sequence

$$F_2 \longrightarrow F_1 \longrightarrow X \longrightarrow 0$$

with $F_1, F_2$ finitely generated free $T$-modules. Since $\otimes$ is a right exact functor and $Y$ is an injective $S$-module, so that the functor $\mathrm{Hom}_S(*, Y)$ is exact, we obtain a commutative diagram

$$
\begin{array}{ccccccc}
F_2 \otimes_T \mathrm{Hom}_S(M, Y) & \longrightarrow & F_1 \otimes_T \mathrm{Hom}_S(M, Y) & \longrightarrow & X \otimes_T \mathrm{Hom}_S(M, Y) & \longrightarrow & 0 \\
\downarrow {\scriptstyle \tau_2} & & \downarrow {\scriptstyle \tau_1} & & \downarrow {\scriptstyle \tau} & & \\
\mathrm{Hom}_T(\mathrm{Hom}_S(F_2, M), Y) & \longrightarrow & \mathrm{Hom}_T(\mathrm{Hom}_S(F_1, M), Y) & \longrightarrow & \mathrm{Hom}_T(\mathrm{Hom}_S(X, M), Y) & \longrightarrow & 0
\end{array}
$$

with exact rows. It follows from lemma 4.5.1 that $\tau_2, \tau_1$ are isomorphisms. Therefore $\tau$ is also an isomorphism by [103, corollary 1.2.9].

2. The proof of this statement is similar to the previous one. $\square$

At the end of this section we consider the important notion of injective cogenerator and some of its main properties. The notion of a cogenerator for the category $\text{Mod}_r A$ is dual to the notion of a generator for $\text{Mod}_r A$ as examined in [100, section 10.6].

**Definition 4.5.3.** A right module $C$ over a ring $A$ is said to be a **cogenerator** for the category $\text{Mod}_r A$ if for every right $A$-module $M$ there exists a monomorphism $\sigma : M \hookrightarrow C^{(I)}$ for some set $I$, where $C^{(I)}$ denotes the product of $I$ copies of $C$, i.e. $C^{(I)} = \prod_{i \in I} C_i$ and $C_i = C$ for all $i \in I$.

The following proposition can be considered as giving equivalent definitions of the notion of a cogenerator. It is dual to [100, Proposition 10.6.8].

**Proposition 4.5.4.** *Let $C$ be a right $A$-module over a ring $A$. Then the following statements are equivalent:*

1. *The module $C$ is a cogenerator for right $A$-modules.*
2. *For each $A$-module $M$ and non-zero element $x \in M$ there is $\varphi \in \text{Hom}_A(M, C)$ such that $\varphi(x) \neq 0$.*
3. *$\text{Hom}_A(M, C) \neq 0$ for any module $M \neq 0$.*
4. *For each $A$-module $M$*

$$\bigcap_{\varphi \in \text{Hom}_A(M,C)} \text{Ker}(\varphi) = 0.$$

5. *For any $A$-modules $M$, $N$ and every non-zero homomorphism $f \in \text{Hom}_A(N, M)$ there is a homomorphism $\varphi \in \text{Hom}_A(M, C)$ such that $\varphi f \neq 0$.*

*Proof.*
$2 \Longrightarrow 3$ and $3 \Longrightarrow 2$ are clear.

$4 \Longrightarrow 5$. Let $f \in \text{Hom}_A(N, M)$ and $f \neq 0$. Then there exists an $n \in N$ such that $0 \neq f(n) \in M$. Therefore there is a homomorphism $\varphi \in \text{Hom}_A(M, C)$ such that $f(n) \neq \text{Ker}(\varphi)$, i.e. $\varphi f(n) \neq 0$, which implies that $\varphi f \neq 0$.

$5 \Longrightarrow 4$. Assume that

$$N = \bigcap_{\varphi \in \text{Hom}_A(M,C)} \text{Ker}(\varphi) \neq 0.$$

Taking into account that $N$ is a submodule in $M$ we have an embedding $\alpha : N \hookrightarrow M$. Since $\alpha \neq 0$ there exists a $\varphi \in \text{Hom}_A(M, C)$ such that $\varphi \alpha \neq 0$, which implies that $\varphi(N) \neq 0$, i.e. $N \nsubseteq \text{Ker}(\varphi)$. A contradiction.

$1 \Longrightarrow 4$. Let $\sigma : M \longrightarrow C^{(I)}$ be a monomorphism and $\pi_i : C^{(I)} \hookrightarrow C_i$ be the $i$-th projection, then $\pi_i \sigma \in \text{Hom}_A(M, C)$. Taking into account that

$$X = \bigcap_{\varphi \in \text{Hom}_A(M,C)} \text{Ker}(\varphi) \subseteq \bigcap_{i \in I} \text{Ker}(\pi_i \sigma) = \text{Ker}(\sigma) = 0,$$

we obtain $X = 0$.

$4 \Longrightarrow 1$. Assume that $X = \bigcap\limits_{\varphi \in \mathrm{Hom}_A(M,C)} \mathrm{Ker}(\varphi) = 0$ for each $A$-module $M$. Let

$$Y = \prod\limits_{\varphi \in \mathrm{Hom}_A(M,C)} C_\varphi,$$

where $C_\varphi = C$ for all $\varphi \in \mathrm{Hom}_A(M, C)$. Consider the homomorphism $f : M \longrightarrow Y$ which is constructed in the following way: $f(m) = (\varphi(m))$ for all $\varphi \in \mathrm{Hom}_A(M, C)$. Suppose $m \in \mathrm{Ker}(f)$, then $\varphi(m) = 0$ for all $\varphi \in \mathrm{Hom}_A(M, C)$. So $m \in X = 0$, i.e. $f$ is a monomorphism.

$5 \Longrightarrow 2$. Let $0 \neq x \in M$ and consider the homomorphism $f : A \longrightarrow M$ giving by $f(1) = x$. So $f \neq 0$ and therefore there exists a non-zero homomorphism $\varphi : A \longrightarrow C$ such that $\varphi f \neq 0$. So $\varphi(x) = \varphi f(1) \neq 0$.

$2 \Longrightarrow 1$. Let $M$ be an arbitrary $A$-module and $0 \neq x \in M$. Then there exists a homomorphism $\varphi_x \in \mathrm{Hom}_A(M, C)$ such that $\varphi_x(x) \neq 0$. Consider the homomorphism

$$\varphi = (\varphi_x) : M \longrightarrow \prod\limits_{0 \neq x \in M} E.$$

Since $\varphi(x) \neq 0$ for any $0 \neq x \in M$, $\varphi$ is a monomorphism. $\square$

**Definition 4.5.5.** An injective right $A$-module $C$ is called an **injective cogenerator** for $\mathrm{Mod}_r A$ if it is a cogenerator for $\mathrm{Mod}_r A$.

**Example 4.5.6.** Since $\mathbf{Q}/\mathbf{Z}$ is an injective group, and by [100, lemma 5.4.8] $\mathrm{Hom}_{\mathbf{Z}}(G, \mathbf{Q}/\mathbf{Z}) \neq 0$ for any Abelian group $G$, $\mathbf{Q}/\mathbf{Z}$ is an injective cogenerator for Abelian groups.

The following lemma can be considered as a generalization of [103, theorem 1.2.11] when instead of all modules we only consider one injective cogenerator.

**Proposition 4.5.7.** *Let a right $A$-module $C$ be an injective cogenerator for $\mathrm{Mod}_r A$. Then a sequence*

$$0 \longrightarrow X \xrightarrow{\varphi} M \xrightarrow{\psi} Y \longrightarrow 0 \tag{4.5.8}$$

*is exact if and only if the induced sequence*

$$0 \longrightarrow \mathrm{Hom}_A(Y, C) \xrightarrow{\bar{\psi}} \mathrm{Hom}_A(M, C) \xrightarrow{\bar{\varphi}} \mathrm{Hom}_A(X, C) \longrightarrow 0 \tag{4.5.9}$$

*is exact.*

*Proof.*

$\Longrightarrow$. Since $C$ is an injective module, exactness of the sequence (4.5.8) implies exactness of the sequence (4.5.9) by [103, theorem 1.2.11] and [103, proposition 1.7.4].

$\Longleftarrow$. Let the sequence (4.5.9) be exact, i.e. $\mathrm{Im}(\bar{\psi}) = \mathrm{Ker}(\bar{\varphi})$, $\bar{\psi}$ be a monomorphism, and $\bar{\varphi}$ be an epimorphism. We will show that $\mathrm{Im}(\varphi) = \mathrm{Ker}(\psi)$, that $\varphi$ is a monomorphism, and that $\psi$ is an epimorphism.

Suppose that $\text{Im}(\varphi) \neq \text{Ker}(\psi)$ and $\text{Im}(\varphi) \not\subset \text{Ker}(\psi)$, then we can choose an element $0 \neq m \in \text{Im}(\varphi) \setminus \text{Ker}(\psi)$. Therefore $x = \psi(m) \neq 0$ and there exists an element $0 \neq x \in X$ such that $m = \varphi(x)$. Since $C$ is a cogenerator and $x = \psi(m) \neq 0$, by proposition 4.5.4(2) there is a homomorphism $0 \neq f \in \text{Hom}_A(Y, C)$ such that $f(x) = f(\psi(m)) \neq 0$. Taking into account that $m = \varphi(x)$ we obtain that $f(\psi(\varphi(m))) \neq 0$, i.e. $f\psi\varphi \neq 0$. On the other hand, since (4.5.9) is an exact sequence which is induced by (4.5.8), we have that $\bar{\varphi}\bar{\psi} = 0$ and $f\psi\varphi = \bar{\varphi}\bar{\psi}(f) = 0$. A contradiction.

Therefore we can assume that $\text{Im}(\varphi) \neq \text{Ker}(\psi)$ and $\text{Im}(\varphi) \subset \text{Ker}(\psi)$. So we can choose an element $n \in \text{Ker}(\psi) \setminus \text{Im}(\varphi)$, which means that $\psi(n) = 0$ and $n \notin \text{Im}(\varphi)$. So $n + \text{Im}(\varphi)$ is a non-zero element in $M/\text{Im}(\varphi)$. Since $C$ is a cogenerator there exists a non-zero homomorphism $f \in \text{Hom}(M/\text{Im}\varphi, C)$ such that $f(n + \text{Im}\varphi) \neq 0$. Then we have a commutative diagram:

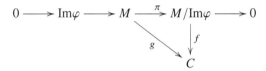

where $\pi$ is the natural projection and $g = f\pi$. Hence $g(n) = f\pi(n) = f(n + \text{Im}\varphi) \neq 0$. So $\bar{\varphi}(g(n)) = g\varphi(n) = f\pi\varphi(n) = 0$, since $\varphi(n) \in \text{Im}\varphi$. Therefore $g \in \text{Ker}(\bar{\varphi}) = \text{Im}(\bar{\psi})$, i.e. $g = \bar{\psi}(h)$, where $h \in \text{Hom}_A(Y, C)$. Since $\psi(n) = 0$, we obtain that $g(n) = \bar{\psi}(h(n)) = h(\psi(n)) = 0$. A contradiction, since $g(n) \neq 0$. So $\text{Im}(\varphi) = \text{Ker}(\psi)$.

In a similar way one can show that $\text{Ker}\varphi = 0$ and $\text{Im}\psi = Y$. $\square$

**Lemma 4.5.10.** *A right injective A-module $C$ is a cogenerator for the category $\text{Mod}_r A$ if and only if for each simple right A-module $S$ the module $C$ contains a direct summand which is isomorphic to an injective hull $E(S)$.*

*Proof.*

$\Longrightarrow$. Let $C$ be a cogenerator for right $A$-modules, and let $E(S)$ be the injective hull of a simple right $A$-module $S$ with its natural embedding $\alpha : S \to E(S)$. Since $C$ is a cogenerator,

$$\bigcap_{\varphi \in \text{Hom}_A(E(S), C)} \text{Ker}(\varphi) = 0.$$

Therefore there exists a homomorphism $\varphi : E(S) \longrightarrow C$ such that $S \not\subset \text{Ker}\varphi$. Since $S$ is a simple right $A$-module and $S \cap \text{Ker}\varphi \subseteq S$, $S \cap \text{Ker}\varphi = 0$. On the other hand, $S$ is an essential module in $E(S)$ and $\text{Ker}\varphi \subset E(S)$, hence $S \cap \text{Ker}\varphi \neq 0$. This contradiction shows that $\text{Ker}\varphi = 0$, i.e. $\varphi$ is a monomorphism. Therefore there is a commutative diagram:

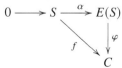

where $f = \varphi\alpha : S \to C$ is a monomorphism. Since $C$ is injective module, $C = C_1 \oplus C_2$, where $C_1 \simeq E(S)$, by [100, Proposition 5.3.6].

$\Longleftarrow$. Let $M$ be a right $A$-module, and let $0 \neq m \in M$. Then $mA \subseteq M$ is a finitely generated $A$-module. Therefore it contains a maximal submodule $N$. So $S = mA/N$ is a simple right $A$-module, therefore we have an embedding $\alpha : S \hookrightarrow E(S)$ and the injective hull $E(S)$ is isomorphic to a direct summand of $C$. Suppose $N \neq mA$, i.e. $m \notin N$. Then we can consider the following commutative diagram

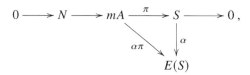

where $\pi$ is the natural projection. Then $\alpha\pi(m) = \alpha(m + N) \neq 0$. So $\alpha\pi \neq 0$. Let $\beta : mA \longrightarrow M$ be the natural embedding. Since $E(S)$ is an injective module and $\beta$ is a monomorphism there results a commutative diagram

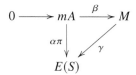

where $\gamma\beta = \alpha\pi$. Since $E(S)$ is isomorphic to a direct summand of $C$, there is an isomorphism $f : E(S) \oplus X \longrightarrow C$ and a monomorphism $\delta : E(S) \longrightarrow E(S) \oplus X$. Giving rise to a commutative diagram

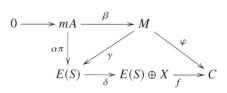

where $\varphi = f\delta\gamma$. So $\varphi(m) = f\delta\gamma(m) = f\delta\alpha\pi(m) \neq 0$, which implies $\mathrm{Hom}(M, C) \neq 0$, i.e. $C$ is a cogenerator.

If $mA = N$, i.e. $mA = S$ is a simple module, we can use the previous arguments when we use $\alpha$ instead of $\alpha\pi$. $\square$

**Proposition 4.5.11.** *Let $S_1, S_2, \ldots, S_n$ be representatives of the isomorphism classes of all simple right $A$-modules. If $E(S_i)$ is an injective hull of $S_i$ for $i = 1, 2, \ldots, n$, then $C = \bigoplus_{i=1}^{n} E(S_i)$ is an injective cogenerator for $\mathrm{Mod}_r A$.*

*Proof.* This result follows immediately from lemma 4.5.10. $\square$

**Corollary 4.5.12.** *If A is a local ring with Jacobson radical R then the injective hull E(A/R) is an injective cogenerator for* Mod$_r$ *A.*

## 4.6 Pure Submodules and Pure Exact Sequences

In this section we consider the important concepts of a pure submodule which can be considered as a generalization of the idea of a direct summand, and a pure exact sequences which can be considered as a generalization of the idea of a split exact sequence.

**Definition 4.6.1.** (P.M. Cohn [45]). Let $M$ be a right $A$-module. A submodule $L$ of $M$ is said to be **pure** if for the natural embedding $i : L \hookrightarrow M$ the induced sequence

$$0 \to L \otimes_A X \xrightarrow{i \otimes 1_X} M \otimes_A X$$

is exact for any left $A$-module $X$.

A **pure monomorphism** is a monomorphism $L \longrightarrow M$ whose image is a pure submodule of $M$.

The concept of a pure submodule is closely connected with concept of a pure exact sequence which is a natural generalization of split exact sequences.

**Definition 4.6.2.** A short exact sequence $0 \to L \longrightarrow M \longrightarrow N \to 0$ of right $A$-modules is called **pure exact** if for every left $A$-module $X$ the induced sequence

$$0 \to L \otimes_A X \longrightarrow M \otimes_A X \longrightarrow N \otimes_A X \to 0 \qquad (4.6.3)$$

is exact.

**Remarks 4.6.4.**

1. It is obvious that every split sequence is pure exact.
2. Since any module is a direct limit of finitely presented modules and tensor products commute with direct limits, in definition 4.6.2 one can assume that a left module $X$ is finitely presented.

**Example 4.6.5.** Since every split sequence is pure exact, every direct summand of a right $A$-module $M$ is pure in $M$. In particular, every subspace of a vector space over a field is pure.

Below we will show the close relationships between pure submodules, pure exact sequences and flat modules.

**Lemma 4.6.6.** *A submodule L of a right A-module M is pure if and only if the natural sequence*

$$0 \to L \longrightarrow M \longrightarrow M/L \to 0 \qquad (4.6.7)$$

*is pure exact.*

*Proof.*

$\Longrightarrow$. If $L$ is a pure submodule of $A$ then the induced sequence to (4.6.7)

$$0 \to L \otimes_A X \longrightarrow M \otimes_A X \longrightarrow M/L \otimes_A X \to 0$$

is exact since the functor of tensor product is right exact. So the sequence (4.6.7) is pure exact.

$\Longleftarrow$. This immediately follows from definitions 4.6.1 and 4.6.2. $\square$

**Proposition 4.6.8.** *A module $M$ is flat if and only if each submodule of $M$ is pure.*

*Proof.* This follows immediately from [103, proposition 4.6.1]. $\square$

**Proposition 4.6.9.** *An $A$-module $F$ is flat if and only if every exact sequence*

$$0 \to L \longrightarrow M \longrightarrow F \to 0 \tag{4.6.10}$$

*is pure exact.*

*Proof.* Applying the functor $\mathrm{Tor}^A(-, X)$ to the exact sequence (4.6.10) we obtain the exact sequence:

$$\mathrm{Tor}_1^A(F, X) \longrightarrow L \otimes_A X \longrightarrow M \otimes_A X \longrightarrow F \otimes_A X \to 0$$

Now taking into account that $F$ is a flat module if and only if $\mathrm{Tor}_1^A(F, X) = 0$ for every left $A$-module $X$ by [103, theorem 1.6.6], we get the exact sequence $0 \to L \otimes_A X \longrightarrow M \otimes_A X \longrightarrow F \otimes_A X \to 0$, i.e. the sequence (4.6.10) is pure exact, as required. $\square$

Recall that a ring $A$ is von Neumann regular if for every $a \in A$ there exists an $x \in A$ such that $a = axa$. There exists many different conditions which equivalently definite von Neumann regular rings. One of these conditions is given by M. Auslander and M. Harada (see [103, theorem 6.2.35]) which says that a ring $A$ is von Neumann regular if and only if all right (left) $A$-modules are flat. Taking into account this result and proposition 4.6.8 we immediately obtain the following characterization of von Neumann regular rings in terms of pure submodules.

**Proposition 4.6.11.** *A ring $A$ is von Neumann regular if and only if any submodule of every $A$-module $M$ is pure.*

Write $\bar{Q} = Q/Z$, which is an injective cogenerator for $Z$-modules. Recall that $X^* = \mathrm{Hom}_Z(X, \bar{Q})$ is the character module of $X$.

The following theorem gives equivalent characterizations of a pure exact sequence.

**Theorem 4.6.12.** *Let $0 \to L \longrightarrow M \longrightarrow N \to 0$ be a short exact sequence of right $A$-modules. The following statements are equivalent.*

   1. *The sequence is pure exact.*

2. *The induced sequence of character modules*

$$0 \longrightarrow N^* \longrightarrow M^* \longrightarrow L^* \longrightarrow 0 \qquad (4.6.13)$$

*is split exact.*

3. *The induced sequence (4.6.13) is pure exact.*
4. *For every finitely presented right A-module X, the induced sequence*

$$0 \longrightarrow \operatorname{Hom}_A(X, L) \longrightarrow \operatorname{Hom}_A(X, M) \longrightarrow \operatorname{Hom}_A(X, N) \longrightarrow 0 \quad (4.6.14)$$

*is exact.*

5. *For every commutative diagram*

$$
\begin{array}{ccc}
A^m & \xrightarrow{\ f_1\ } & A^n \\
\varphi_1 \downarrow & & \downarrow \varphi_2 \\
0 \longrightarrow L & \xrightarrow{\ f\ } & M
\end{array}
$$

*with $m, n \in \mathbf{N}$, there exists $h : A^n \longrightarrow L$ such that $\varphi_1 = h f_1$.*

6. *Any finite set of linear equations over A with constants in $f(L)$ which is solvable in M is solvable in $f(L)$, i.e. for any matrix $(a_{ij}) \in M_{m \times n}(A)$ and any finite family $(u_j)_{j=1}^m$ of elements of $f(L)$, if there exist elements $(y_i)_{i=1}^n$ in M such that*

$$\sum_{i=1}^n a_{ij} y_i = u_j \quad \text{for } j = 1, \ldots, m$$

*then there also exist elements $(z_i')_{i=1}^n$ in $f(L)$ satisfying the same equations, i.e.*

$$\sum_{i=1}^n a_{ij} z_i' = u_j \quad \text{for } j = 1, \ldots, m$$

*Proof.*

$1 \Longleftrightarrow 2$. By definition of an exact sequence $0 \to L \longrightarrow M \longrightarrow N \to 0$ is pure exact if and only if the induced sequence (4.6.3) is exact for every left $A$-module $X$. Since any $A$-module can be considered as $\mathbf{Z}$-module, and $\mathbf{Q}/\mathbf{Z}$ is an injective cogenerator of $\mathbf{Z}$-modules, we obtain from proposition 4.5.7 that sequence (4.6.3) is exact if and only if the sequence

$$0 \to (N \otimes_A X)^* \longrightarrow (M \otimes_A X)^* \longrightarrow (L \otimes_A X)^* \to 0$$

is exact for every left $A$-module $X$. Applying the adjoint isomorphism

$$(Y \otimes_A X)^* = \operatorname{Hom}_{\mathbf{Z}}(Y \otimes_A X, \bar{\mathbf{Q}}) \simeq \operatorname{Hom}_A(X, \operatorname{Hom}_{\mathbf{Z}}(Y, \bar{\mathbf{Q}})) = \operatorname{Hom}_A(X, Y^*)$$

to this sequence, there results that the sequence (4.6.3) is exact if and only if the sequence

$$0 \to \operatorname{Hom}_A(X, N^*) \longrightarrow \operatorname{Hom}_A(X, M^*) \longrightarrow \operatorname{Hom}_A(X, L^*) \to 0$$

is exact for every left $A$-module $X$, which happens if and only if $L^*$ is projective and so the sequence (4.6.13) is a split exact sequence.

$2 \Longrightarrow 3$. This is trivial.

$3 \Longleftrightarrow 4$. Since the sequence (4.6.13) is pure exact, the induced sequence

$$0 \to X \otimes N^* \longrightarrow X \otimes M^* \longrightarrow X \otimes L^* \to 0$$

is exact for any right $A$-module $X$. Suppose that $X$ is a f.p. $A$-module. Since $\mathbf{Q}/\mathbf{Z}$ is an injective $\mathbf{Z}$-module, applying theorem 4.5.2(1) we obtain an exact sequence:

$$0 \to (\mathrm{Hom}_A(X, N))^* \longrightarrow (\mathrm{Hom}_A(X, M))^* \longrightarrow (\mathrm{Hom}_A(X, L))^* \to 0$$

So the sequence (13.6.14) is also exact by proposition 4.5.7, since $\mathbf{Q}/\mathbf{Z}$ is an injective cogenerator.

All these arguments are reversible.

$3 \Longrightarrow 1$. Suppose that the sequence (4.6.13) of left $A$-modules is pure exact. Then taking into account statement 4 of this theorem we obtain that for any f.p. right $A$-module $X$ the sequence

$$0 \longrightarrow \mathrm{Hom}_A(X, N^*) \longrightarrow \mathrm{Hom}_A(X, M^*) \longrightarrow \mathrm{Hom}_A(X, L^*) \longrightarrow 0$$

is also exact. Now applying adjoint isomorphisms

$$\mathrm{Hom}_A(X, Y^*) \simeq \mathrm{Hom}_A(X \otimes Y, \mathbf{Q}/\mathbf{Z}) = (X \otimes Y)^*$$

there results the exact sequence

$$0 \longrightarrow (X \otimes N)^* \longrightarrow (X \otimes M)^* \longrightarrow (X \otimes L)^* \longrightarrow 0$$

which implies the exactness of the sequence

$$0 \longrightarrow X \otimes N \longrightarrow X \otimes M \longrightarrow X \otimes L \longrightarrow 0$$

which follows from proposition 4.5.7.

$4 \Longrightarrow 5$. Let $X$ be a f.p. $A$-module, and $\varphi_3 : X \longrightarrow N$. Then there exists a commutative diagram

$$
\begin{array}{ccccccccc}
A^m & \xrightarrow{f_1} & A^n & \xrightarrow{f_2} & X & \longrightarrow & 0 \\
\varphi_1 \downarrow & & \varphi_2 \downarrow & & \varphi_3 \downarrow & & \\
0 \longrightarrow & L & \xrightarrow{f} & M & \xrightarrow{g} & N & \longrightarrow & 0
\end{array}
$$

**Diag. 4.6.15.**

Since by assumption $\mathrm{Hom}_A(X, M) \longrightarrow \mathrm{Hom}_A(X, N)$ is an epimorphism there exists $\psi : X \longrightarrow M$ such that $g\psi = \varphi_3$. So by the Homotopy Lemma 3.2.14 there exists a homomorphism $h : A^n \longrightarrow L$ such that $\varphi_1 = hf_1$:

$$
\begin{array}{ccccccccc}
A^m & \xrightarrow{f_1} & A^n & \xrightarrow{f_2} & X & \longrightarrow & 0 \\
\varphi_1 \downarrow & {}^{h}\swarrow \varphi_2 \downarrow & & {}^{\psi}\swarrow & \varphi_3 \downarrow & & \\
0 \longrightarrow & L & \xrightarrow{f} & M & \xrightarrow{g} & N & \longrightarrow & 0
\end{array}
$$

$5 \Longrightarrow 4$. Let $X$ be a f.p. $A$-module, and $\varphi_3 : X \longrightarrow N$. Consider the commutative diagram 4.6.15. Suppose that there exists a homomorphism $h : A^n \longrightarrow L$ such that $\varphi_1 = h f_1$. Then by the Homotopy Lemma 3.2.14 there exists a homomorphism $\psi : X \longrightarrow M$ such that $g\psi = \varphi_3$, i.e. $\operatorname{Hom}_A(X, M) \longrightarrow \operatorname{Hom}_A(X, N)$ is an epimorphism.

$4 \Longrightarrow 6$. Suppose the sequence (4.6.14) is exact for every f.p. $A$-module $X$ and there exist elements $y_i \in M$ such that $\sum_{i=1}^{n} a_{ij} y_i = u_j \in f(L)$ where $a_{ij} \in A$ for $i = 1, \ldots, n$ and $j = 1, \ldots, m$. Then we can define a finitely presented module $X$ with $n$ generators $x_1, \ldots, x_n$ and $m$ relations $\sum_{i=1}^{n} a_{ij} x_i = 0$. Let $\varphi : X \longrightarrow M$ be a module homomorphism such that $\varphi(x_i) = g(y_i)$ for each generator $x_i$ of $X$. Since sequence (4.6.14) is exact there exists a homomorphism $\psi : X \longrightarrow M$ such that $g\psi = \varphi$. Then elements $z_i = y_i - \psi(x_i)$ satisfy the given system of equations:

$$\sum_{i=1}^{n} a_{ij} z_i = \sum_{i=1}^{n} a_{ij} y_i - \psi \left( \sum_{i=1}^{n} a_{ij} x_i \right) = u_j$$

and $z_i \in f(L)$ since $g(z_i) = g(y_i) - g\psi(x_i) = g(y_i) - \varphi(x_i) = 0$.

$6 \Longrightarrow 4$. Suppose $X$ is a f.p. $A$-module and $\varphi \in \operatorname{Hom}_A(X, N)$. Since $X$ is a f.p. $A$-module, there exist elements $(x_i)_{i=1}^{n}$ which are generators in $X$ and which satisfy some finite system of relations $\sum_{i=1}^{n} a_{ij} x_i = 0$, where $a_{ij} \in A$ for all $i = 1, \ldots, n$ and $j = 1, \ldots, m$. Since $g$ is an epimorphism, for each $i = 1, \ldots, n$ there exists an element $y_i \in M$ such that $g(y_i) = \varphi(x_i)$. Since $\operatorname{Ker} g = \operatorname{Im} f$ and

$$g \left( \sum_{i=1}^{n} a_{ij} y_i \right) = \varphi \left( \sum_{i=1}^{n} a_{ij} x_i \right) = 0,$$

$\sum_{i=1}^{n} a_{ij} y_i = u_j \in f(L)$ for $j = 1, \ldots, m$. Then by assumption there exist elements $z_i \in f(L)$ such that $\sum_{i=1}^{n} a_{ij} z_i = u_j$ and $g(z_i) = 0$. So one can define the mapping $\psi : X \longrightarrow M$ such that $\psi(x_i) = y_i - z_i$ for each $i$. Obviously this is a module homomorphism of $X$ into $M$. Since $g\psi(x_i) = g(y_i) - g(z_i) = \varphi(x_i)$ for each generator of $X$, $g\psi = \varphi$, i.e. $\operatorname{Hom}_A(X, M) \longrightarrow \operatorname{Hom}_A(X, N)$ is an epimorphism, i.e. the sequence (4.6.14) is exact. $\square$

**Proposition 4.6.16.**

1. *If $f : L \longrightarrow M$ and $g : M \longrightarrow N$ are pure monomorphisms, then $gf : L \longrightarrow N$ is also a pure monomorphism.*
2. *If $f : L \longrightarrow M$ and $g : M \longrightarrow N$ are homomorphisms and $gf : L \longrightarrow N$ is a pure monomorphism, then $f$ is also a pure monomorphism.*

*Proof.*

1.  Consider the commutative diagram

$$
\begin{array}{ccc}
L & \xrightarrow{\ f\ } & M \\
1_L \big\| & & \big\downarrow g \\
L & \xrightarrow{\ fg\ } & M
\end{array}
$$

**Diag. 4.6.17.**

Since $f, g$ are monomorphisms, so is $fg$. Using lemma 3.2.1 this diagram can be completed by kernels and cokernels to the following commutative diagram

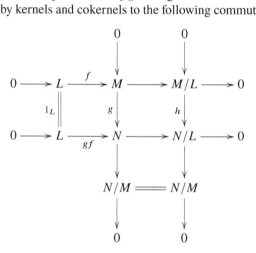

where $h$ is a monomorphism by [103, lemma 1.2.11]. Suppose that $X$ is a f.p. $A$-module. Applying to this diagram the functor $\mathrm{Hom}(X, -)$ and taking into account that $f$ and $g$ are pure monomorphisms, we obtain the induced commutative diagram:

$$
\begin{array}{ccccccccc}
& & 0 & & & & 0 & & \\
& & \downarrow & & & & \downarrow & & \\
0 \longrightarrow & \mathrm{Hom}(X, L) & \longrightarrow & \mathrm{Hom}(X, M) & \longrightarrow & \mathrm{Hom}(X, M/L) & \longrightarrow & 0 \\
& \big\| & & \downarrow & & \downarrow & & \\
0 \longrightarrow & \mathrm{Hom}(X, L) & \longrightarrow & \mathrm{Hom}(X, N) & \longrightarrow & \mathrm{Hom}(X, N/L) & & \\
& & & \downarrow & & \downarrow & & \\
& & & \mathrm{Hom}(X, N/M) & = & \mathrm{Hom}(X, N/M) & & \\
& & & \downarrow & & & & \\
& & & 0 & & & &
\end{array}
$$

Since $f$ and $g$ are pure monomorphisms, the first row and the second column of this diagram are exact sequences. Then the second row is also exact, i.e. $gf$ is a pure monomorphism by theorem 4.6.12.

2. Since $gf$ is a monomorphism, $f$ is also a monomorphism and we can complete diagram 4.6.17 by kernels and cokernels to the following commutative diagram:

$$
\begin{array}{ccccccccc}
0 & \longrightarrow & L & \overset{f}{\longrightarrow} & M & \longrightarrow & M/L & \longrightarrow & 0 \\
& & \parallel{\scriptstyle 1_L} & & \downarrow{\scriptstyle g} & & \downarrow{\scriptstyle h} & & \\
0 & \longrightarrow & L & \underset{gf}{\longrightarrow} & N & \longrightarrow & N/L & \longrightarrow & 0
\end{array}
$$

Suppose that $X$ is a f.p. $A$-module. Applying to this diagram the functor $\mathrm{Hom}(X, -)$ and taking into account that $fg$ is a pure monomorphism, we obtain the induced commutative diagram:

$$
\begin{array}{ccccccccc}
0 & \longrightarrow & \mathrm{Hom}(X, L) & \longrightarrow & \mathrm{Hom}(X, M) & \longrightarrow & \mathrm{Hom}(X, M/L) & & \\
& & \parallel & & \downarrow & & \downarrow & & \\
0 & \longrightarrow & \mathrm{Hom}(X, L) & \underset{gf}{\longrightarrow} & \mathrm{Hom}(X, N) & \longrightarrow & \mathrm{Hom}(X, N/L) & \longrightarrow & 0
\end{array}
$$

Since the second row in this diagram is an exact sequence, the first row is also exact. $\square$

## 4.7 Almost Split Sequences over Semiperfect Rings

In this section we show the existence of almost split sequences for some classes of modules over semiperfect rings, proved by M. Auslander [17].

**Lemma 4.7.1.** *Let $A$ be a semiperfect ring, and $M$ a finitely presented stable right $A$-module. Suppose that $0 \longrightarrow X \longrightarrow Y \longrightarrow Z \longrightarrow 0$ is a short exact sequence of right $A$-modules. Then there exists a natural homomorphism $\delta$ such that the sequence*

$$
0 \to \mathrm{Hom}_A(M, X) \to \mathrm{Hom}_A(M, Y) \to \mathrm{Hom}_A(M, Z) \overset{\delta}{\longrightarrow} X \otimes_A \mathrm{Tr}M \to
$$

$$
\to Y \otimes_A \mathrm{Tr}M \to Z \otimes_A \mathrm{Tr}M \to 0
$$

*is exact.*

*Proof.* Let $P_1 \overset{f_1}{\longrightarrow} P_0 \overset{f_0}{\longrightarrow} M \longrightarrow 0$ be a minimal projective resolution of $M$, where $P_0, P_1$ are finitely generated projective modules. Then, by [103, proposition 4.5.9], $\mathrm{Hom}_A(P_i, N) \simeq N \otimes_A P_i^*$ for $i = 1, 2$ and each right $A$-module $N$. Therefore the cokernel of the homomorphism $\mathrm{Hom}_A(P_0, N) \to \mathrm{Hom}_A(P_1, N)$ is isomorphic to $N \otimes_A \mathrm{Tr}M$. Thus, taking into account that the functor Hom is left exact in each variable and that the $P_0$, $P_1$ are projective modules we have the following commutative diagram:

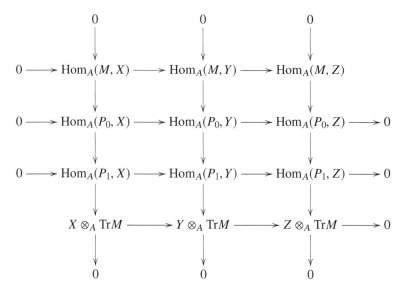

Then, the snake lemma (theorem 3.2.5) supplies the required sequence. □

To formulate the main results of this section we use the important notion of local duality which was first introduced by M. Auslander and M. Bridger in [13].

**Definition 4.7.2.** Let $M$ be a finitely presented right $A$-module with endomorphism ring $S = \text{End}_A M$, which is a local ring with unique maximal ideal $J = \text{rad}(S)$. Let $Q$ be the injective hull of the $S$-module $S/J$. Then the left $A$-module

$$M^+ = \text{Hom}_S(M, Q) \tag{4.7.3}$$

is called the **local dual** of $M$.

**Lemma 4.7.4.** [3, Lemma 2.3.1]. *Let $M$ be a finitely presented right $A$-module with a local endomorphism ring $S$, and let $Q$ be the injective hull of the $S$-module $S/J$, where $J$ is the Jacobson radical of $S$. Then $M^+$ is a strongly indecomposable left $A$-module and*

$$\text{End}_A(M^+) \simeq \text{End}_S(Q). \tag{4.7.5}$$

*Proof.* Let $\varphi \in M^+$, and $a \in A$, then $a\varphi \in M^+$ as well, where $a\varphi : M \to Q$ is defined by the following rule: $a\varphi(m) = \varphi(ma)$.

By the adjoint isomorphism (see [103, proposition 1.3.1]) we have

$$\text{End}_A M^+ = \text{Hom}_A(\text{Hom}_S(M, Q), \text{Hom}_S(M, Q)) \simeq$$

$$\simeq \text{Hom}_S(M \otimes_A \text{Hom}_S(M, Q), Q).$$

Since $Q$ is an injective left $S$-module and $M$ is a finitely presented right $A$-module,

$$M \otimes_A \mathrm{Hom}_S(M, Q) \simeq \mathrm{Hom}_S(\mathrm{Hom}_A(M, M), Q),$$

by theorem 4.5.2. So

$$\mathrm{End}_A(M^+) \simeq \mathrm{Hom}_S(\mathrm{Hom}_S(\mathrm{Hom}_A(M, M), Q), Q) \simeq$$

$$\mathrm{Hom}_S(\mathrm{Hom}_S(S, Q), Q) \simeq \mathrm{End}_S(Q).$$

Finally, $\mathrm{End}_S(Q)$ is a local ring, by lemma 4.4.3. $\square$

**Definition 4.7.6.** A homomorphism $f : L \longrightarrow M$ is said to **factor through homomorphism** $h$ if it can be written as a composition of homomorphisms $f = gh$ or $f = hg$.

**Proposition 4.7.7.** (**The Factor Theorem**). *Let $L$, $M$ and $N$ be right $A$-modules and let $f : L \longrightarrow M$ be a homomorphism of $A$-modules. If $g : L \longrightarrow N$ is a homomorphism with $\mathrm{Ker} g \subseteq \mathrm{Ker} f$ then there exists a unique homomorphism $h : N \longrightarrow M$ such that $f = hg$.*

*Proof.* Since $g$ is an epimorphism, for any $x \in N$ there exists an $y \in L$ such that $g(y) = x$. We define $h(x) = f(y)$. This is a module homomorphism which is well defined. Indeed, if also $y_1 \in L$ with $g(y_1) = x$ then $g(y - y_1) = 0$, i.e. $y - y_1 \in \mathrm{Ker} g$. Since $\mathrm{Ker} g \subseteq \mathrm{Ker} f$, $f(y) - f(y_1) = f(y - y_1) = 0$, i.e. $f(y) = f(y_1)$. So that $f(y) = hg(y)$, i.e. $f = hg$. The uniqueness of $h$ follows immediately from the fact that $g$ is an epimorphism. $\square$

Below we now give statement and proof of the main result of this section which was first proved by M. Auslander [17, theorem 3.9] for an arbitrary rings. Here the proof of this theorem is given for a particular case for semiperfect rings following W. Zimmermann [236] and L. Angeleri Hügel [3].

**Theorem 4.7.8.** (M. Auslander [17, theorem 3.9]). *Let $A$ be a semiperfect ring. For each finitely presented non-projective right $A$-module $N$ with a local endomorphism ring there exists an almost split sequence*

$$0 \to (\mathrm{Tr}N)^+ \xrightarrow{f} M \xrightarrow{g} N \to 0, \tag{4.7.9}$$

*in* $\mathrm{Mod}_r A$.

*Proof.* Since $\mathrm{End}_A N$ is a local ring, the ring $T = \mathrm{End}_A(\mathrm{Tr}N)$ is also local, by theorem 4.3.17(3). Let $U$ be the injective hull of $T/J$, where $J$ is the Jacobson radical of $T$, and let $L = \mathrm{Hom}_T(\mathrm{Tr}N, U) = (\mathrm{Tr}N)^+$. Then, by lemma 4.7.4, $L$ is a right $A$-module with a local endomorphism ring $S = \mathrm{End}_A L \simeq \mathrm{End}_T U$.

Let $\pi : P \longrightarrow N$ be a projective cover of $N$, and consider the exact sequence

$$0 \longrightarrow K \xrightarrow{i} P \xrightarrow{\pi} N \longrightarrow 0.$$

Our goal is to construct a morphism $\alpha : K \longrightarrow L$ such that the pushout diagram for $(i, \alpha)$

$$
\begin{array}{ccccccccc}
0 & \longrightarrow & K & \xrightarrow{\ i\ } & P & \xrightarrow{\ \pi\ } & N & \longrightarrow & 0 \\
& & \downarrow{\scriptstyle \alpha} & & \downarrow{\scriptstyle \beta} & & \parallel{\scriptstyle 1_N} & & \\
0 & \longrightarrow & L & \xrightarrow{\ f\ } & M & \xrightarrow{\ g\ } & N & \longrightarrow & 0
\end{array}
$$

yields an almost split sequence for $L$ in $\mathrm{Mod}_r A$.

First we show that if the morphism $\alpha$ satisfies the following properties:

a. $\alpha$ does not factor through $i$,
b. The composition $s\alpha$ factors through $i$ for any $s \in \mathrm{rad}S$,

then the sequence

$$0 \to L \xrightarrow{\ f\ } M \xrightarrow{\ g\ } N \to 0 \tag{4.7.10}$$

constructed above is an almost split sequence in $\mathrm{Mod}_r A$.

First we show that $f : L \longrightarrow M$ is a left almost split homomorphism. If there existed $\bar{f} : M \longrightarrow L$ such that $\bar{f}f = 1_L$, then $\alpha = \bar{f}\beta i$, which contradicts condition a. So $f$ is not a split monomorphism.

Now we will show that for any homomorphism $h : L \to X$ which is not a split monomorphism there exists a homomorphism $h_1 : M \longrightarrow X$ such that $h = fh_1$.

Assume that $h : L \to X$ is not a split monomorphism. Consider the pushout diagram for $(f, h)$

$$
\begin{array}{ccccccccc}
0 & \longrightarrow & L & \xrightarrow{\ f\ } & M & \xrightarrow{\ g\ } & N & \longrightarrow & 0 \\
& & \downarrow{\scriptstyle h} & & \downarrow{\scriptstyle \beta_1} & & \parallel{\scriptstyle 1_N} & & \\
0 & \longrightarrow & X & \xrightarrow{\ f_1\ } & Y & \xrightarrow{\ g_1\ } & N & \longrightarrow & 0
\end{array}
$$

Suppose $\varphi \in \mathrm{Hom}_A(X, L)$, then $\varphi h \in \mathrm{rad}(S)$, and so $\varphi h \alpha$ factors through $i$ by condition b, i.e. there exists a $\gamma : P \longrightarrow L$ such that $\varphi h \alpha = \gamma i$. Since $(M, \beta, f)$ is a pushout for a pair $(\alpha, i)$, there exists a homomorphism $t : M \longrightarrow L$ such the following diagram

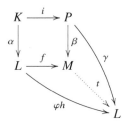

is commutative, i.e. $\varphi h = ft$ and $\gamma = t\beta$. Since $(Y, \beta_1, f_1)$ is a pushout for a pair $(h, f)$, there exists a homomorphism $\psi : Y \longrightarrow L$ such the following diagram

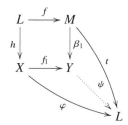

is commutative, i.e. $\varphi = \psi f_1$. This shows that the sequence

$$0 \to \mathrm{Hom}_A(N, L) \to \mathrm{Hom}_A(Y, L) \to \mathrm{Hom}_A(X, L) \to 0$$

is exact.

Since $L = \mathrm{Hom}_T(\mathrm{Tr}N, U)$, for any right $A$-module $B$ we have $\mathrm{Hom}_A(B, L) \simeq \mathrm{Hom}_T(B \otimes_A \mathrm{Tr}N, U)$ by the adjoint isomorphism (see [103, proposition 1.3.1]). So we obtain an exact sequence:

$$0 \to \mathrm{Hom}_T(X \otimes_A \mathrm{Tr}N, U) \longrightarrow \mathrm{Hom}_T(Y \otimes_A \mathrm{Tr}N, U) \longrightarrow$$
$$\longrightarrow \mathrm{Hom}_T(N \otimes_A \mathrm{Tr}N, U) \longrightarrow 0$$

Taking into account that $U$ is an injective cogenerator by corollary 4.5.12, we obtain that the sequence

$$0 \to X \otimes_A \mathrm{Tr}N \to Y \otimes_A \mathrm{Tr}N \to N \otimes_A \mathrm{Tr}N \to 0$$

is also exact by proposition 4.5.7. Then, by lemma 4.7.1, there results a long exact sequence

$$0 \to \mathrm{Hom}_A(N, X) \to \mathrm{Hom}_A(N, Y) \xrightarrow{\mathrm{Hom}_A(N, g_1)} \mathrm{Hom}_A(N, N) \xrightarrow{\delta}$$
$$\xrightarrow{\delta} X \otimes_A \mathrm{Tr}N \to Y \otimes_A \mathrm{Tr}N$$

Hence $\mathrm{Hom}_A(N, g_1)$ is an epimorphism by the snake lemma, and so $g_1$ is a split epimorphism. Therefore the sequence

$$0 \to X \xrightarrow{f_1} Y \xrightarrow{g_1} N \to 0$$

is split exact, and so there exists $f_2 : Y \longrightarrow X$ such $f_2 f_1 = 1_l$, which implies that $h = f_2 \beta_1 f$. So $h$ factors through $f$. Thus, $f$ is a left almost split homomorphism. Since $\mathrm{End}_A N$ is a local ring, (4.7.9) is an almost split sequence by theorem 4.4.14.

Now we proceed to show the existence of a morphism $\alpha$ which satisfies conditions a and b.

By lemma 4.7.1, there exists a long exact sequence

$$0 \to \mathrm{Hom}_A(N, K) \to \mathrm{Hom}_A(N, P) \xrightarrow{\mathrm{Hom}_A(N,\pi)} \mathrm{Hom}_A(N, N) \xrightarrow{\delta}$$

$$\xrightarrow{\delta} K \otimes_A \mathrm{Tr}N \to P \otimes_A \mathrm{Tr}N.$$

Then $\mathrm{Ker}(\delta) = \mathrm{Im}(\mathrm{Hom}_A(N, \pi)) = \mathcal{P}_A(N, N)$. So, by theorem 4.3.17, there is a commutative diagram

$$
\begin{array}{ccccccc}
0 & \longrightarrow & T/\mathcal{P}(\mathrm{Tr}N, \mathrm{Tr}N) & \longrightarrow & K \otimes_A \mathrm{Tr}N & \xrightarrow{i\otimes 1} & P \otimes_A \mathrm{Tr}N \\
 & & \downarrow{\simeq} & \nearrow{\overline{\delta}} & & & \\
 & & \mathrm{End}_A N / \mathcal{P}(N, N) & & & &
\end{array}
$$

with exact row.

Applying the left exact functor $\mathrm{Hom}_T(*, U_T)$ to this sequence, and taking into account the adjoint isomorphism and the isomorphism $S \simeq \mathrm{End}_T U$, there results an exact sequence of left $S$-modules:

$$\mathrm{Hom}_A(P, L) \to \mathrm{Hom}_A(K, L) \xrightarrow{\widetilde{\delta}} \mathrm{Hom}_A(T/\mathcal{P}(\mathrm{Tr}N, \mathrm{Tr}N), U) \to 0$$

Since $\mathrm{Tr}N$ is not projective, $T/\mathcal{P}(\mathrm{Tr}N, \mathrm{Tr}N) \neq 0$. Hence $\mathrm{Hom}_A(T/\mathcal{P}(\mathrm{Tr}N, \mathrm{Tr}N), U)$ is a non-zero submodule of $\mathrm{Hom}_T(T, U) \simeq U$. And so, by lemma 4.4.3, it contains a simple module $V \simeq T/J$. Since $T/J$ is a division ring we can choose a generator $\alpha$ of $V_T$. Let $\mu \in \mathrm{Hom}_A(K, L)$ be a preimage of $\alpha$ under $\widetilde{\delta}$, i.e. $\widetilde{\delta}(\mu) = \alpha$. Then $\alpha$ satisfies condition a, since $\widetilde{\delta}(\mu) \neq 0$. Moreover, taking into account that $\mathrm{soc}U \cdot \mathrm{rad}(\mathrm{End}_T U) = 0$, one sees that $\alpha$ satisfies also condition b. So the sequence (4.7.9) is indeed almost split. $\square$

Note that theorem 4.7.8 states an existence of almost split sequences for semiperfect rings in the category of all modules. W. Zimmermann in [237] generalized this theorem for the category of finitely presented modules. The proof of this theorem will be done below, at the end of this section.

**Proposition 4.7.11.** (W.Zimmermann [236, Proposition 3]). *Let $A$ be a semiperfect ring, and $N$ a finitely presented non-projective $A$-module with local endomorphism ring. Let*

$$0 \to L_1 \xrightarrow{f_1} M_1 \xrightarrow{g_1} N \to 0 \qquad (4.7.12)$$

*be an almost split sequence in* $\mathrm{mod}_r A$, *and*

$$0 \to (\mathrm{Tr}N)^+ \xrightarrow{f} M \xrightarrow{g} N \to 0 \qquad (4.7.13)$$

*the almost split sequence in* $\mathrm{Mod}_r A$ *which is constructed by theorem 4.7.8. Then there exists a pure monomorphism $\alpha : L_1 \to (\mathrm{Tr}N)^+$ and a homomorphism $\beta : M_1 \to M$*

*such that the diagram*

$$0 \longrightarrow L_1 \xrightarrow{f_1} M_1 \xrightarrow{g_1} N \longrightarrow 0$$

$$\alpha \downarrow \qquad \beta \downarrow \qquad 1_N \|$$

$$0 \longrightarrow (\mathrm{Tr}N)^+ \xrightarrow{f} M \xrightarrow{g} N \longrightarrow 0$$

**Diag. 4.7.14.**

*is commutative.*

*Proof.* Since (4.7.12) and (4.7.13) are almost split sequences, $g$ is a right almost split epimorphism, $g_1$ is not a split epimorphism. Therefore there is a homomorphism $\beta : M_1 \longrightarrow M$ such that $g\beta = g_1$. Since $L_1 = \mathrm{Ker} f_1$ and $(\mathrm{Tr}N)^+ = \mathrm{Ker} f$, by lemma 3.2.1 there exists a homomorphism $\alpha : L_1 \longrightarrow (\mathrm{Tr}N)^+$ such that the diagram 4.7.14 is commutative.

Since $\alpha$ does not factors through $f_1$, $\alpha$ is a monomorphism. Indeed, suppose $\alpha$ is not a monomorphism and $K \neq 0$ is a finitely generated submodule of $\mathrm{Ker}\alpha$. Then by proposition 4.7.7 for a canonical epimorphism $\varphi : L_1 \rightarrow L_1/K$ there is a homomorphism $h : L_1/K \rightarrow (\mathrm{Tr}N)^+$ such that $\alpha = h\varphi$. On the other hand, since $\varphi$ is not a split monomorphism, $L_1/K$ is a f.p. module, and $f_1$ is an almost split monomorphism, there is a homomorphism $\psi : M_1 \longrightarrow L_1/K$ such that $\varphi = \psi f_1$. So that $\alpha = h\psi f_1$, i.e. $\alpha$ factors through $f_1$.

We show now that $\alpha$ is a pure monomorphism using the equivalent condition for purity given by theorem 4.6.12(4). Let $X \in \mathrm{mod}_r A$ and $\tau \in \mathrm{Hom}_A(X, K)$, where $K = \mathrm{Coker}\alpha$. By lemma 3.1.18 there exists a commutative diagram

$$0 \longrightarrow L_1 \xrightarrow{u_1} Y \xrightarrow{v_1} X \longrightarrow 0$$

$$1_{L_1} \| \qquad \tau_1 \downarrow \qquad \tau \downarrow$$

$$0 \longrightarrow L_1 \xrightarrow{\alpha} (\mathrm{Tr}N)^+ \xrightarrow{v} K \longrightarrow 0$$

with exact rows.

From the Homotopy lemma 3.2.14 there exists a homomorphism $\gamma : X \longrightarrow (\mathrm{Tr}N)^+$ such that $\gamma = \tau v$ if and only if there exists a homomorphism $\beta : Y \longrightarrow L_1$ such that $\beta u_1 = 1_{L_1}$, i.e. the diagram

$$0 \longrightarrow L_1 \xrightarrow{u_1} Y \xrightarrow{v_1} X \longrightarrow 0$$

$$1_{L_1} \| \quad \beta \quad \tau_1 \downarrow \quad \gamma \quad \tau \downarrow$$

$$0 \longrightarrow L_1 \xrightarrow{\alpha} (\mathrm{Tr}N)^+ \xrightarrow{v} K \longrightarrow 0$$

commutes. In this case the top row of this diagram is a split exact sequence, and we are done.

Suppose that the top row of this diagram is not split. Then $u_1$ is not a split monomorphism and therefore there exists a homomorphism $h : M_1 \longrightarrow Y$ such that the diagram

is commutative, i.e. $u_1 = hf_1$. Then $\alpha = \tau_1 u_1 = \tau_1 h f_1$, i.e. $\alpha$ factors through $h$ that is not the case. So that $\alpha$ is a pure monomorphism. $\square$

**Theorem 4.7.15.** (W. Zimmermann [237, Theorem 1]). *Let A be a semiperfect ring, and N a finitely presented non-projective A-module with a local endomorphism ring. Then there exists an almost split sequence*

$$0 \to L \to M \to N \to 0, \qquad (4.7.16)$$

*in $\mathrm{mod}_r A$ if and only if $(\mathrm{Tr}N)^+$ contains a finitely presented non-zero pure submodule with a local endomorphism ring. Up to isomorphism, the later is uniquely determined.*

*Proof.* From proposition 4.7.11 it follows that if $0 \to L \to M \to N \to 0$ is an almost split sequence in $\mathrm{mod}_r A$ then there exists a pure embedding $\alpha : L \to (\mathrm{Tr}N)^+$.

Conversely, assume that $N$ is a finitely presented non-projective $A$-module and $(\mathrm{Tr}N)^+$ contains a finitely presented non-zero pure submodule $L$ with a local endomorphism ring. Let $\alpha : L \to (\mathrm{Tr}N)^+$ be a pure monomorphism. By theorem 4.7.8 for the f.p. module $N$ there is an almost split sequence $0 \to (\mathrm{Tr}N)^+ \xrightarrow{f_1} M_1 \xrightarrow{g_1} N \to 0$ in $\mathrm{Mod}_r A$.

If $\alpha$ is an isomorphism, $\beta$ is also an isomorphism by [103, corollary 1.2.7]. So that

$$0 \longrightarrow (\mathrm{Tr}N)^+ \longrightarrow M_1 \longrightarrow N \to 0$$

is an almost split sequence in $\mathrm{mod}_r A$, and we are done.

Assume that $\alpha$ is not an isomorphism. Write $K = \mathrm{Coker}(\alpha)$ and so there is an exact sequence $0 \to L \xrightarrow{\alpha} (\mathrm{Tr}N)^+ \xrightarrow{\pi} K \to 0$. Since $\pi$ is not a split monomorphism and $f_1$ is left almost split, there is a map $h : M_1 \to K$ such that $h f_1 = \pi$. Let $M = \mathrm{Ker}h$ and $\beta : M \to M_1$ be the natural embedding. Then there is a homomorphism $f : L \longrightarrow M$ such that $\beta f = f_1 \alpha$. Since $f_1$, $\alpha$ and $\beta$ are monomorphisms, $f$ is also a monomorphism. Let $g = g_1 \beta : M \longrightarrow N$. Then $g$ is an epimorphism and $gf = g_1 \beta f = g_1 f_1 \alpha = 0$. Thus we obtain a commutative diagram with exact rows and columns:

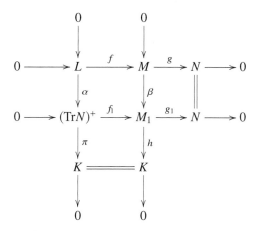

Since $L$ and $N$ are finitely presented modules, $M$ is also finitely presented. So we need only show that the first row in this diagram is an almost split sequence in $\mathrm{mod}_r A$. It does not split because the second row is not split. Since $N$ and $L$ have local endomorphism rings, it suffices to show that $g$ is right almost split in $\mathrm{mod}_r A$. Let $X \in \mathrm{mod}_r A$ and $\varphi : X \to N$ a morphism which is not a split epimorphism. Since $g_1$ is a right almost split in $\mathrm{Mod}_r A$ there is $\varphi_1 : X \to M_1$ such that $g_1 \varphi_1 = \varphi$. By the purity of $\alpha$ there is a $\varphi_2 : X \to (\mathrm{Tr}N)^+$ such that $h\varphi_1 = \pi\varphi_2 = h f_1 \varphi_2$. Therefore there is a $\varphi_3 : X \to M$ such that $\beta\varphi_3 = \varphi_1 - f_1\varphi_2$, and we have $\varphi = g_1\varphi_1 = g_1(\beta\varphi_3 + f_1\varphi_2) = g_1\beta\varphi_3 = g\varphi_3$. So $g$ is right almost split and $\mathrm{End}_A L$ is a local ring. Then the sequence $0 \to L \longrightarrow M \longrightarrow N \to 0$ is almost split by theorem 4.4.14. Because almost split sequences are uniquely determined up to isomorphism by proposition 4.4.18, a submodules $L$ is also uniquely determined up to isomorphism. $\square$

**Example 4.7.17.** (W. Zimmermann [237]).
This is an example of a ring $A$ which is an Artinian ring, but neither an Artin algebra, nor a ring of finite representation type, which admits almost split sequences in the category $\mathrm{mod}_r A$.

Let $k$ be a field of characteristic $0$, $F = k((X))$ be the field of formal Laurent series in the variable $X$ over $k$, i.e. the field of fraction of the power series ring $k[[X]]$. Let ${}_F N_F$ be a bimodule over $F$, which as a left $F$-vector space has a basis $\{x, y\}$, and as a right $F$-vector space is defined by the relations $x\alpha = \alpha x$ and $y\alpha = \alpha y + \alpha' x$, where $\alpha'$ is the derivative with respect to $X$ of $\alpha \in F$. Obviously, $\{x, y\}$ are also a right basis of $N$. Consider the triangular matrix ring:

$$A = \begin{bmatrix} F & {}_F N_F \\ 0 & F \end{bmatrix}.$$

Then $A$ is an Artinian ring, which is not an Artin algebra, because its center is equal to $k \cdot 1$. And as follows from [101, theorem 2.7.1], $A$ has infinite representation type.

## 4.8 Linkage and Duality of Modules over Semiperfect Rings

This section presents a short introduction to the theory of linkage of modules over semiperfect Noetherian rings using two types of functors: syzygy and transpose. These functors play a fundamental role in the homological theory of Noetherian rings and were first introduced and studied by M. Auslander and M. Bridger [13], and O. Iyama [104]. The theory of linkage of modules over a wide class of rings was first considered by A. Martsinkovsky and J.R. Strooker [146]. In this section some results of this theory are given based on the paper [146].

All rings considered in this section are assumed to be semiperfect and Noetherian, and all modules are finitely generated.

In section 4.4 we considered the transpose functor

$$\text{Tr} : \underline{\text{mod}}_r A \longrightarrow \underline{\text{mod}}_l A,$$

which is a duality functor between the projective stable categories of right and left modules over $A$. Now consider another duality functor, which is called the syzygy functor.

**Definition 4.8.1.** Let $M \in \text{mod}_r A$ and $P \xrightarrow{f} M \longrightarrow 0$ a projective cover of $M$. We write $\Omega M = \text{Ker} f$ and call it as the **syzygy module** of $M$.

By proposition 3.1.26, the syzygy module is defined uniquely up to stable equivalence, i.e. the projective equivalence class of this module is uniquely defined. If $A$ is a semiperfect ring then this module is defined uniquely up to isomorphism, by corollary 4.2.2.

Recall that a module $M$ is called **stable** if it has no projective summands.

**Proposition 4.8.2.** *Let $A$ be a semiperfect ring, $M$ a stable finitely presented $A$-module. Then $M^* = \text{Hom}_A(M, A)$ is isomorphic to $\Omega^2 \text{Tr}(M)$.*

*Proof.* Let $P_1 \xrightarrow{f} P_0 \longrightarrow M \to 0$ be a minimal projective resolution of $M$. Applying to this sequence the duality operator $^*$, we obtain the exact sequence

$$0 \to M^* \to P_0^* \xrightarrow{f^*} P_1^* \xrightarrow{\varphi} \text{Tr}(M) \to 0, \qquad (4.8.3)$$

which, by proposition 4.3.5, is a minimal projective resolution for $\text{Tr}(M)$. Then $\Omega\text{Tr}(M) = \text{Ker}\varphi$, and the sequence (4.8.3) is really a combination (Yoneda splice) of two exact sequences:

$$0 \to \Omega\text{Tr}(M) \to P_1^* \to \text{Tr}(M) \to 0$$

$$0 \to M^* \to P_0^* \to \Omega\text{Tr}(M) \to 0,$$

where $P_1^*$ is a projective cover of $\text{Tr}(M)$, and $P_0^*$ is a projective cover of $\Omega\text{Tr}(M)$. Since $P_0$ is a finitely generated projective module, $P_0^*$ is also projective, by [103, proposition 4.5.9]. Therefore, by definition, $M^* \simeq \Omega(\Omega\text{Tr}M) = \Omega^2\text{Tr}(M)$. $\square$

Write the composite of $\Omega$ and Tr as

$$\lambda = \Omega \text{Tr}.$$

This operator was introduced by A. Martsinkowsky and J.R. Strooker in [146]. Using this operator they defined a notion of linkage of modules.

Since $A$ is a semiperfect ring, from corollary 4.2.2 it follows that $\lambda M$ is determined uniquely up to isomorphism for any finitely generated $A$-module $M$.

**Lemma 4.8.4.** *Let $A$ be a semiperfect Noetherian ring, $M$ a finitely generated right $A$-module with minimal projective resolution*

$$P_1 \to P_0 \to M \to 0.$$

*Then $M$ is stable if and only if $P_0^* \to \lambda M$ is a projective cover.*

The proof of this statement follows immediately from proposition 4.3.5.

**Definition 4.8.5.** Let $A$ be a semiperfect Noetherian ring. A finitely generated right $A$-module $M$ and a left $A^{op}$-module $N$ are said to be **horizontally linked** if $M \simeq \lambda N$ and $N \simeq \lambda M$. It follows that $M$ is **horizontally linked** (to $\lambda M$) if and only if $M \simeq \lambda^2 M$.

Obviously, if $M$ is a f.g. $A$-module and $M$ and $N$ are horizontally linked then $N$ is also finitely generated. It follows from definition 4.8.5 that a projective module is horizontally linked if and only if it is isomorphic to the zero module, since the transpose of a projective module over a semiperfect ring is zero.

Note also the following simple fact. Suppose $M$ is a finitely generated $A$-module over a semiperfect Noetherian ring $A$ having no non-zero projective direct summands, and that the projective dimension of Tr$M$ is equal to one. If $P_1 \xrightarrow{\varphi} P_0 \longrightarrow M \to 0$ is a minimal projective resolution for the module $M$ then, since $M$ is a stable module, $P_0^* \xrightarrow{\varphi^*} P_1^* \longrightarrow \text{Tr}(M) \to 0$, is a minimal projective resolution for Tr$(M)$ by proposition 4.3.5(2). Since proj.dim$(\text{Tr}(M)) = 1$, $P_0^* = 0$. So from the sequence (4.3.2) it follows that $M^* = 0$. Therefore $M$ is horizontally linked if and only if $M$ isomorphic to the zero module.

**Proposition 4.8.6.** *Let $A$ be a semiperfect Noetherian ring. Then each horizontally linked $A$-module $M$ is stable.*

*Proof.* Suppose that an $A$-module $M$ is finitely generated and horizontally linked, i.e. $M \cong \lambda^2 M$. Assume that $M = M_1 \oplus P$, where $M_1$ is stable and $P$ is projective. Let $P_1 \to P_0 \to M_1 \to 0$ be a minimal projective resolution of $M_1$. Since $M_1$ is stable, from proposition 4.3.5 it follows that

$$0 \to M_1^* \to P_0^* \xrightarrow{f^*} P_1^* \xrightarrow{\varphi} \text{Tr}(M_1) \to 0$$

is a minimal projective resolution for Tr$(M_1)$. This means that

$$0 \to M_1^* \to P_0^* \to \lambda M_1 \to 0$$

is a projective cover for $\lambda M_1$. Now we can apply this construction again and obtain an epimorphism $P_0^{**} \to \lambda^2 M_1 \to 0$. Since $P_0$ is a f.g. projective module, $P_0^{**} \cong P_0$. Since $\text{Tr}(M) = \text{Tr}(M_1)$, we get $\lambda^2 M_1 \cong \lambda^2 M \cong M = M_1 \oplus P$. Suppose that $Q$ is a projective cover of $\lambda^2 M$, then $Q \cong P_0 \oplus P$ and $Q \cong P_0^{**} \cong P_0$. Since $A$ is a semiperfect ring, $P = 0$, i.e. $M$ is stable. $\square$

**Proposition 4.8.7.** *If $M$ is a horizontally linked $A$-module, then $\lambda M$ is also horizontally linked, and, in particular, $\lambda M$ is stable.*

*Proof.* Since $M$ is a horizontally linked $A$-module, $M \simeq \lambda^2 M$. Therefore $\lambda^2(\lambda M) \simeq \lambda(\lambda^2 M) \simeq \lambda M$. $\square$

**Remark 4.8.8.** The proof of proposition 4.8.7 shows, in particular, that the definition of a horizontal linkage is symmetric in the sense that $M$ is horizontally linked to $\lambda M$ if and only if $\lambda M$ is horizontally linked to $M$.

Write $T_k = \text{Tr}\Omega^{k-1} M$ for $k > 0$. This functor was first introduced by M. Auslander and M. Bridger in [13] to define and study torsionless (i.e. semireflexive) and reflexive modules. O.Iyama used this functor in [104] to study the duality between some full subcategories defined by (reduced) grade (which, in turn, has to do with vanishing of higher $\mathcal{E}xt$ groups).

**Proposition 4.8.9.** *For any $k > 0$ and an $A$-module $M$ the module $T_k M$ is horizontally linked if and only if $\text{Ext}_A^k(M, A) = 0$.*

*Proof.* From the definition it follows that for any $k > 0$ there is an exact sequence

$$0 \to \text{Ext}_A^k(M, A) \to T_k M \to \lambda^2 T_k M \to 0. \qquad (4.8.10)$$

So, the statement immediately follows from this sequence and the definition of linkage 4.8.5. $\square$

**Proposition 4.8.11.** (M. Auslander). *Let $M$ be a finitely generated right $A$-module. Then there is an exact sequence*

$$0 \to \text{Ext}_{A^{op}}^1(\text{Tr}M, A) \to M \xrightarrow{\delta_M} M^{**} \to \text{Ext}_{A^{op}}^2(\text{Tr}M, A) \to 0. \qquad (4.8.12)$$

*Proof.* Let $M$ be a finitely generated $A$-module. Let

$$P_1 \xrightarrow{f_1} P_0 \xrightarrow{f_0} M \to 0 \qquad (4.8.13)$$

be a minimal projective resolution of $M$ which can be seen as obtained as the composition (Yoneda splice) of the two projective covers: $0 \to K \xrightarrow{i_1} P_0 \xrightarrow{f_0} M \to 0$ and $P_1 \xrightarrow{\pi_1} K \to 0$, where $K = \text{Ker}(f_0)$. Setting $f_1 = i_1\pi_1$ there results a commutative diagram:

$$P_1 \xrightarrow{\ f_1\ } P_0 \xrightarrow{\ f_0\ } M \longrightarrow 0$$

with $\pi_1 : P_1 \to K$, $i_1 : K \to P_0$.

Applying the duality functor $^*$ to the exact sequence (4.8.13) yields a commutative diagram with exact row:

$$0 \longrightarrow M^* \xrightarrow{\ f_0^*\ } P_0^* \xrightarrow{\ f_1^*\ } P_1^* \xrightarrow{\ \varphi\ } \mathrm{Tr}M \longrightarrow 0 \,,$$

with $\pi_2 : P_0^* \to C$, $i_2 : C \to P_1^*$.

where $C = \mathrm{Coker}(f_0^*)$. Applying the duality functor $^*$ to the exact sequence

$$0 \to C \xrightarrow{\ i_2\ } P_1^* \xrightarrow{\ \varphi\ } \mathrm{Tr}M \to 0$$

we obtain the long exact sequence

$$0 \to (\mathrm{Tr}M)^* \xrightarrow{\ \varphi^*\ } P_1^{**} \xrightarrow{\ i_2^*\ } C^* \to \mathrm{Ext}^1_{A^{op}}(\mathrm{Tr}M, A) \to \mathrm{Ext}^1_{A^{op}}(P_1^*, A) \to$$

$$\to \mathrm{Ext}^1_{A^{op}}(C, A) \to \mathrm{Ext}^2_{A^{op}}(\mathrm{Tr}M, A) \to \mathrm{Ext}^2_{A^{op}}(P_1^*, A). \qquad (4.8.14)$$

Since $M$ is a finitely generated module over a Noetherian ring, $P_0$, $P_1$ are finitely generated projective modules so that $P_0^*$, $P_1^*$ are also projective, and hence $\mathrm{Ext}^1_{A^{op}}(P_1^*, A) = \mathrm{Ext}^2_{A^{op}}(P_1^*, A) = 0$. Therefore (4.8.14) immediately implies that

$$\mathrm{Ext}^1_{A^{op}}(C, A) \simeq \mathrm{Ext}^2_{A^{op}}(\mathrm{Tr}M, A). \qquad (4.8.15)$$

On the other hand, applying the duality functor $^*$ to the exact sequence

$$0 \to M^* \xrightarrow{\ f_0^*\ } P_0^* \xrightarrow{\ \pi_2\ } C \to 0$$

gives the long exact sequence

$$0 \to C^* \xrightarrow{\ \pi_2^*\ } P_0^{**} \xrightarrow{\ f_0^{**}\ } M^{**} \to \mathrm{Ext}^1_{A^{op}}(C, A) \to 0,$$

since $\mathrm{Ext}^1_{A^{op}}(P_0^*, A) = 0$. Therefore one can consider the commutative diagram

$$
\begin{array}{ccccccccc}
0 & \longrightarrow & K & \xrightarrow{\ i_1\ } & P_0 & \xrightarrow{\ f_0\ } & M & \longrightarrow & 0 \\
 & & \downarrow{\scriptstyle g} & & \downarrow{\scriptstyle \delta_{P_0}} & & \downarrow{\scriptstyle \delta_M} & & \\
0 & \longrightarrow & C^* & \xrightarrow{\ \pi_2^*\ } & P_0^{**} & \xrightarrow{\ f_0^{**}\ } & M^{**} & \longrightarrow & \mathrm{Ext}^1_{A^{op}}(C, A) \longrightarrow 0
\end{array}
$$

Diag. 4.8.16.

where $g$ is a homomorphism induced by homomorphisms $\delta_{P_0}$ and $\delta_M$. Since $\delta_{P_0}$ is an isomorphism by [103, proposition 4.5.9], from the snake lemma we have that

$$\mathrm{Ker}(\delta_M) \simeq \mathrm{Coker}(g). \tag{4.8.17}$$

Taking into account (4.8.15) we obtain:

$$\mathrm{Coker}(\delta_M) \simeq \mathrm{Ext}^1_{A^{op}}(C, A) \simeq \mathrm{Ext}^2_{A^{op}}(\mathrm{Tr}M, A). \tag{4.8.18}$$

Since diagram 4.8.16 is commutative, $\delta_{P_0} i_1 = \pi_2^* g$. Since $f_1 = i_1 \pi_1$, we obtain that

$$\delta_{P_0} f_1 = (\delta_{P_0} i_1)\pi_1 = (\pi_2^* g)\pi_1. \tag{4.8.19}$$

On the other hand, from the commutative diagram

$$
\begin{array}{ccc}
P_1 & \xrightarrow{f_1} & P_0 \\
\downarrow{\scriptstyle \delta_{P_1}} & & \downarrow{\scriptstyle \delta_{P_0}} \\
P_1^{**} & \xrightarrow{f_1^{**}} & P_0^{**}
\end{array}
$$

it follows that $\delta_{P_0} f_1 = f_1^{**} \delta_{P_1}$. So

$$\delta_{P_0} f_1 = \pi_2^* i_2^* \delta_{P_1}, \tag{4.8.20}$$

since $f_1^* = i_2 \pi_2$. Comparing (4.8.19) with (4.8.20), and taking into account that $\pi_1^*$ is a monomorphism, we obtain that $g\pi_1 = i_2^* \delta_{P_1}$. Hence $\mathrm{Im}(i_2^* \delta_{P_1}) \subseteq \mathrm{Im}(g)$ and so there is an induced commutative diagram:

$$
\begin{array}{ccccccccc}
P_1 & \xrightarrow{\delta_{P_1}} & P_1^{**} & \xrightarrow{i_2^{**}} & C^* & \longrightarrow & \mathrm{Ext}^1_{A^{op}}(\mathrm{Tr}M, A) & \longrightarrow & 0 \\
\downarrow{\scriptstyle \pi_1} & & \downarrow{\scriptstyle \mathrm{id}} & & \| & & \downarrow{\scriptstyle h} & & \\
0 \longrightarrow K & \xrightarrow{\quad g \quad} & & & C^* & \longrightarrow & \mathrm{Coker} g & \longrightarrow & 0 \\
\downarrow & & & & & & & & \\
0 & & & & & & & &
\end{array}
$$

Then from the snake lemma it follows that $h$ is an isomorphism, and so $\mathrm{Coker} g \simeq \mathrm{Ext}^1_{A^{op}}(\mathrm{Tr}M, A)$. Hence from (4.8.17) we obtain that $\mathrm{Ker}(\delta_M) \simeq \mathrm{Ext}^1_{A^{op}}(\mathrm{Tr}M, A)$, as required. $\square$

**Definition 4.8.21.** The left $A$-module $\mathrm{Ker}(\delta_M) = \mathrm{Ext}^1_{A^{op}}(\mathrm{Tr}M, A)$ is called the **1-torsion submodule** of $M$.

If $A$ is a commutative domain, then the 1-torsion submodule is the usual torsion submodule. But in general this is not the case. From the existence of the exact sequence (4.8.12) we immediately obtain the following result.

**Corollary 4.8.22.** *Let M be a finitely generated right A-module. Then M is torsionless if and only if* $\mathrm{Ext}^1_{A^{op}}(\mathrm{Tr}M, A) = 0$.

**Lemma 4.8.23.** *Let M be a finitely generated stable A-module. Then* $\mathrm{Im}(\delta_M) \simeq \lambda^2 M$.

*Proof.* Let $P_1 \to P_0 \xrightarrow{\varphi} M \to 0$ be a minimal projective resolution of $M$. Since $P_0^{**} \simeq P_0$ and $\varphi$ is an epimorphism, from the commutative diagram

$$
\begin{array}{ccc}
P_0 & \xrightarrow{\varphi} & M \\
\delta_{P_0} \downarrow \simeq & & \downarrow \delta_M \\
P_0^{**} & \xrightarrow{\varphi^{**}} & M^{**}
\end{array}
$$

it follows that $\mathrm{Im}(\delta_M) = \mathrm{Im}(\varphi^{**})$. Applying the functor $*$ to a minimal projective resolution of $M$ yields two exact sequences

$$0 \to M^* \xrightarrow{\varphi^*} P_0^* \to P_1^* \to \mathrm{Tr}(M) \to 0$$

$$0 \to M^* \xrightarrow{\varphi^*} P_0^* \to \lambda M \to 0.$$

Let $Q \xrightarrow{\psi} M^* \to 0$ be a projective cover of $M^*$. Set $\alpha = \varphi^* \psi$. Since $M$ is a stable module, the sequence $Q \xrightarrow{\alpha} P_0^* \to \lambda M \to 0$ is a minimal projective resolution. Applying to this sequence the functor $*$ yields a commutative diagram with exact row:

$$
\begin{array}{ccccccccc}
0 & \longrightarrow & (\lambda M)^* & \longrightarrow & P_0^{**} & \xrightarrow{\alpha^*} & Q^* & \longrightarrow & \mathrm{Tr}(\lambda M) & \longrightarrow & 0 \\
& & & & & \searrow \varphi^* \quad \nearrow \psi^* & & & & & \\
& & & & & M^{**} & & & & &
\end{array}
$$

By construction, $\mathrm{Im}(\alpha^*) \simeq \Omega\mathrm{Tr}(\lambda M) = \lambda^2 M$. On the other hand, since $\psi^*$ is a monomorphism, $\mathrm{Im}(\alpha^*) \simeq \mathrm{Im}(\varphi^{**}) \simeq \mathrm{Im}(\delta_M)$. Hence $\lambda^2 M \simeq \mathrm{Im}(\delta_M)$, as required. $\square$

**Corollary 4.8.24.** *Let M be a finitely generated stable right A-module. Then there is a short exact sequence*

$$0 \to \mathrm{Ext}^1_{A^{op}}(\mathrm{Tr}M, A) \to M \to \lambda^2 M \to 0. \tag{4.8.25}$$

**Theorem 4.8.26.** *Let M be a finitely generated right A-module. Then M is horizontally linked if and only if it is stable and torsionless.*

*Proof.* If $M$ is horizontally linked, then $M \simeq \lambda^2 M$, by the definition. By proposition 4.8.6, $M$ is stable. So from corollary 4.8.24 it follows that $\mathrm{Ext}^1_{A^{op}}(\mathrm{Tr}M, A) = 0$, and so $M$ is torsionless, by corollary 4.8.22.

The converse statement immediately follows from corollary 4.8.24. $\square$

**Corollary 4.8.27.** *A stable reflexive module is horizontally linked.*

**Corollary 4.8.28.** *A finitely generated A-module M is horizontally linked if and only if M is a stable syzygy module.*

*Proof.* If $M$ is horizontally linked, then $M$ is stable, by proposition 4.8.6. From the definition it follows that $M \simeq \lambda^2 M = \Omega \mathrm{Tr} \Omega \mathrm{Tr} M$, which shows that $M$ is a syzygy module.

The converse statement follows from [103, corollary 4.5.6] and theorem 4.8.26. $\square$

**Lemma 4.8.29.** *Let M be a finitely presented right A-module such that $\delta_M : M \to M^{**}$ is an epimorphism, then $\mathrm{Ext}^1_{A^{op}}(\lambda M, A) = 0$.*

*Proof.* From proposition 4.8.11 it follows that $\mathrm{Ext}^1_{A^{op}}(\lambda M, A) \simeq \mathrm{Ext}^1_{A^{op}}(\mathrm{Tr} M, A)$. $\square$

## 4.9 Notes and References

The notion of the transpose Tr of a module was first introduced by M. Auslander at the Stockholm International Congress of Mathematicians in 1962 in the study of commutative Noetherian regular local rings [12]. The general theory of the Auslander-Bridger translation was developed by M. Auslander and M. Bridger in [13].

Huang Zhaoyong in [238] generalized the notion of the Auslander-Bridger transpose to that the transpose with respect to a faithfully balanced self-orthogonal bimodule $_A\omega_A$ over an Artinian algebra $A$ and obtained some properties about dual modules with respect to $_A\omega_A$.

The notion of almost split sequences was first introduced and studied by M. Auslander and I. Reiten (see [15] and [20]), and therefore these sequences are often called Auslander-Reiten sequences. The existence of these sequences was first proved by M. Auslander in [16] for Artin algebras, and by M. Auslander and I. Reiten for Artin rings of finite representation type in [20]. Theorem 4.7.8 for semiperfect rings was proved by M. Auslander in [17]. We here present a proof of this theorem following [236]. In [237] W. Zimmermann studied the existence of Auslander-Reiten sequences for other classes of Artin rings, in particular, he gave the example of an Artin ring $A$ which is neither an Artin algebra nor of finite representation type, but which admits Auslander-Reiten sequences in the category $\mathrm{mod}_r A$ of finitely generated $A$-modules.

The concept of linkage (liaison) for algebraic varieties, notably curves, dates from the late 19-th century and early 20-th century. A major break-through came with paper of C. Peskine and L. Szpiro in 1974 [173] who reduced it (and generalized it) to be a certain relation between ideals in local rings. These remarks are a condensed version of material in the introduction of [146]. In this paper by A. Martsinkowsky and J.R. Strooker 'linkage' was greatly generalized to deal with modules over a wide class of rings, including non-commutative semiperfect Noetherian rings. This theory uses two types of functors: syzygy and transpose, which play a fundamental role in the homological algebra of Noetherian rings. The origins go back to M. Auslander and M. Bridger [13], and O. Iyama [104].

# CHAPTER 5

# Representations of Primitive Posets

This chapter is devoted to finite partially ordered sets (posets, in short) and their representations, which play an important role in representation theory in general. Representations of posets were first introduced and studied by L.A. Nazarova and A.V. Roiter [163] in 1972 in connection with problems of representations of finite dimensional algebras. M.M. Kleiner characterized posets of finite type [129] and described pairwise non-isomorphic indecomposable representations [130]. He proved a theorem which gives a criterium for posets to be of finite representation type.

Recall that a finite poset $\mathcal{P}$ is called **primitive** if it is a cardinal sum of linearly ordered sets $L_1, \ldots, L_m$. It is then denoted by $\mathcal{P} = L_1 \sqcup \cdots \sqcup L_m$.

This chapter gives the proof of a criterium for primitive posets to be of finite representation type following [30].

To prove this criterium we use only the trichotomy lemma which was proved by P. Gabriel and A.V. Roiter in [82], the Kleiner lemma about the representations of a pair of finite posets proved by M. Kleiner in [129] and the main (so-called) construction, considered in section 5.5. Note that this construction (in some form) was introduced by L.A. Nazarova and A.V. Roiter in [163].

## 5.1 Representations of Finite Posets

Let's recall the main notions and definitions of the representation theory of posets.

Let $(\mathcal{P}, \leq)$ be a finite poset, where $\mathcal{P}$ is a finite set and $\leq$ is a partial order relation. We denote by $x < y$ the strict order, i.e. the relation "$x \leq y$ and $x \neq y$".

The **height** of a poset $\mathcal{P}$ is the largest cardinality of sub-posets of $\mathcal{P}$ which are chains, and the **width** of $\mathcal{P}$ is the largest cardinality of sub-posets of $\mathcal{P}$ which are antichains. In other words, the height of a poset $\mathcal{P}$ is the length of the longest chain in $\mathcal{P}$, and the width of a poset $\mathcal{P}$ is the maximal number of pairwise incomparable elements of $\mathcal{P}$. We denote the height and the width of $\mathcal{P}$ by $h(\mathcal{P})$ and $w(\mathcal{P})$ respectively.

If $\mathcal{P}$ is a finite poset, then a chain $C$ and an antichain $A$ of $\mathcal{P}$ have at most one element in common. Hence the least number of antichains, whose union is $\mathcal{P}$, is not less then the size $h(\mathcal{P})$ of the largest chain in $\mathcal{P}$. In fact, there is a partition of $\mathcal{P}$ into $h(\mathcal{P})$ antichains. There is also some kind of dual statement which is known as the Dilworth theorem.

**Theorem 5.1.1.** (Dilworth [51].) *For a poset $\mathcal{P}$ with finite width $w(\mathcal{P})$, the minimal number of disjoint chains that together contain all elements of $\mathcal{P}$ is equal to $w(\mathcal{P})$.*

In order to visualize a poset $\mathcal{P}$ we will use its diagram. Let $x$ and $y$ be distinct elements of $\mathcal{P}$. We say that $y$ **covers** $x$ if $x < y$ but there is no element $z$ such that $x < z < y$. Recall that the **diagram** of a poset $\mathcal{P} = (p_1, \ldots, p_n)$ is the directed graph whose vertex set is $\mathcal{P}$ and the set of edges is given by the set of covering pairs $(p_i, p_j)$ of $\mathcal{P}$, moreover, there is an edge from a vertex $p_i$ up to a vertex $p_j$ if and only if $p_j$ covers $p_i$.

For example, the diagram below

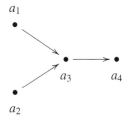

represents a poset $(\mathcal{P}, \leq)$ with 4 elements $\{a_1, a_2, a_3, a_4\}$ and relations $a_2 < a_3 < a_4$ and $a_1 < a_3$.

The diagram of a poset is often called its **Hasse diagram**. Usually it is drawn in the plane in such a way that if $y$ covers $x$ then the point representing $y$ is drawn higher than the point representing $x$. In this case we draw the Hasse diagram without arrows. For example, the Hasse diagram below

represents the poset $(\mathcal{P}, \leq)$, where $\mathcal{P} = \{a_1, a_2, a_3\}$ with one relation $a_2 < a_3$.

**Definition 5.1.2.** Let $(X, \leq_1)$ and $(Y, \leq_2)$ be any two (disjoint) posets. The **cardinal sum** $X \sqcup Y$ (or **disjoint union** $X \cup Y$) of $X$ and $Y$ is the set of all $x \in X$ and $y \in Y$. The relations $x \leq_1 x_1$ and $y \leq_2 y_1$ $(x, x_1 \in X; y, y_1 \in Y)$ have the same meanings and there are no another relations in $X \sqcup Y$.

In the published literature one can found a few different definitions of representations of a finite poset $(\mathcal{P}, \leq)$, where $\mathcal{P} = \{1, 2, \ldots, n\}$ over a division ring $K$.

In this chapter we present two of them: in the language of vector spaces and the matrix language, and the relationships between them.

Historically the first definition of representations of a poset $\mathcal{P}$ over a division ring $K$ was given in the matrix language by A.V. Roiter and L.A. Nazarova in 1972 (see [163])[1].

**Definition 5.1.3.** Let $(\mathcal{P}, \preceq)$ be a finite poset with $n$ elements. A **matrix representation** of $\mathcal{P}$ over a division ring $K$ is an arbitrary block matrix

$$\mathbf{A} = \left[ \mathbf{A}_1 \,\|\, \mathbf{A}_2 \,\|\, \ldots \,\|\, \mathbf{A}_n \right] \tag{5.1.4}$$

partitioned horizontally into $n$ (vertical) blocks (also called strips), where $A_i$ is a $(d_0 \times d_i)$-matrix with entries in $K$ for $i = 1, 2, ..., n$. Moreover, some of these matrices can be empty. In that case we are in fact dealing with a representation of a sub-poset. We will denote this matrix representation as follows: $\mathbf{A} = (\{\mathbf{A}_i\}_{i=1}^n)$.

The vector

$$\mathbf{d} = \underline{\dim}(\mathbf{A}) = (d_0; d_1, \ldots, d_{n-1}, d_n), \tag{5.1.5}$$

where $d_0$ is the number of rows of a matrix $\mathbf{A}$ and $d_i$ is the number of columns of the matrix $\mathbf{A}_i$ ($i = 1, 2, \ldots, n$), is called the **coordinate vector** (or **dimension vector**) of $\mathbf{A}$, and the number

$$d = d(\mathbf{A}) = d_0 + \sum_{i=1}^n d_i \tag{5.1.6}$$

is called the **dimension** of $\mathbf{A}$.

Sometimes it is useful to consider also $\widetilde{d_0}$, the number of non-zero rows. In part of the literature $\widetilde{d_0}$ is written as $d_{n+1}$.

**Remark 5.1.7.** Note that in the definition of matrix representations 5.1.3 we admit the case when some of matrices $\mathbf{A}_i$ or even all of them are empty. One of the reasons to introduce these matrix representations is that it is naturally to consider matrix representations of the following form

$$\left[ \frac{\mathbf{A}}{\mathbf{O}} \right] \quad \text{or} \quad \left[ \, \mathbf{A} \,\|\, \mathbf{O} \, \right]$$

to be decomposable.

Another reason to introduce these empty matrices is that there is no full analogy between matrices and linear spaces without these matrices. Indeed, no nonempty matrix corresponds to the unique linear mapping from $m$-dimensional vector space to

---

[1]As L.A. Nazarova and A.V. Roiter write: "The definition (of a matrix representation) presented appears to be totally absurd because the order relation on $\mathcal{P}$ is totally unaccounted for. However the order relation will be taken into account when defining the similarity of representations". And the similarity classes of representations are the main object of study.

the 0-dimensional vector space, and to the unique linear mapping from 0-dimensional vector space to the $m$-dimensional vector space.

We now introduce 'ideal' matrices (following the terminology of L.A. Nazarova and A.V. Roiter [163]) which have one of the following forms:

1. $\Im_{m,0}$ is an $(m \times 0)$-sized empty matrix with $m$ rows and 0 columns. It corresponds to the unique linear mapping from $m$-dimensional vector space to the 0-dimensional vector space.

2. $\Im_{0,m}$ is a $(0 \times m)$-sized empty matrix with 0 rows and $m$ columns. It corresponds to the unique linear mapping from 0-dimensional vector space to the $m$-dimensional vector space.

   In particular, $\Im_{0,0} = \pi_0$ is a $0 \times 0$ empty matrix with zero rows and zero columns. It corresponds to the unique linear mapping from 0-dimensional vector space to the 0-dimensional vector space.

The operation of multiplication with these matrices is defined as follows:

$$\Im_{0,m}\Im_{m,0} = \Im_{0,0}, \quad \Im_{0,m}\mathbf{A}_{m,n} = \Im_{0,n}$$

$$\Im_{n,0}\Im_{0,m} = 0, \quad \mathbf{A}_{m,n}\Im_{n,0} = \Im_{m,0}$$

where $\mathbf{A}_{m,n} \in M_{m \times n}(K)$, and the operation of direct sum with these matrices is defined as follows:

$$\Im_{m,0} \oplus \Im_{n,0} = \Im_{m+n,0} \quad \Im_{0,m} \oplus \Im_{0,n} = \Im_{0,m+n}$$

$$\Im_{m,0} \oplus \Im_{0,n} = \Im_{0,n} \oplus \Im_{m,0} = \mathbf{O}_{m,n}.$$

In particular, $\Im_{0,1} \oplus \Im_{1,0} = (0)$, the $1 \times 1$ zero matrix consisting of a single zero.

$$\mathbf{A}_{m,n} \oplus \Im_{k,0} = \left[ \begin{array}{c} \mathbf{A}_{m,n} \\ \hline \mathbf{O}_{k,n} \end{array} \right] \quad \Im_{k,0} \oplus \mathbf{A}_{m,n} = \left[ \begin{array}{c} \mathbf{O}_{k,n} \\ \hline \mathbf{A}_{m,n} \end{array} \right]$$

$$\mathbf{A}_{m,n} \oplus \Im_{0,k} = \left[ \begin{array}{c|c} \mathbf{A}_{m,n} & \mathbf{O}_{m,k} \end{array} \right] \quad \Im_{0,k} \oplus \mathbf{A}_{m,n} = \left[ \begin{array}{c|c} \mathbf{O}_{m,k} & \mathbf{A}_{m,n} \end{array} \right]$$

where $\mathbf{A}_{m,n} \in M_{m \times n}(K)$ and $\mathbf{O}_{k,n}$ is a $k \times n$ zero matrix.

In particular, $\mathbf{A}_{m,n} \oplus \Im_{0,0} = \mathbf{A}_{m,n}$.

Using these matrices we introduce the following 'ideal' representations of a finite poset $\mathcal{P}$ with $n$ elements:

(i) $\mathbf{I}_0^n$ is an empty matrix which has zero rows and all $n$ strip matrices are empty of the following form

$$\mathbf{I}_0^n = \left[ \begin{array}{c|c|c|c} \Im_{0,0} & & \cdots & \Im_{0,0} \end{array} \right] \tag{5.1.8}$$

with dimension vector

$$\underline{\dim}(\mathbf{I}_0^n) = (0; 0, \ldots, 0).$$

(ii) $\mathbf{I}_n^0$ is the strip matrix which has one row and $n$ strips of which all are empty matrices of the following form

$$\mathbf{I}_n^0 = \left[\; \mathfrak{I}_{1,0} \;\middle\|\; \cdots \;\middle\|\; \mathfrak{I}_{1,0} \;\right] \tag{5.1.9}$$

with dimension vector

$$\underline{\dim}(\mathbf{I}_n^0) = (1; 0, \ldots, 0).$$

(iii) $\mathbf{J}_n^i$ is the strip matrix with zero rows and $n$ strips which all strip matrices are empty except the $i$-th strip matrix which is equal to $\mathfrak{I}_{0,1}$ of the following form

$$\mathbf{J}_n^i = \left[\; \mathfrak{I}_{0,0} \;\middle\|\; \cdots \;\middle\|\; \mathfrak{I}_{0,1} \;\middle\|\; \cdots \;\middle\|\; \mathfrak{I}_{0,0} \;\right] \tag{5.1.10}$$

with dimension vector

$$\underline{\dim}(\mathbf{J}_n^i) = (0; 0, \ldots, 0, 1, 0, \ldots, 0)$$

i.e. $d_0 = 0$, $d_j = 0$ for $j \neq i$ and $d_i = 1$.

Let $\mathbf{B} = (\{\mathbf{B}_i\}_{i=1}^n)$ be another matrix representation of the poset $(\mathcal{P}, \leq)$:

$$\mathbf{B} = \left[\; \mathbf{B}_1 \;\middle\|\; \mathbf{B}_2 \;\middle\|\; \cdots \;\middle\|\; \mathbf{B}_n \;\right] \tag{5.1.11}$$

with dimension vector

$$\mathbf{d}' = \underline{\dim}(\mathbf{B}) = (d_0'; d_1', \ldots, d_{n-1}', d_n')$$

A morphism of matrix representations $X : \mathbf{A} \to \mathbf{B}$ is given by a pair of matrices $(\mathbf{\Phi}_0, \mathbf{\Phi})$ such that

$$\mathbf{\Phi}_0 \mathbf{A}_j = \sum_{i \leq j} \mathbf{B}_i \mathbf{\Phi}_{ij} \tag{5.1.12}$$

where $\mathbf{\Phi}_0$ is a $d_0' \times d_0$-matrix and

$$\mathbf{\Phi} = \begin{bmatrix} \mathbf{\Phi}_{11} & \mathbf{\Phi}_{12} & \cdots & \mathbf{\Phi}_{1n} \\ \mathbf{\Phi}_{21} & \mathbf{\Phi}_{22} & \cdots & \mathbf{\Phi}_{2n} \\ \vdots & \vdots & \ddots & \vdots \\ \mathbf{\Phi}_{n1} & \mathbf{\Phi}_{n2} & \cdots & \mathbf{\Phi}_{nn} \end{bmatrix}, \tag{5.1.13}$$

where $\mathbf{\Phi}_{ij}$ is a $(d_i' \times d_j')$-matrix for $i, j = 1, \ldots, n$; moreover, $\mathbf{\Phi}_{ij}$ can be $\neq 0$ if $i < j$ in $\mathcal{P}$, otherwise $\mathbf{\Phi}_{ij}$ is a zero matrix. Renumbering (if it is necessary) the labels in the poset $\mathcal{P}$ we can consider that the matrix $\mathbf{\Phi}$ is upper triangular.

If $Y : \mathbf{A} \to \mathbf{B}$ is a morphism of matrix representations given by $Y = (\mathbf{\Psi}_0, \mathbf{\Psi})$ then there is a composition of the morphisms $X$ and $Y$: the morphism $Z : \mathbf{B} \to \mathbf{C}$ which is given in the following way: $Z = Y \circ X = (\mathbf{\Psi}_0 \mathbf{\Phi}_0, \mathbf{\Psi}\mathbf{\Phi})$.

Since the product of matrices is associative, compositions of morphisms is also associative.

For any matrix representation $\mathbf{A} = (\{\mathbf{A}_i\}_{i=1}^n)$ with dimension vector $\underline{\dim}(\mathbf{A}) = (d_0; d_1, \ldots, d_{n-1}, d_n)$ the identity morphism $1_{\mathbf{A}}$ is given by a pair of identity matrices $(\mathbf{I}_{d_0}, \mathbf{I}_m)$, where $m = \sum_{i=1}^n d_i$.

Thus, matrix representations of a poset $\mathcal{P}$ over a division ring $K$ with morphisms given above form a category of matrix representations which is denoted by $\mathrm{repM}_K(\mathcal{P})$.

The morphism $X = (\mathbf{\Phi}_0, \mathbf{\Phi})$ is an **isomorphism** of matrix representations $\mathbf{A}$ and $\mathbf{B}$ if $\mathbf{\Phi}_0$ and $\mathbf{\Phi}$ are nonsingular matrices. In this case each $\mathbf{\Phi}_{ii}$ is a nonsingular matrix for $i = 1, \ldots, n$.

In is easy to prove the following fact.

**Lemma 5.1.14.** *If two matrix representations* $\mathbf{A}$ *and* $\mathbf{B}$ *are isomorphic then*

$$\underline{\dim}(\mathbf{A}) = \underline{\dim}(\mathbf{B})$$

Recall that an **elementary row operation** over a division ring $K$ for an arbitrary matrix is one of the followings:

(i) Interchanging two rows.
(ii) Multiplying a row (on the left) by a non-zero scalar (i.e. an element of $K$).
(iii) Replacing a row by itself plus a scalar multiple of another row.

Analogously there can be introduced the **elementary column operations**. Here the scalar multiplications of columns are on the right.

Let $\mathbf{A}$ be a matrix representation of a poset $(\mathcal{P}, \leq)$. The following elementary operations:

(a) Elementary row operations on the whole matrix $\mathbf{A}$.
(b) Elementary column operations within each vertical strip $\mathbf{A}_i$.
(c) Additions of scalar multiples of columns of a strip $\mathbf{A}_i$ to columns of a strip $\mathbf{A}_j$ if $i \prec j$ holds in $\mathcal{P}$

are called the **admissible elementary operations** for $\mathbf{A}$. A composition of admissible elementary operations for $\mathbf{A}$ is called an **admissible operation** for $\mathbf{A}$.

Note that by means of a sequence of operations (b) and (c) an arbitrary linear combination of columns of $\mathbf{A}_i$ can be added to a column of $\mathbf{A}_j$ if $i \prec j$ in $\mathcal{P}$.

Thus an **isomorphism** of matrix representations $\varphi : \mathbf{A} \to \mathbf{B}$ given by a pair of nonsingular matrices $(\mathbf{\Phi}_0, \mathbf{\Phi})$ can be given by a set of admissible operations (a), (b),

(c) and (i), (ii), (iii) by which one can obtain the matrix $\mathbf{B}$ from the matrix $\mathbf{A}$, and vice versa.

The **direct sum** of two matrix representations $\mathbf{A}$ and $\mathbf{B}$ is a matrix representation $\mathbf{A} \oplus \mathbf{B}$ which is equal to the following block matrix:

$$\mathbf{A} \oplus \mathbf{B} = \left[\begin{array}{c|c||c|c||c|c|c} \mathbf{A}_1 & \mathbf{O} & \mathbf{A}_2 & \mathbf{O} & \ddots & \mathbf{A}_n & \mathbf{O} \\ \hline \mathbf{O} & \mathbf{B}_1 & \mathbf{O} & \mathbf{B}_2 & \ddots & \mathbf{O} & \mathbf{B}_n \end{array}\right], \tag{5.1.15}$$

i.e. $(\mathbf{A} \oplus \mathbf{B})_i = \mathbf{A}_i \oplus \mathbf{B}_i$.

It is easy to verify that

$$\underline{\dim}(\mathbf{A} \oplus \mathbf{B}) = \underline{\dim}(\mathbf{B}) + \underline{\dim}(\mathbf{B}). \tag{5.1.16}$$

**Definition 5.1.17.** A matrix representation $\mathbf{A}$ is called **decomposable** if it is equivalent to a direct sum of at least two matrix representations $\mathbf{A}_1$ and $\mathbf{A}_2$ with $d(\mathbf{A}_1) \neq 0$ and $d(\mathbf{A}_2) \neq 0$. Otherwise it is called **indecomposable**.

Note that $\mathbf{I}_n^0$ and $\mathbf{J}_n^i$ are non-zero indecomposable representations of a finite poset $\mathcal{P}$ with $n$ elements.

**Remark 5.1.18.** If

$$\mathbf{A} = \left[\begin{array}{c||c||c|c} \mathbf{A}_1 & \mathbf{A}_2 & \ldots & \mathbf{A}_n \end{array}\right]$$

is a matrix representation of a poset $\mathcal{P}$ with dimension vector $\underline{\dim}(\mathbf{A}) = (d_0; d_1, \ldots, d_n)$, then

$$1) \qquad \mathbf{A} \oplus \mathbf{I}_n^0 = \left[\begin{array}{c||c||c|c} \mathbf{A}_1' & \mathbf{A}_2' & \ldots & \mathbf{A}_n' \end{array}\right]$$

where

$$\mathbf{A}_i' = \left[\begin{array}{c} \mathbf{A}_1 \\ \hline 0 \ldots 0 \end{array}\right] \quad \text{if } \mathbf{A}_i \neq \mathfrak{J}_{d_0,0}$$

and

$$\mathbf{A}_i' = \mathfrak{J}_{d_0+1,0} \quad \text{if } \mathbf{A}_i = \mathfrak{J}_{d_0,0}.$$

Moreover,

$$\underline{\dim}(\mathbf{A} \oplus \mathbf{I}_n^0) = (d_0 + 1; d_1, \ldots, d_n).$$

$$2) \qquad \mathbf{A} \oplus \mathbf{J}_n^i = \left[\begin{array}{c||c||c|c} \mathbf{A}_1' & \mathbf{A}_2' & \ldots & \mathbf{A}_n' \end{array}\right]$$

where

$$\mathbf{A}'_j = \mathbf{A}_j \quad \text{for} \quad j \neq i$$

and

$$\mathbf{A}'_i = \begin{bmatrix} \mathbf{A}_i & \begin{matrix} 0 \\ 0 \\ \vdots \\ 0 \end{matrix} \end{bmatrix} \quad \text{if } \mathbf{A}_i \neq \mathfrak{J}_{d_0,0}, \quad \text{and if } \mathbf{A}_i = \mathfrak{J}_{d_0,0} \text{ then } \mathbf{A}'_i = \begin{bmatrix} 0 \\ 0 \\ \vdots \\ 0 \end{bmatrix}.$$

Moreover,

$$\underline{\dim}(\mathbf{A} \oplus \mathbf{J}_{i,1}) = (d_0; d_1, \dots, d_i + 1, \dots, d_n).$$

Note that $\mathbf{A} \oplus \mathbf{I}_0^n = \mathbf{A}$, i.e. $\mathbf{I}_0^n$ is a zero element with respect to taking direct sum.

Thus taking direct sums $\mathbf{A} \oplus \mathbf{I}_n^0$ and $\mathbf{A} \oplus \mathbf{J}_n^i$ serves to insert zero rows and zero columns.

**Proposition 5.1.19.** *With respect to the operation of taking direct sums the category of all matrix representations (including "ideal" representations) is an additive $K$-linear category* $\mathrm{repM}_K(\mathcal{P})$ *whose zero object is the representation* $\mathbf{I}_0^n$ *defined by* (5.1.8).

**Remark 5.1.20.** The monoid of zeroes matrix representations.

Let $\mathcal{P} = \{1, 2, \dots, n\}$ be a poset.

Write $m\mathbf{I}_n^0 = \mathbf{I}_n^0 \oplus \mathbf{I}_n^0 \oplus \cdots \oplus \mathbf{I}_n^0$, the direct sum of $m$ copies of $\mathbf{I}_n^0$, i.e.

$$m\mathbf{I}_n^0 = \begin{bmatrix} \mathfrak{J}_{m,0} & \Big\| & \cdots & \Big\| & \mathfrak{J}_{m,0} \end{bmatrix} \tag{5.1.21}$$

with dimension vector

$$\underline{\dim}(m\mathbf{I}_n^0) = (m; 0, \dots, 0).$$

Analogously, write $m\mathbf{J}_n^i = \mathbf{J}_n^i \oplus \mathbf{J}_n^i \oplus \cdots \oplus \mathbf{J}_n^i$, the direct sum of $m$ copies of $\mathbf{J}_n^i$, i.e.

$$m\mathbf{J}_n^i = \begin{bmatrix} \mathfrak{J}_{0,0} & \Big\| & \cdots & \Big\| & \mathfrak{J}_{0,m} & \Big\| & \cdots & \Big\| & \mathfrak{J}_{0,0} \end{bmatrix} \tag{5.1.22}$$

with dimension vector

$$\underline{\dim}(m\mathbf{J}_n^i) = (0; 0, \dots, m, 0, \dots, 0).$$

Consider the set $\mathrm{Rep}_0(\mathcal{P})$ of 'ideal' matrix representations of $\mathcal{P}$ of the following form

$$\mathbf{M} = \mathbf{I}_0^n \oplus d_0 \mathbf{I}_n^0 \oplus \sum_{i=1}^{n} d_i \mathbf{J}_n^i, \tag{5.1.23}$$

with dimension vector

$$\underline{\dim}(\mathbf{M}) = (d_0; d_1, \dots, d_n) \qquad (5.1.24)$$

where $d_0, d_i \in \mathbf{N} \cup \{0\}$ for all $i \in \mathcal{P}$.

Then the set $\mathrm{Rep}_0(\mathcal{P})$ forms a monoid with respect to the operation of taking direct sums. The identity in this monoid is an 'ideal' matrix representation $\mathbf{I}_0^n$. This monoid is finitely generated with $n + 1$ generators $\mathbf{I}_n^0$ and $\mathbf{J}_n^i$, and it depends only on the size of $\mathcal{P}$ and not on the ordering.

Consider the monoid

$$\mathcal{D}_n = (\mathbf{N} \cup \{0\})^{n+1} = \{(d_0; d_1, \dots, d_n) \; : \; d_i \in \mathbf{N} \cup \{0\}\}$$

with operation of addition on vectors and the identity element of $\mathcal{D}_n$ is the zero vector $(0, 0, \dots, 0)$. Then $\underline{\dim}$ defined by (5.1.24) on $\mathrm{Rep}_0(\mathcal{P})$ is an isomorphism from the monoid $\mathrm{Rep}_0(\mathcal{P})$ of all zeros representations to the monoid $\mathcal{D}_n$ of all possible dimension vectors.

It is interesting that elements of the monoid $\mathrm{Rep}_0(\mathcal{P})$ are also ordinary zeroes matrix representations, i.e. representations $(\{\mathbf{A}_i\}_{i=1}^n)$ where all nonempty matrices $\mathbf{A}_i$ are zero. Consider such a zeroes representation $\mathbf{A} = (\{\mathbf{A}_i\}_{i=1}^n)$ with dimension vector $\underline{\dim}(\mathbf{M}) = (d_0; d_1, \dots, d_n)$. We now show that this representation can be decomposed into a finite direct sum of basis elements of $\mathrm{Rep}_0(\mathcal{P})$. Write $\mathcal{P} = \mathcal{P}_1 \cup \mathcal{P}_2$, where $i \in \mathcal{P}_1$ if and only if $\mathbf{A}_i = \mathfrak{I}_{d_0,0}$ and $j \in \mathcal{P}_2$ if and only if $\mathbf{A}_j$ is a $d_0 \times d_j$ zero matrix. Then

$$\mathbf{A} = \mathbf{B} \oplus \bigoplus_{k \in \mathcal{P}_2} \mathbf{C}^{(k)},$$

where $\mathbf{B} = (\{\mathbf{B}_i\}_{i=1}^n)$ and $\mathbf{C}^{(k)} = (\{\mathbf{C}_i^{(k)}\}_{i=1}^n)$ with $\mathbf{B}_i = \mathfrak{I}_{d_0,0}$ for all $i \in \mathcal{P}$ and

$$\mathbf{C}_i^{(k)} = \begin{cases} \mathfrak{I}_{0,d_i} & \text{if } i = k \text{ and } \mathbf{A}_i = \mathbf{O}_{d_0 \times d_i} \\ \mathfrak{I}_{0,0} & \text{otherwise} \end{cases}$$

Then it is easy to see that $\mathbf{B} = d_0 \mathbf{I}_n^0$ and $\mathbf{C}^{(k)} = d_k \mathbf{J}_n^k$. Thus $\mathbf{A} \in \mathrm{Rep}_0(\mathcal{P})$.

**Remark 5.1.25.** Note that D. Simson [191] gave an equivalent interpretation of the category of matrix representations $\mathrm{repM}_K(\mathcal{P})$ as a category $\mathrm{Mat}_K^{\mathrm{ad}}(\mathcal{P})$ in the following way.

Let $\mathcal{P} = \{1, 2, \dots, n\}$. The objects of the category $\mathrm{Mat}_K^{\mathrm{ad}}(\mathcal{P})$ are systems

$$V = (V_1, V_2, \dots, V_n, V_{n+1}, A_1, A_2, \dots, A_n)$$

where the $V_i$ are the finite dimensional $K$-linear vector spaces and $A_i : V_i \to V_{n+1}$ are $K$-linear vector morphism mappings for $i = 1, \dots, n$. If $W = (W_i, B_i)$ is another object of the category $\mathrm{Mat}_K^{\mathrm{ad}}(\mathcal{P})$ then a morphism $f : V \to W$ is defined as a pair $(\varphi, \varphi_{n+1})$ of two $K$-linear spaces such that the diagram

$$V_1 \oplus \cdots \oplus V_n \xrightarrow{(A_i)} V_{n+1}$$

$$\varphi \downarrow \qquad\qquad \downarrow \varphi_{n+1}$$

$$W_1 \oplus \cdots \oplus W_n \xrightarrow[(B_i)]{} W_{n+1}$$

**Diag. 5.1.26.**

is commutative, where $\varphi$ is a $K$-linear map with upper triangular matrix

$$\mathbf{\Phi} = \begin{pmatrix} \varphi_{11} & \varphi_{12} & \cdots & \varphi_{1n} \\ 0 & \varphi_{22} & \cdots & \varphi_{2n} \\ \vdots & \vdots & \ddots & \vdots \\ 0 & 0 & \cdots & \varphi_{nn} \end{pmatrix}$$

where $\varphi_{ij} : V_j \to W_i$ is a $K$-linear map and $\varphi_{ij} = 0$ if $i \not\leq j$ in $\mathcal{P}$. The direct sum of two objects $V$ and $W$ is

$$V \oplus W = (U_1, \ldots, U_n, U_{n+1}, C_1, \ldots, C_n)$$

where $U_i = (V \oplus W)_i = V_i \oplus W_i$, $U_{n+1} = (V \oplus W)_{n+1} = V_{n+1} \oplus W_{n+1}$, $C_i = A_i \oplus B_i$. The zero element in $\mathrm{Mat}_K^{\mathrm{ad}}(\mathcal{P})$ is an element $V$ such that all $V_i = 0$ and all $A_i$ are trivial linear maps. An object $U$ of this category is called **decomposable** if it can be written as a direct sum $U = V \oplus W$ of two non-zero objects $V$ and $W$. Otherwise $U$ is called **indecomposable**.

It is easy to show that these categories are equivalent. In this case to any matrix representation $(\{\mathbf{A}_i\}_{i=1}^n)$ with dimension vector $\underline{\dim}(\mathbf{A}) = (d_0; d_1, \ldots, d_{n-1}, d_n)$ we let correspond the object $V$ with $V_i = K^{d_i}$ for $i = 1, \ldots, n$, $V_{n+1} = K^{d_0}$ and the $K$-linear map $A_i : V_i \to V_{n+1}$ is induced by the matrix $\mathbf{A}_i$ with respect to the standard basis in $K^{d_i}$ and $K^{d_0}$.

Another definition of representations of a finite poset over a division ring was given by P. Gabriel in [78] in the language of vector spaces.

**Definition 5.1.27.** A (space) representation of a poset $\mathcal{P}$ over a division ring $K$ is a collection of finite dimensional $K$-spaces $V = (V_0, V_1, \ldots, V_n)$ such that $V_i \subseteq V_0$ and $V_i \subseteq V_j$ if and only if $i \leq j$ in $\mathcal{P}$. If all $K$-vector spaces are finite dimensional over $K$, $V$ is called a **finite dimensional** representation.

Note, that some of these spaces or all of them can be zero.

The **dimension vector** of a space representation $V$ is the vector

$$d = \underline{\dim}V = (d_0; d_1, \ldots, d_n) \in (\mathbf{N} \cup \{0\})^{n+1} \tag{5.1.28}$$

where $d_0 = \dim V_0$, and $d_i = \dim V_i / \sum_{j < i} V_j$.

Let $V = (V_0, V_1, \ldots, V_n)$ and $W = (W_0, W_1, \ldots, W_n)$ be two representations of a poset $P$. A **morphism** $f : V \longrightarrow W$ is a $K$-linear map $V_0 \longrightarrow W_0$ such that $f(V_i) \subseteq W_i$ for all $i \in P$.

Two space representations $V$ and $W$ is said to be **isomorphic** if there is an isomorphic $K$-linear map $f : V \to W$ such that $f(V_i) = W_i$ for all $i \in \mathcal{P}$.

Thus, the set of all space representations with morphisms defined above form a category which is called the **category of space representations of a poset** $\mathcal{P}$ over a division ring $K$ and we will denote it by $\mathrm{RepS}_K(\mathcal{P})$.

The zero element in this category is $V$ with $V_0 = V_i = 0$ for all $i \in P$. This category is additive with respect to direct sums: $V \oplus W = U$, where $U_0 = V_0 \oplus W_0$ and $U_i = V_i \oplus W_i$. A (space) representation $U \in \mathrm{RepS}_K(\mathcal{P})$ is called **indecomposable** if it cannot be written as a direct sum of two non-zero representations, otherwise it called **decomposable**. Representations of the form $V = (V_0, 0, \ldots, 0)$ are called **trivial**.

The category of all finite dimensional representations of $\mathcal{P}$ over $K$ is denoted by $\mathrm{RepS}_K(\mathcal{P})$. We will consider only finite dimensional representations, i.e. the category $\mathrm{repS}_K(\mathcal{P})$.

The category of all indecomposable finite dimensional representations of $\mathcal{P}$ over $K$ will be denoted $\mathrm{indS}(\mathcal{P})_K$.

For any (space) representation $V \in \mathrm{repS}_K(\mathcal{P})$ the ring of endomorphisms $\mathrm{End}_K(V)$ is a finite dimensional $K$-algebra, and so it is a semiperfect ring. Moreover, since the category $\mathrm{repS}_K(\mathcal{P})$ has split idempotents, it is a Krull-Schmidt category. So that any representation $V \in \mathrm{rep}_K(\mathcal{P})$ is a direct sum of indecomposable representations and this decomposition is unique up to isomorphism and to the order of the direct summands. Moreover, a representation $V \in \mathrm{repS}_K(\mathcal{P})$ is indecomposable if and only if $\mathrm{End}_K(V)$ is a local ring.

We now discuss the relationship between the two categories $\mathrm{repM}_K(\mathcal{P})$ and $\mathrm{repS}_K(\mathcal{P})$ of representations of a poset $\mathcal{P}$ over a division ring $K$. We define the map

$$F : \mathrm{repM}_K(\mathcal{P}) \longrightarrow \mathrm{repS}_K(\mathcal{P}) \qquad (5.1.29)$$

in the following way. Let $\mathbf{A} = (\{\mathbf{A}_i\}_{i=1}^n) \in \mathrm{repM}_K(\mathcal{P})$ with dimension vector (5.1.5). Note that any matrix $\mathbf{A}_i \in M_{d_0 \times d_i}(K)$ induces a $K$-linear homomorphism

$$A_i : K^{d_i} \longrightarrow K^{d_0}$$

with respect to the standard basis in $K^{d_i}$ and $K^{d_0}$. Using these homomorphisms we can define a $K$-linear homomorphism

$$\alpha_i = \sum_{j \leq i} A_j : \bigoplus_{j \leq i} K^{d_j} \longrightarrow K^{d_0}.$$

Thus, we define $F(\mathbf{A}) = V = (V_0, V_i)$ by setting

$$V_0 = K^{d_0} \quad \text{and} \quad V_i = \sum_{j \leq i} \mathrm{Im}(A_j) \text{ for all } i \in I. \qquad (5.1.30)$$

Let $\mathbf{B} = (\{\mathbf{B}_i\}_{i=1}^n) \in \mathrm{repM}_K(\mathcal{P})$ be another (matrix) representation with dimension vector $\underline{\dim}(\mathbf{B}) = (d'_0; d'_1, \ldots, d'_{n-1}, d'_n)$. Suppose that a $K$-linear homomorphism

$$B_i : K^{d'_i} \longrightarrow K^{d'_0}$$

is induced by the matrix $\mathbf{B}_i \in M_{d'_0 \times d'_i}(K)$ and

$$\beta_i = \sum_{j \leq i} B_j : \bigoplus_{j \leq i} K^{d'_j} \longrightarrow K^{d'_0}.$$

Let $F(\mathbf{B}) = W$, where

$$W_0 = K^{d'_0} \quad \text{and} \quad W_i = \sum_{j \leq i} \mathrm{Im}(B_j) \quad \text{for all } i \in I. \tag{5.1.31}$$

Let $f : \mathbf{A} \to \mathbf{B}$ be a morphism given by a pair of matrices $(\mathbf{\Phi}_0, \mathbf{\Phi})$. Then we define $F(X) : F(\mathbf{A}) \to F(\mathbf{B})$ as follows: $F(f) = \varphi_0$, where $\varphi_0 : V_0 \to W_0$ is the $K$-linear homomorphism induced by the matrix $\mathbf{\Phi}_0$ with respect to the standard basis in $V_0$ and $W_0$. We also write $\varphi_i : V_i \to W_i$ for the homomorphism induced by the matrix $(\mathbf{\Phi}_{r,s})_{r,s \leq i}$.

We now show that $F$ is a functor. Let $x = \bigoplus_{i \leq j} x_i \in V_i$, then

$$\varphi_0 \alpha_i(x) = \sum_{j \leq i} \varphi A_j(x_j)$$

and

$$\beta_i \varphi_i(x) = \sum_{j \leq i} (\sum_{k \leq j} B_k \Phi_{k,j}(x_j))$$

Taking into account the definition of a homomorphism of matrix representations we obtain that $\varphi_0 \alpha_i(x) = \beta_i \varphi_i(x)$ for each $x \in V_i$, i.e. the following diagram

$$
\begin{array}{ccc}
V_i & \xrightarrow{\alpha_i} & V_0 \\
{\scriptstyle \varphi_i} \downarrow & & \downarrow {\scriptstyle \varphi_0} \\
W_i & \xrightarrow{\beta_i} & W_0
\end{array}
$$

is commutative. So that $\varphi(V_i) \subseteq W_i$ for all $i \in I$, i.e. $F(f) = \varphi_0$ is a morphism of (space) representations $F(\mathbf{A})$ and $F(\mathbf{B})$. Thus $F : \mathrm{repM}_K(\mathcal{P}) \longrightarrow \mathrm{repS}_K(\mathcal{P})$ is a functor.

**Definition 5.1.32.** A functor $F : \mathfrak{C} \to \mathfrak{D}$ is said to be **full** if for each $X, Y \in Ob(\mathfrak{C})$ the morphism mapping $\mathrm{Hom}_{\mathfrak{C}}(X, Y) \to \mathrm{Hom}_{\mathfrak{D}}(F(X), F(Y))$ is surjective. The functor $F$ is said to be **faithful** if this morphism mapping is injective. The functor $F$ is said to be **dense** if for every $Y \in Ob(\mathfrak{D})$ there exists an element $X \in Ob(\mathfrak{C})$ such that $Y \simeq F(X)$.

**Proposition 5.1.33.** (Yu.A. Drozd [62], D. Simson [191]). *Let $F : \mathrm{repM}_K(\mathcal{P}) \to$ $\mathrm{repS}_K(\mathcal{P})$ be the functor defined above as (5.1.29). Then $F$ is full and dense.*

*Proof.* Let $\mathbf{A} = (\{A_i\}_{i=1}^n) \in \mathrm{repM}_K(\mathcal{P})$ and $\mathbf{B} = (\{B_i\}_{i=1}^n) \in \mathrm{repM}_K(\mathcal{P})$ be two matrix representations with dimension vectors $\underline{\dim}(\mathbf{A}) = (d_0; d_1, \ldots, d_n)$ and $\underline{\dim}(\mathbf{B}) = (d_0'; d_1', \ldots, d_n')$ correspondingly. By the definition of the functor $F$ constructed above $F(\mathbf{A}) = (V_0, V_i)$ and $F(\mathbf{B}) = (W_0, W_i)$, where $V_0, V_i$ are given by (5.1.30) and $W_0, W_i$ are given by (5.1.31). We first show that $F$ is a full functor, i.e. for any morphism $g : F(\mathbf{A}) \to F(\mathbf{B})$ there is a morphism $f : \mathbf{A} \to \mathbf{B}$ such that $g = F(f)$. Since $g$ is a $K$-linear morphism in $\mathrm{repS}_K(\mathcal{P})$, it defines a homomorphism of $K$-spaces $g : V_0 \to W_0$ with corresponding matrix $\mathbf{\Phi}_0 \in M_{d_0 \times d_0'}(K)$. From the definition of morphisms in $\mathrm{repS}_K(\mathcal{P})$ it follows that $g(V_i) \subseteq W_i$, or what is the same

$$g(\sum_{j \leq i} \mathrm{Im}(A_j)) \subseteq \sum_{j \leq i} \mathrm{Im}(B_j)$$

Then we can choose homomorphisms $\varphi_{ji} : K^{d_i} \to K^{d_j'}$ such that $\varphi_{ji} = 0$ if $j \not\leq i$ in $\mathcal{P}$, so that the diagram

$$
\begin{array}{ccc}
K^{d_i} & \xrightarrow{\quad A_i \quad} & K^{d_0} \\
{\scriptstyle (\varphi_{ji})_{j \leq i}} \downarrow & & \downarrow {\scriptstyle g} \\
\bigoplus_{j \leq i} K^{d_j'} & \xrightarrow{\quad \sum_{j \leq i} B_j \quad} & K^{d_i'}
\end{array}
$$

is commutative. Then setting $\mathbf{\Phi}$ to be the matrix corresponding to the $K$-linear homomorphism $\mathbf{\Phi} = (\varphi_{ji})$ we obtain that the pair of matrices $(\mathbf{\Phi_0}, \mathbf{\Phi})$ defines a morphism $f : \mathbf{A} \to \mathbf{B}$. Thus, $F(f) = g$, i.e. $F$ is a full functor.

Secondly we show that $F$ is a dense functor, i.e. for any representation $V = (V_0, V_i) \in \mathrm{repS}_K(\mathcal{P})$ there exists a matrix representation $\mathbf{A} = (\{A_i\}_{i=1}^n) \in \mathrm{repM}_K(\mathcal{P})$ such that $F(\mathbf{A}) \simeq V$. We first define another space representation $W = (W_0, W_i)$ in the following way. Let $W_0 = V_0$ and $W_i = V_i$ for each minimal element $i \in \mathcal{P}$. If $i$ is not minimal we will define $W_i$ recursively in the following way. If for all $j < i$ the subspaces $W_j$ are already constructed then we define $W_i$ as a direct summand of the decomposition $V_i = W_i \oplus \sum_{j < i} W_j$. It is clear that $V_i = \bigoplus_{j \leq i} W_j$ and that $W_i \subseteq W_0$ for all $i \in I$. If $\dim W_i = d_i$ then there is a $K$-linear isomorphism $\varphi_i : W_i \to K^{d_i}$ and $\varphi_i^{-1} : K^{d_i} \to W_i$. If $\dim W_0 = d_0$ then there is a $K$-linear homomorphism $\beta_0 : W_0 \to K^{d_0}$ and a natural injection $\beta_i : W_i \to W_0$. So that we obtain a commutative diagram

$$
\begin{array}{ccc}
W_i & \xrightarrow{\quad \beta_i \quad} & W_0 \\
{\scriptstyle \varphi_i^{-1}} \big\Vert {\scriptstyle \varphi_i} & & \downarrow {\scriptstyle \beta_0} \\
K^{d_i} & \xrightarrow{\quad \gamma_i \quad} & K^{d_0}
\end{array}
$$

where $\gamma_i = \beta_0\beta_i\varphi_i^{-1}$. Suppose that a $K$-linear homomorphism $A_i = \beta_0\beta_i = \gamma_i\varphi_i$ : $W_i \rightarrow K^{d_0}$ corresponds to a matrix $\mathbf{A}_i$. Consider a matrix representation $\mathbf{A} = (\{\mathbf{A}_i\}_{i=1}^n)$. We show now that $F(\mathbf{A}) \simeq V$. Indeed,

$$F(\mathbf{A}) = (K^{d_0}, \sum_{j \leq i} \mathrm{Im}(A_j)).$$

Since $\beta_0(V_0) = \beta_0(W_0) = K^{d_0}$ and $V_i = \bigoplus_{j \leq i} W_j$, we obtain a commutative diagram:

$$
\begin{array}{ccccc}
V_i & \xrightarrow{\ \alpha_i\ } & V_0 & \xrightarrow{\ \beta_0\ } & K^{d_0} \\
\| & & \| & & \| \\
\displaystyle\bigoplus_{j \leq i} W_j & \xrightarrow[\bigoplus_i \varphi_i]{\simeq} & \displaystyle\bigoplus_i K^{d_i} & \xrightarrow[\underset{j \leq i}{\Sigma\, A_j}]{} & K^{d_0}
\end{array}
$$

which implies that $\beta_0(V_i) = \sum_{j \leq i} \mathrm{Im}(A_i)$. Thus $F(\mathbf{A}) \simeq V$ as required. $\square$

**Example 5.1.34.**
The following example shows that the functor $F : \mathrm{repM}_K(\mathcal{P}) \rightarrow \mathrm{repS}_K(\mathcal{P})$ defined above is not faithful. Let

$$
\mathbf{A} = \left[\begin{array}{c|c} 1 & 0 \\ 0 & 1 \\ 0 & 0 \end{array}\right] \quad \text{and} \quad \mathbf{B} = \left[\begin{array}{c|c} 1 & 0 \\ 0 & 0 \\ 0 & 0 \end{array}\right]
$$

be two matrix representations of the poset $\mathcal{P} = \{ \bullet \qquad \bullet \}$ without relations. Consider two morphisms of these representations $f : \mathbf{A} \rightarrow \mathbf{B}$ given by a pair of matrices: $\mathbf{X}_0 = \begin{pmatrix} 0 & 0 & 0 \\ 0 & 0 & 0 \\ 0 & 0 & 0 \end{pmatrix}$ and $\mathbf{X} = \begin{pmatrix} 0 & 0 \\ 0 & 0 \\ 0 & 0 \end{pmatrix}$ and $g : \mathbf{A} \rightarrow \mathbf{B}$ given by a pair of matrices: $\mathbf{Y}_0 = \begin{pmatrix} 0 & 0 & 0 \\ 0 & 0 & 0 \\ 0 & 0 & 0 \end{pmatrix}$ and $\mathbf{Y} = \begin{pmatrix} 0 & 0 \\ 1 & 0 \\ 0 & 1 \end{pmatrix}$. Then $F(f) = 0 = F(g)$, nevertheless $f \neq g$. Therefore the functor $F$ is not faithful.

**Example 5.1.35.**
Let

$$
\mathbf{A} = \left[\begin{array}{c|c} 1 & 0 \\ 0 & 1 \\ 0 & 0 \end{array}\right] \quad \text{and} \quad \mathbf{B} = \left[\begin{array}{c|c|c} 1 & 0 & 0 \\ 0 & 0 & 0 \\ 0 & 0 & 0 \end{array}\right]
$$

be two matrix representations of the poset $\mathcal{P} = \{ \bullet \qquad \bullet \}$ without relations. Then these representations are not isomorphic since $\underline{\dim}(\mathbf{A}) = (3, 1, 1) \neq \underline{\dim}(\mathbf{B}) = (3, 2, 1)$. Nevertheless $F(\mathbf{A}) = F(\mathbf{B})$. Therefore an isomorphism of space representations in

$\mathrm{repS}_K(\mathcal{P})$ does not imply an isomorphism of corresponding matrix representations in $\mathrm{repM}_K(\mathcal{P})$.

We now restrict our attention to the subcategory $\mathrm{repM}^0_K(\mathcal{P})$ of $\mathrm{repM}_K(\mathcal{P})$ whose objects are matrix representations which do not contain direct summands isomorphic to the "ideal" representations of the form $\mathbf{J}^i_n$ defined by (5.1.10).

**Definition 5.1.36.** An additive functor $F : C \to \mathcal{D}$ between additive categories is said to be a **representation equivalence** if it is full, dense and reflects isomorphism, i.e. $F(f) \in \mathrm{Mor}\mathcal{D}$ is an isomorphism if and only if $f \in \mathrm{Mor}C$ is an isomorphism. The functor $F$ is said to be **equivalence** if it is full, dense and faithful.

**Proposition 5.1.37.** (Yu.A. Drozd [62], D. Simson [191]). *Let $F : \mathrm{repM}_K(\mathcal{P}) \to \mathrm{repS}_K(\mathcal{P})$ be the functor defined above as (5.1.24). Then the following statements hold.*

1. *Let $\mathbf{A}$ be an indecomposable representation in $\mathrm{repM}_K(\mathcal{P})$. Then $F(\mathbf{A}) = 0$ if and only if $\mathbf{A}$ is isomorphic to one of the "ideal" representations of the form $\mathbf{J}^i_n$ for some $i \in \mathcal{P}$.*
2. *The restriction of the functor $F$ to the subcategory $\mathrm{repM}^0_K(\mathcal{P})$ induces a representation equivalence between categories $\mathrm{repM}^0_K(\mathcal{P})$ and $\mathrm{repS}_K(\mathcal{P})$.*

*Proof.* 1. This follows immediately from the definition of the functor $F$ as (5.1.29).

Note also that $\mathbf{A} = \{\mathbf{A}_i\}_{i \in \mathcal{P}}$ is isomorphic to $\mathbf{J}^i_n$ for some $i \in \mathcal{P}$ if and only if the $A_i$ are injective and $W_i = \mathrm{Im}A_i \cap (\sum_{j < i} \mathrm{Im}(A_j)) = 0$ for all $i \in \mathcal{P}$.

2. Let $\mathbf{A}, \mathbf{B} \in \mathrm{repM}^0_K(\mathcal{P})$ with dimension vectors $\underline{\dim}(\mathbf{A}) = (d_0; d_1, \ldots, d_n)$ and $\underline{\dim}(\mathbf{B}) = (d'_0; d'_1, \ldots, d'_n)$. Suppose that $f : F(\mathbf{A}) \to F(\mathbf{B})$ is an isomorphism in $\mathrm{repS}_K(\mathcal{P})$. Since $F$ is a full functor from $\mathrm{repM}_K(\mathcal{P})$ to $\mathrm{repS}_K(\mathcal{P})$, its restriction to $\mathrm{repM}^0_K(\mathcal{P})$ is also full, i.e. there exists a morphism $g : \mathbf{A} \to \mathbf{B}$ given by a pair of matrices $(\mathbf{\Phi}_0, \mathbf{\Phi})$ such that $F(g) = f = \Phi_0$. Thus $\Phi_0 : K^{d_0} \to K^{d'_0}$ is an isomorphism, which implies that $d_0 = d'_0$. So it remains to prove that the $K$-linear mapping corresponding to the matrix $\mathbf{\Phi}$ is an isomorphism.

Suppose that $F(\mathbf{A}) = (K^{d_0}, \sum_{j \le i} \mathrm{Im}(A_j))$ and $F(\mathbf{B}) = (K^{d_0}, \sum_{j \le i} \mathrm{Im}(B_j))$.

Since $\mathbf{A}, \mathbf{B} \in \mathrm{repM}^0_K(\mathcal{P})$, all $A_i$ and $B_i$ are injective for $i \in \mathcal{P}$. Denote the restriction of $f$ to $\sum_{j \le i} \mathrm{Im}(A_j))$ by $f_i$.

Suppose that $i$ is minimal in $\mathcal{P}$. Then since $A_i$ and $B_i$ are injective from the definition of the functor $F$ and the homomorphism $g : \mathbf{A} \to \mathbf{B}$ we have the commutative diagram:

$$
\begin{array}{ccc}
K^{d_i} & \xrightarrow{\;A_i\;} & \mathrm{Im}(A_i) \\
\Big\downarrow{\scriptstyle \Phi_{ii}} & & \Big\downarrow{\scriptstyle f_i} \\
K^{d'_i} & \xrightarrow[\;B_i\;]{} & \mathrm{Im}(B_i)
\end{array}
$$

Since $A_i$ and $B_i$ are both injective, their restrictions in $\mathrm{Im}(A_i)$ and $\mathrm{Im}(B_i)$ are isomorphisms. Because $f$ is injective on $K^{d_0}$, its restriction $f_i$ is also injective. By definition of a homomorphism of matrix representations we have that $f A_i = B_i$, so $f_i$ is surjective. So that $f_i$ is an isomorphism, which implies that $\Phi_{ii}$ is also an isomorphism for all $i$ which are minimal in $\mathcal{P}$.

If $i$ is not a minimal element in $\mathcal{P}$ we use an induction argument. Assume that $\Phi_{jj}$ is an isomorphism for all $j \prec i$. Then from the the definition of the functor $F$ and the homomorphism $g : \mathbf{A} \to \mathbf{B}$ we have the commutative diagram:

$$
\begin{array}{ccccc}
K^{d_i} & \xrightarrow{A_i} & \sum_{j \le i} \mathrm{Im}(A_j) & \xhookrightarrow{\quad} & K^{d_0} \\
{\scriptstyle (\Phi_{ij})_{j \le i}}\Big\downarrow & & {\scriptstyle f_i}\Big\downarrow & & {\scriptstyle \Phi_0}\Big\downarrow \\
\bigoplus_{j \le i} K^{d'_j} & \xrightarrow{(B_j)_{j \le i}} & \sum_{j \le i} \mathrm{Im}(B_j) & \xhookrightarrow{\quad} & K^{d'_0}
\end{array}
$$

Since all $A_i$ and $B_i$ are injective we have that $\mathrm{Im}(A_i) \cap (\sum_{j \prec i} \mathrm{Im}(A_j)) = 0$ and $\mathrm{Im}(B_i) \cap (\sum_{j \prec i} \mathrm{Im}(A_j)) = 0$ for all $i \in \mathcal{P}$. So that we have that $\sum_{j \le i} \mathrm{Im}(A_j) = \mathrm{Im}(A_i) \oplus \sum_{j \prec i} \mathrm{Im}(A_j)$ and $\sum_{j \le i} \mathrm{Im}(B_j) = \mathrm{Im}(B_i) \oplus \sum_{j \prec i} \mathrm{Im}(B_j)$. Since $f_i$ is an isomorphism, its restriction $f_{ii} : \mathrm{Im}(A_i) \to \mathrm{Im}(B_i)$ is also an isomorphism. Thus the diagram

$$
\begin{array}{ccc}
K^{d_i} & \xrightarrow{A_i} & \mathrm{Im}(A_i) \\
{\scriptstyle \Phi_{ii}}\Big\downarrow & & {\scriptstyle f_{ii}}\Big\downarrow \\
K^{d'_i} & \xrightarrow{B_i} & \mathrm{Im}(B_i)
\end{array}
$$

is commutative and as was shown above it follows that $\Phi_{ii}$ is an isomorphism for all $i \in \mathcal{P}$. Since the matrix $\Phi$ is upper triangular, then this implies that $\Phi$ is an isomorphism, as required. $\square$

**Definition 5.1.38.** A matrix representation $\mathbf{A}$ is called **sincere** if it is indecomposable and all coordinates of the vector $\mathbf{d}(\mathbf{A})$ are non-zero. A matrix representation $\mathbf{A}$ will be called **ideal** (following the terminology L.A. Nazarova and A.V. Roiter [163]) if some (possibly all) of the coordinates of the vector $\mathbf{d}(\mathbf{A})$ are zero.

We shall denote by $n(\mathcal{P}, K)$ the cardinal number of pairwise non-isomorphic indecomposable representations of the poset $\mathcal{P}$ over a division ring $K$.

**Definition 5.1.39.** A partially ordered set $\mathcal{P}$ is said to be of **finite representation type** (or **finite type**, in short) over a division ring $K$ if it has finitely many pairwise non-isomorphic indecomposable representations, i.e. $n(\mathcal{P}, K) < \infty$. A poset $\mathcal{P}$ is said

to be of **infinite representation type** (or **infinite type**, in short) over a division ring
$K$ if it has infinitely many pairwise non-isomorphic indecomposable representations,
i.e. $n(\mathcal{P}, K) = \infty$.

**Remark 5.1.40.** From proposition 5.1.37 it follows that the categories $\mathrm{repM}_K(\mathcal{P})$ and
$\mathrm{repS}_K(\mathcal{P})$ differ only by a finite number of indecomposable representations. Thus
their representation type is the same, and there is no difference if we would study
the representation type of a finite poset in the language of matrix representations or
space representations.

M. Kleiner in [129] proved the following theorem which gives a criterium for
posets to be of finite representation type.

**Theorem 5.1.41.** (M. Kleiner [129]). *A finite partially ordered set $\mathcal{P}$ is of finite
representation type over a field $k$ if and only if $\mathcal{P}$ does not contain as a full subposet
any poset from the following list*:

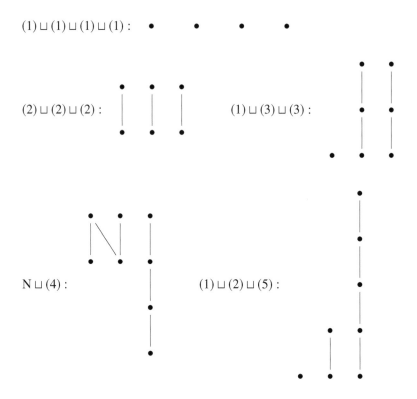

Here the $n$-element chain is denoted by $(n)$ and the symbol N denotes the 4-element poset $\{a_1 \prec a_2 \succ a_3 \prec a_4\}$. The posets in the list of the Kleiner theorem are called the **critical subposets**.

## 5.2 Main Canonical Forms of Matrix Problems

Recall the following important and well-known definitions and facts from elementary linear algebra.

**Definition 5.2.1.** A matrix $\mathbf{A}$ is an **echelon matrix** (or in echelon form) if the following conditions hold:

1. All zero rows of $\mathbf{A}$ are at the bottom.
2. Each leading (i.e. left-most) non-zero entry in a row is strictly to the right of the leading non-zero entry in the row above.

An upper-triangular matrix with non-zero diagonal entries is a special case of an echelon matrix, but an echelon matrix is not necessarily upper triangular. For example, the matrix

$$\mathbf{A} = \begin{pmatrix} 0 & 1 & 0 & 0 \\ 0 & 0 & 0 & 1 \end{pmatrix}$$

is an echelon matrix, but it is not upper triangular.

**Definition 5.2.2.** A matrix is said to be in **row canonical form** (or in **reduced row echelon form**) if it is in echelon form, each leading non-zero entry is equal to 1 and it is the only non-zero entry in its column.

A row canonical form is simpler than an echelon form and it is unique in the sense that it is independent of the algorithm used. Moreover, it gives the unique form of an ordered basis for the row space.

**Definition 5.2.3.** A matrix $\mathbf{A}$ is said to be in **normal form** if:

$$\mathbf{A} = \left[ \begin{array}{c|c} \mathbf{I} & \mathbf{O} \\ \hline \mathbf{O} & \mathbf{O} \end{array} \right], \tag{5.2.4}$$

where $\mathbf{I}$ is an identity matrix.

From standard linear algebra the following three statements are well-known. The first theorem states that any matrix over a field $k$ by means of elementary row operations can be reduced to an echelon form. This row reduction algorithm is often called **Gaussian elimination**. The second theorem says that any echelon matrix over a field $k$ by means of elementary row operations can be reduced to a row canonical form. This row reduction algorithm is called **Gauss-Jordan elimination**. The third theorem states that any matrix $\mathbf{A}$ over a field $k$ by means of elementary row and column operations can be reduced to normal form. It is easy to prove that these three theorems also hold for matrices over a division ring $K$ with row operations involving only multiplications by scalars on the left, and column operations involving only multiplications by scalar on the right.

**Definition 5.2.5.** Let $\mathcal{P} = \{p_1, p_2, \ldots, p_n\}$ be a finite poset. An indecomposable matrix representation of $\mathcal{P}$ with dimension vector $\mathbf{d} = (d_0; d_1, \ldots, d_n)$ is called **elementary** if it has one of the following forms:

1. $\mathbf{V}_n^i$ is the strip matrix with one row and $n$ strips which all strip matrices are empty except the $i$-th strip matrix which is equal to (1):

$$\mathbf{V}_n^i = \left[ \begin{array}{c|c|c|c|c} \Im_{1,0} & \cdots & 1 & \cdots & \Im_{1,0} \end{array} \right] \tag{5.2.6}$$

with dimension vector

$$\underline{\dim}(\mathbf{V}_n^i) = (1; 0, \ldots, 0, 1, 0, \ldots, 0)$$

i.e. $d_0 = 1$, $d_j = 0$ for $j \neq i$ and $d_i = 1$.

2. $\mathbf{V}_n^{i,j}$ is the strip matrix with one row and $n$ strips which all strip matrices are empty except the $i$-th ans $j$-th strip matrices which are equal to (1):

$$\mathbf{V}_n^{i,j} = \left[ \begin{array}{c|c|c|c|c|c|c} \Im_{1,0} & \cdots & 1 & \Im_{1,0} & \cdots & 1 & \cdots & \Im_{1,0} \end{array} \right] \tag{5.2.7}$$

with dimension vector

$$\underline{\dim}(\mathbf{V}_n^{i,j}) = (1; 0, \ldots, 0, 1, 0, \ldots, 0, 1, 0, \ldots, 0)$$

i.e. $d_0 = 1$, $d_k = 0$ for $k \neq i, j$ and $d_i = d_j = 1$.

(3) The matrix representation $\mathbf{I}_n^0$ whose matrix has one row and all strip matrices are empty.

(4) The matrix representation $\mathbf{J}_n^i$ for which all strip matrices are empty except the $i$-th strip matrix which has one zero column and zero rows.

**Lemma 5.2.8.** *Let $C_n$ be a chain. Then a matrix representation $\mathbf{R}$ of $C_n$ over a division ring $K$ can be reduced to the following canonical form:*

$$\mathbf{R} = \left[ \begin{array}{c|c|c|c|c|c|c|c|c} I & O & O & O & O & O & \cdots & O & O \\ \hline O & O & I & O & O & O & \cdots & O & O \\ \hline O & O & O & O & I & O & \cdots & O & O \\ \hline \cdots & \cdots & \cdots & \cdots & \cdots & \cdots & \ddots & \cdots & \cdots \\ \hline O & O & O & O & O & O & \cdots & I & O \\ \hline O & O & O & O & O & O & \cdots & O & O \end{array} \right]. \tag{5.2.9}$$

*Therefore $n(C_n, K) = 2n + 1$ and all indecomposable representations are elementary of the forms $\mathbf{V}_n^i$ and $\mathbf{I}_n^0$, $\mathbf{J}_n^i$ for $i = 1, \ldots, n$.*

*Proof.* Let $C_n = \{p_1 \leq p_2 \leq \ldots \leq p_n\}$ be a chain, and $\mathbf{R}$ a matrix representation of $C_n$. Then the matrix $\mathbf{R}$ is partitioned into $n$ vertical strips:

$$\mathbf{R} = [\ \overbrace{\mathbf{A}_1}^{p_1} \ \|\ \overbrace{\mathbf{A}_2}^{p_2} \ \|\ \cdots \ \|\ \overbrace{\mathbf{A}_n}^{p_n}\ ],$$

where the matrix $\mathbf{A}_i$ corresponds to the element $p_i$ of the poset $C_n$. Let

$$\mathbf{d}(\mathbf{R}) = (d_0; d_1, \ldots, d_n)$$

be the dimension vector of $\mathbf{R}$.

Thus, the following operations are admissible on the matrix $\mathbf{R}$:

a. Elementary row operations on the whole matrix $\mathbf{R}$.

b. Elementary column operations within each vertical strip $\mathbf{A}_i$.

c. Additions of columns of a strip $\mathbf{A}_i$ to columns of a strip $\mathbf{A}_j$ if $i < j$.

The lemma is proved by induction on the number $n$. If $n = 1$ then any matrix by row and column operations can be reduced to the normal form (5.2.4). Assume that $n > 1$ and the statement is true for all $k < n$. Write $\widetilde{d_0}$ for the number of non-zero rows of the matrix $\mathbf{A}$.

If $\widetilde{d_0} = 0$, the matrix $\mathbf{A}$ is equivalent to matrices $\mathbf{J}_n^i$ ($i = 1, 2, \ldots, n$) or $\mathbf{I}_n^0$ and we are done. We have in fact a zeroes matrix representation as discussed in remark 5.1.20.

Suppose that $\widetilde{d_0} \neq 0$. If $d_1 = 0$ or $\mathbf{A}_1$ is a zero matrix, then $\mathbf{A}$ can be considered as a matrix representation of the subposet $C_{n-1} = \{p_2 \leq \ldots \leq p_n\}$ and the result follows from the inductive assumption.

Thus we can assume that $d_1 > 0$ and that there are non-zero elements in $\mathbf{A}_1$. Using the elementary operations the matrix $\mathbf{A}_1$ can be reduced to the normal form (5.2.4). Adding multiples of the columns of the first strip to the other strips of $\mathbf{A}$, all the entries in the first upper horizontal strip can be made zero. Therefore, this yields the following equivalent form:

$$\mathbf{R} \simeq \left[ \begin{array}{cc|c|c|c} \mathbf{I} & \mathbf{O} & \mathbf{O} & \cdots & \mathbf{O} \\ \mathbf{O} & \mathbf{O} & \mathbf{A}_{21} & \cdots & \mathbf{A}_{n1} \end{array} \right] = \left[ \begin{array}{cc|c} \mathbf{I} & \mathbf{O} & \mathbf{O} \\ \mathbf{O} & \mathbf{O} & \mathbf{R}_1 \end{array} \right]$$

for some matrices $\mathbf{A}_{21}, \ldots, \mathbf{A}_{n1}$.

Since the matrix

$$\mathbf{R}_1 = [\ \overbrace{\mathbf{A}_{21}}^{p_2} \ \|\ \overbrace{\mathbf{A}_{31}}^{p_3} \ \|\ \cdots \ \|\ \overbrace{\mathbf{A}_{n1}}^{p_n}\ ],$$

corresponds to a matrix representation of the chain $C_{n-1} = \{p_2 \leq p_3 \leq \ldots \leq p_n\}$, by the induction hypothesis, it can be reduced to the form (5.2.9). So the matrix $\mathbf{R}$ can be reduced to this form as well. $\square$

**Lemma 5.2.10.** *Let $\mathcal{P}$ be a poset of the form*

$$
\begin{array}{cc}
1 & 2 \\
\bullet & \bullet
\end{array}
$$

*then the matrix representation $\mathbf{R}$ of $\mathcal{P}$ over a division ring $K$ can be reduced to the following form:*

$$
\mathbf{R} = \left[\begin{array}{cc|cc||cc}
I & 0 & 0 & 0 & I & 0 \\
\hline
0 & I & 0 & 0 & 0 & 0 \\
\hline
0 & 0 & 0 & I & 0 & 0 \\
\hline
0 & 0 & 0 & 0 & 0 & 0
\end{array}\right]. \tag{5.2.11}
$$

*Therefore, $n(\mathcal{P}) = 6$ and all indecomposable representations are elementary of the forms $\mathbf{V}_2^i$, $\mathbf{I}_2^0$, $\mathbf{J}_2^i$ for $i = 1, 2$ and $\mathbf{V}_2^{1,2}$.*

*Proof.* Consider a matrix representation $\mathbf{R}$ of $\mathcal{P}$:

$$
\mathbf{R} = \left[\ \mathbf{X} \parallel \mathbf{Y}\ \right],
$$

where the matrix $\mathbf{X}$ corresponds to the element 1 and the matrix $\mathbf{Y}$ corresponds to the element 2. Let

$$
\mathbf{d}(\mathbf{R}) = (d_0; d_1, d_2)
$$

be the dimension vector of $\mathbf{R}$.

If $\widetilde{d_0} = 0$, the matrix $\mathbf{R}$ is a zeroes matrix representation and we are done.

Suppose that $\widetilde{d_0} \neq 0$. If $d_1 = 0$ or $\mathbf{X}$ is a zero matrix, then $\mathbf{R}$ can be considered as a matrix representation of the subposet $C_1 = \{p_2\}$ and the result follows from lemma 5.2.8.

Thus we can assume that $d_1 > 0$ and there are non-zero elements in $\mathbf{X}$. By elementary operations on the matrix $\mathbf{X}$, it can be reduced to the normal form (5.2.4), whence

$$
\mathbf{R} \simeq \left[\begin{array}{cc||c}
I & 0 & Y_1 \\
\hline
0 & 0 & Y_2
\end{array}\right] = \mathbf{R}_1.
$$

Now use row operations on $\mathbf{Y}_2$ and column operations on $\left[\dfrac{\mathbf{Y}_1}{\mathbf{Y}_2}\right]$ to reduce $\mathbf{Y}_2$ to normal form. This gives the matrix

$$
\mathbf{R} = \left[\begin{array}{cc||c|c}
I & 0 & Y_{11} & Y_{12} \\
\hline
0 & 0 & I & 0 \\
\hline
0 & 0 & 0 & 0
\end{array}\right].
$$

Then with row operations $\mathbf{Y}_{11}$ can be made zero. Next use row operations on the first block row and column operations on $\mathbf{Y}_{12}$ to reduce $\mathbf{Y}_{12}$ to normal form. This spoils the left top block $\mathbf{I}$ but only to the extent of replacing it with an invertible matrix. So we get

$$\left[\begin{array}{cc|c|c|c|c}\mathbf{M}_{11} & \mathbf{M}_{12} & \mathbf{O} & \mathbf{O} & \mathbf{I} & \mathbf{O} \\ \hline \mathbf{M}_{21} & \mathbf{M}_{22} & \mathbf{O} & \mathbf{O} & \mathbf{O} & \mathbf{O} \\ \hline \mathbf{O} & \mathbf{O} & \mathbf{O} & \mathbf{I} & \mathbf{O} & \mathbf{O} \\ \hline \mathbf{O} & \mathbf{O} & \mathbf{O} & \mathbf{O} & \mathbf{O} & \mathbf{O}\end{array}\right].$$

Finally, because $\mathbf{M} = \left[\begin{array}{c|c}\mathbf{M}_{11} & \mathbf{M}_{12} \\ \hline \mathbf{M}_{21} & \mathbf{M}_{22}\end{array}\right]$ is invertible we can use column transformations on the first strip to bring back $\mathbf{M}$ to its canonical form $\mathbf{I}$ giving the final form

$$\left[\begin{array}{c|c|c|c|c|c}\mathbf{I} & \mathbf{O} & \mathbf{O} & \mathbf{O} & \mathbf{I} & \mathbf{O} \\ \hline \mathbf{O} & \mathbf{I} & \mathbf{O} & \mathbf{O} & \mathbf{O} & \mathbf{O} \\ \hline \mathbf{O} & \mathbf{O} & \mathbf{O} & \mathbf{I} & \mathbf{O} & \mathbf{O} \\ \hline \mathbf{O} & \mathbf{O} & \mathbf{O} & \mathbf{O} & \mathbf{O} & \mathbf{O}\end{array}\right]$$

as required. □

**Remark 5.2.12.** As P. Gabriel and A.V. Roiter write in their book "Representations of finite dimensional algebras", Encyclopaedia of Mathematics 73, Algebra VIII. Springer, 1992, Ch.1, the case of a 3-element poset with no comparable elements can be handled similarly. But things break down for posets of 4 (or more) incomparable elements. Then the

$$\left[\begin{array}{c|c|c|c}1 & 0 & 1 & \lambda \\ \hline 0 & 1 & 1 & 1\end{array}\right] \tag{5.2.13}$$

form a one parameter family of pairwise non-isomorphic indecomposable matrix representations.

**Lemma 5.2.14.** *Let $C_n$ be a chain. Then all indecomposable representations of the poset $\mathcal{P} = (1) \sqcup C_n$ over a division ring $K$ are elementary of the forms $\mathbf{V}_{n+1}^i$, $\mathbf{V}_{n+1}^{1,i}$, $\mathbf{J}_{n+1}^i$ for $i = 1, \ldots, n + 1$, and $\mathbf{I}_{n+1}^0$. Moreover $n(\mathcal{P}) = 3n + 4$.*

*Proof.* Let $\mathcal{P} = (1) \sqcup C_n$. Consider a matrix representation $\mathbf{R}$ of $\mathcal{P}$:

$$\mathbf{R} = \left[\begin{array}{c||c}\mathbf{X} & \mathbf{Y}\end{array}\right]$$

where the matrix $\mathbf{X}$ corresponds to the sub-poset $(1)$ and the matrix $\mathbf{Y}$ corresponds to the chain $C_n$. Let

$$\mathbf{d}(\mathbf{R}) = (d_0; d_1, \ldots, d_n, d_{n+1})$$

be the coordinate vector of $\mathbf{R}$.

If $\widetilde{d_0} = 0$, the matrix $\mathbf{R}$ is a zeroes matrix representation and we are done.

Suppose that $\widetilde{d_0} \neq 0$. If $d_1 = 0$ or $\mathbf{X}$ is a zero matrix, then $\mathbf{R}$ can be considered as a matrix representation of the subposet $C_n$ and the result follows from lemma 5.2.8.

So we can assume that $d_1 > 0$ and there are non-zero elements in $\mathbf{X}$. By elementary operations the matrix $\mathbf{X}$ can be reduced to normal form, whence

$$\mathbf{R} \simeq \left[\begin{array}{c|c||c}\mathbf{I} & \mathbf{O} & \mathbf{Y}_1 \\ \hline \mathbf{O} & \mathbf{O} & \mathbf{Y}_2\end{array}\right] = \mathbf{R}_1.$$

Considering the matrix $\mathbf{Y}_2$ as a representation of the chain $C_{n-1}$, by lemma 5.2.8, it can be reduced by means of elementary row operations of the matrix $\mathbf{R}_1$ and elementary column operations of the matrix $\mathbf{Y} = \left[\dfrac{\mathbf{Y}_1}{\mathbf{Y}_2}\right]$ to the canonical form (5.2.9). By means of the identity matrices $\mathbf{I}$ in $\mathbf{Y}_2$, all elements that lie above in the matrix $\mathbf{Y}_1$ can be made zero. Therefore we obtain the following matrix representation:

$$
\mathbf{R} \simeq
\left[
\begin{array}{cc|c|ccc|cc}
\mathbf{I} & \mathbf{O} & \mathbf{Y}_{11} & \ldots & \mathbf{O} & \mathbf{Y}_{1n} \\
\hline
\mathbf{O} & \mathbf{I} & \mathbf{O} & \ldots & \mathbf{O} & \mathbf{O} \\
\vdots & \vdots & \vdots & \ddots & \vdots & \vdots \\
\mathbf{O} & \mathbf{O} & \mathbf{O} & \ldots & \mathbf{I} & \mathbf{O} \\
\mathbf{O} & \mathbf{O} & \mathbf{O} & \ldots & \mathbf{O} & \mathbf{O}
\end{array}
\right]
=
\left[
\begin{array}{c|c}
\mathbf{U}_1 & \mathbf{O} \\
\hline
\mathbf{O} & \mathbf{U}_2
\end{array}
\right]
$$

which is a direct sum of two matrix representations

$$
\mathbf{U}_1 = \left[\ \mathbf{I}\ \|\ \mathbf{Y}_{11}\ |\ \ldots\ |\ \mathbf{Y}_{1n}\ \right]
$$

and

$$
\mathbf{U}_2 =
\left[
\begin{array}{cc|c|c}
\mathbf{I} & \mathbf{O} & \ldots & \mathbf{O} \\
\mathbf{O} & \mathbf{I} & \ldots & \mathbf{O} \\
\vdots & \vdots & \ddots & \vdots \\
\mathbf{O} & \mathbf{O} & \ldots & \mathbf{I} \\
\mathbf{O} & \mathbf{O} & \ldots & \mathbf{O}
\end{array}
\right].
$$

The representation $\mathbf{U}_2$ is a direct sum of elementary representations of the forms $\mathbf{V}_n^i$ for $i = 1, 2, \ldots, n$ and $\mathbf{I}_n^0$. Since the representation

$$
\left[\ \mathbf{Y}_{11}\ |\ \ldots\ |\ \mathbf{Y}_{1n}\ \right]
$$

can be considered as a representation of the chain $C_n$, it is equivalent to a direct sum of elementary representations of the forms $\mathbf{V}_n^i$, $\mathbf{J}_n^i$ for $i = 1, 2, \ldots, n$ and $\mathbf{I}_n^0$, by lemma 5.2.8. Therefore an elementary sub-representation of $\mathbf{U}_1$ is one of the forms $\mathbf{V}_{n+1}^i$, $\mathbf{V}_{n+1}^{1,i}$, $\mathbf{J}_{n+1}^i$ for $i = 1, \ldots, n+1$ and $\mathbf{I}_{n+1}^0$, as required. $\square$

## 5.3 Trichotomy Lemma

Let $(\mathcal{P}, \preceq)$ be a finite poset with $n$ elements. In this section by a representation $V$ of a poset $\mathcal{P}$ we understand a matrix representation with dimension vector $\mathbf{d} = (d_0; d_1, \ldots, d_n)$.

**Definition 5.3.1.** Let $V$ be a representation of a poset $\mathcal{P}$ with dimension vector $\mathbf{d} = (d_0; d_1, \ldots, d_n)$. The subset of $\mathcal{P}$ formed of those $p_i \in \mathcal{P}$ for which $d_i > 0$ is called the **support** of $V$. A representation $V$ is said to **present in** $p_i \in \mathcal{P}$ if $d_i > 0$; i.e. if $p_i$ is in the support of $V$.

**Definition 5.3.2.** Two posets $X$ and $Z$ can be combined to form an **ordinal sum**, denoted by $X < Z$, that is a disjoint sum of posets $X \cup Z$ with additional relation $x < z$ for all $x \in X$ and for all $z \in Z$.

**Definition 5.3.3** ([82]). A **trichotomy of a finite poset** $\mathcal{P}$ is a triple $(X, Y, Z)$ formed by disjoint sub-posets with union $\mathcal{P}$ such that the following conditions are satisfied:

   a. $X \neq \emptyset$, and $Z \neq \emptyset$.
   b. $X < Z$, i.e. $x < z$ for all $x \in X$ and all $z \in Z$.
   c. $w(Y) \leq 1$; ($Y$ is a chain or empty).

In this case $\mathcal{P}$ is written as $\mathcal{P} = \{X < Z\} \sqcup Y$.

**Lemma 5.3.4.** (**Trichotomy Lemma** [82, lemma 5.1]). *Let $(X, Y, Z)$ be a trichotomy of a poset $\mathcal{P}$, i.e. $\mathcal{P} = \{X < Z\} \sqcup Y$, where $X, Z \neq \emptyset$, and $w(Y) \leq 1$. If $R$ is a representation of $\mathcal{P}$, then $R \cong R_1 \oplus R_2$, where the support of $R_1$ is contained in $X \sqcup Y$, and the support of $R_2$ is contained in $Y \sqcup Z$.*

*Proof.* Let $\mathcal{P}$ be a cardinal sum of posets $X$, $Y$, $Z$ such that $X < Z$ and $w(Y) \leq 1$. If $Y = \emptyset$, the statement is obvious. Suppose that $Y = \{a_1 < a_2 < \ldots < a_s\}$ is a chain. Consider the corresponding matrix representation $\mathbf{R}$ of $\mathcal{P}$

$$\mathbf{R} = \left[\ \overbrace{\mathbf{X}}^{X}\ \|\ \overbrace{\mathbf{Y}}^{Y}\ \|\ \overbrace{\mathbf{Z}}^{Z}\ \right],$$

where the columns of $\mathbf{X}$, $\mathbf{Y}$, $\mathbf{Z}$ are assigned to the posets $X$, $Y$ and $Z$ respectively.

By elementary row operations on the matrix $\mathbf{R}$, the matrix $\mathbf{R}$ can be reduced to the form

$$\mathbf{R} \simeq \left[\begin{array}{c|c|c} \mathbf{X}_1 & \mathbf{Y}_1 & \mathbf{Z}_1 \\ \hline \mathbf{O} & \mathbf{Y}_2 & \mathbf{Z}_2 \end{array}\right] = \mathbf{R}_1$$

where $\left[\dfrac{\mathbf{X}_1}{\mathbf{O}}\right]$ is the row canonical form of the matrix $\mathbf{X}$ and the matrix $\mathbf{X}_1$ has no zero rows.

Considering the matrix $\mathbf{Y}_2$ as a representation of the chain $Y$, by lemma 5.2.8 it can be reduced by means of elementary row operations in the matrix $\mathbf{R}_1$ and elementary column operations in the matrix $\mathbf{Y} = \left[\dfrac{\mathbf{Y}_1}{\mathbf{Y}_2}\right]$ to the canonical form (5.2.9).

By means of all identity matrices $\mathbf{I}$ in $\mathbf{Y}_2$, all the elements that lie above it in the matrix $\mathbf{Y}_1$ can be made zero. Therefore we obtain the following matrix representation:

$$\mathbf{R} \simeq \left[\begin{array}{c|c|c|c|c|c|c} \mathbf{X}_1 & \mathbf{O} & \mathbf{Y}_{11} & \cdots & \mathbf{O} & \mathbf{Y}_{1s} & \mathbf{Z}_{11} \\ \hline \mathbf{O} & \mathbf{I} & \mathbf{O} & \cdots & \mathbf{O} & \mathbf{O} & \mathbf{Z}_{21} \\ \hline \cdots & \cdots & \cdots & \ddots & \cdots & \cdots & \cdots \\ \hline \mathbf{O} & \mathbf{O} & \mathbf{O} & \cdots & \mathbf{I} & \mathbf{O} & \mathbf{Z}_{2s} \\ \mathbf{O} & \mathbf{O} & \mathbf{O} & \cdots & \mathbf{O} & \mathbf{O} & \mathbf{Z}_{2s+1} \end{array}\right].$$

Taking into account that the matrix $\mathbf{X}_1$ is in row canonical form without zero rows, and $X < Z$, the matrix $\mathbf{Z}_{11}$ can be made zero. Hence we obtain the following representation:

$$
\mathbf{R} \simeq
\left[
\begin{array}{c||c|c|c|c|c||c}
\mathbf{X}_1 & \mathbf{O} & \mathbf{Y}_{11} & \cdots & \mathbf{O} & \mathbf{Y}_{1s} & \mathbf{O} \\
\hline
\mathbf{O} & \mathbf{I} & \mathbf{O} & \cdots & \mathbf{O} & \mathbf{O} & \mathbf{Z}_{21} \\
\hline
\cdots & \cdots & \cdots & \ddots & \cdots & \cdots & \cdots \\
\hline
\mathbf{O} & \mathbf{O} & \mathbf{O} & \cdots & \mathbf{I} & \mathbf{O} & \mathbf{Z}_{2s} \\
\hline
\mathbf{O} & \mathbf{O} & \mathbf{O} & \cdots & \mathbf{O} & \mathbf{O} & \mathbf{Z}_{2s+1}
\end{array}
\right]
=
\left[
\begin{array}{c|c}
\mathbf{U}_1 & \mathbf{O} \\
\hline
\mathbf{O} & \mathbf{U}_2
\end{array}
\right],
$$

which is a direct sum of the representations

$$
\mathbf{U}_1 =
\left[\; \overbrace{\mathbf{X}_1}^{X} \;\|\; \overbrace{\mathbf{Y}_{11} \;|\; \cdots \;|\; \mathbf{Y}_{1s}}^{Y} \;\right]
$$

and

$$
\mathbf{U}_2 =
\left[
\begin{array}{c|c|c|c||c}
\multicolumn{3}{c}{\overbrace{\hphantom{\mathbf{I} \quad \mathbf{O} \quad \mathbf{O}}}^{Y}} & & \overbrace{}^{Z} \\
\mathbf{I} & \mathbf{O} & \cdots & \mathbf{O} & \mathbf{Z}_{21} \\
\hline
\cdots & \cdots & \ddots & \cdots & \cdots \\
\hline
\mathbf{O} & \mathbf{O} & \cdots & \mathbf{I} & \mathbf{Z}_{2s} \\
\hline
\mathbf{O} & \mathbf{O} & \cdots & \mathbf{O} & \mathbf{Z}_{2s+1}
\end{array}
\right].
$$

The lemma is proved. $\square$

**Proposition 5.3.5.** ( [82, section 5.2]). *If the sub-poset $Y$ of $\mathcal{P}$ consisting of all elements of $\mathcal{P}$ incomparable with element $p \in \mathcal{P}$ is a chain, then each indecomposable representation of $\mathcal{P}$ which is present at $p$ is elementary.*

*Proof.* Let $Y$ be the set of all elements of $\mathcal{P}$ incomparable with $p \in \mathcal{P}$. Write $X = \{q \in \mathcal{P} : q < p\}$, $U = \{q \in \mathcal{P} : p < q\}$. Suppose that $Y$ is a chain. Then $\mathcal{P} = X \sqcup Y \sqcup \{p\} \sqcup U$ and $X < U$. Therefore we can apply lemma 5.3.4 to the sets $X, Y, Z$, where $Z = \{p\} \cup U$. From this lemma it follows that the support of a representation present at $p$ is contained in $\mathcal{P}_1 = Y \sqcup Z = Y \sqcup \{p\} \sqcup U$. Since $\mathcal{P}_1 = \{\{p\} < U\} \sqcup Y$, this lemma can again be applied for $X_1 = \{p\}$, $Y_1 = Y$ and $Z_1 = U$. Then it follows that the support of a representation present at $p$ is contained in $\mathcal{P}_2 = \{p\} \sqcup Y$. Since $Y$ is a chain, from lemma 5.2.14 it follows that all indecomposable representations of $\mathcal{P}_2$ are elementary, as required. $\square$

**Corollary 5.3.6.** ([82, section 5.2]). *Any indecomposable representation of a poset $\mathcal{P}$ of width $\leq 2$ is elementary. Therefore, all posets of width $\leq 2$ are of finite representation type.*

## 5.4 The Kleiner Lemma

The idea of a representation of a pair of finite posets over a field was introduced by M.M. Kleiner in [129]. The pairs of posets of finite type and of tame type were described by M.M. Kleiner in [129] and [131] respectively.

Let $\mathcal{P} = (\{p_1, p_2, \ldots, p_m\}, \preceq_1)$ and $\mathcal{Q} = (\{q_1, q_2, \ldots, q_n\}, \preceq_2)$ be finite posets. A **representation of a pair of posets** $(\mathcal{P}, \mathcal{Q})$ over a division ring $K$ is a matrix $\mathbf{A}$ that is partitioned into $mn$ blocks $\mathbf{A}_{ij}$:

$$
\mathbf{A} = \left[ \begin{array}{c|c|c|c}
\mathbf{A}_{11} & \mathbf{A}_{12} & \cdots & \mathbf{A}_{1n} \\
\hline
\mathbf{A}_{21} & \mathbf{A}_{22} & \cdots & \mathbf{A}_{2n} \\
\hline
\vdots & \vdots & \ddots & \vdots \\
\hline
\mathbf{A}_{m1} & \mathbf{A}_{m2} & \cdots & \mathbf{A}_{mn}
\end{array} \right].
$$

If the dimensions a block $\mathbf{A}_{ij}$ are equal to $u_i \times v_j$, then the integer vector $\mathbf{d} = (u_1, u_2, \ldots, u_m; v_1, v_2, \ldots, v_n)$ is called the **dimension vector** of the partitioned matrix $\mathbf{A}$.

Let

$$
\mathbf{B} = \left[ \begin{array}{c|c|c|c}
\mathbf{B}_{11} & \mathbf{B}_{12} & \cdots & \mathbf{B}_{1n} \\
\hline
\mathbf{B}_{21} & \mathbf{B}_{22} & \cdots & \mathbf{B}_{2n} \\
\hline
\vdots & \vdots & \ddots & \vdots \\
\hline
\mathbf{B}_{m1} & \mathbf{B}_{m2} & \cdots & \mathbf{B}_{mn}
\end{array} \right]
$$

be another matrix representation of a pair of posets $(\mathcal{P}, \mathcal{Q})$.

**Definition 5.4.1.** A representation $\mathbf{A}$ is **isomorphic** to a representation $\mathbf{B}$ of a pair of poset $(\mathcal{P}, \mathcal{Q})$ if $\mathbf{A}$ can be reduced to $\mathbf{B}$ by the following operations:

(a) Elementary row operations within each horizontal strip $\mathbf{A}'_i = \{\mathbf{A}_{i1}, \mathbf{A}_{i2}, \ldots, \mathbf{A}_{in}\}$ for $i = 1, 2, \ldots, m$.

(b) Elementary column operations within each vertical strip $\mathbf{A}''_j = \{\mathbf{A}_{1j}, \mathbf{A}_{2j}, \ldots, \mathbf{A}_{mj}\}$ for $i = 1, 2, \ldots, n$.

(c) Additions of columns of a vertical strip $\{\mathbf{A}_{1j}, \mathbf{A}_{2j}, \ldots, \mathbf{A}_{mj}\}$ to columns of a vertical strip $\{\mathbf{A}_{1k}, \mathbf{A}_{2k}, \ldots, \mathbf{A}_{mk}\}$ if $j \preceq_1 k$ in $\mathcal{P}$.

(d) Additions of rows of a horizontal strip $\{\mathbf{A}_{i1}, \mathbf{A}_{i2}, \ldots, \mathbf{A}_{in}\}$ to rows of a horizontal strip $\{\mathbf{A}_{j1}, \mathbf{A}_{j2}, \ldots, \mathbf{A}_{jn}\}$ if $i \preceq_2 j$ in $\mathcal{Q}$.

In a similar way as for one poset, one can introduce the notion of a direct sum of matrix representations of a pair of posets, and the notion of an indecomposable matrix representation.

**Lemma 5.4.2.** (**The Kleiner lemma** [129]). *Let $(X, Y)$ be a pair of finite posets $X$ and $Y$, where $X = C_{m+1}$ is a chain of length $m + 1$. Then $(X, Y)$ is of finite representation type if and only if the poset $C_m \sqcup Y$ is of finite representation type.*

*Proof.* Consider the cardinal sum of posets $C_m \sqcup Y$, where $C_m = \{x_1 \leq_1 x_2 \leq_1 \ldots \leq_1 x_m\}$ is a chain of length $m$, and $Y = \{y_1, y_2, \ldots, y_n\}$ is a poset with partially order relation $\leq_2$. Let $\mathbf{R}$ be a matrix representation of $C_m \sqcup Y$. Then it is partitioned into $m + n$ vertical strips:

$$\mathbf{R} = \left[\; \overbrace{\mathbf{A}_1 \mid \ldots \mid \mathbf{A}_m}^{C_m} \;\|\; \overbrace{\mathbf{B}_1 \mid \ldots \mid \mathbf{B}_n}^{Y} \;\right]$$

where the matrices $\mathbf{A}_i$ correspond to the elements $x_i$ of the poset $C_m$, and the matrices $\mathbf{B}_j$ correspond to the elements $y_j$ of the poset $Y$.

Considering the matrix

$$\mathbf{A} = \left[\; \mathbf{A}_1 \mid \ldots \mid \mathbf{A}_m \;\right]$$

as a representation of $C_m$, one can reduce it by means of elementary row operations in the matrix $\mathbf{R}$ and elementary column operations inside each $\mathbf{A}_i$, and by means of addition of the columns of matrices $\mathbf{A}_i$ to columns of matrices $\mathbf{A}_j$ if $i \leq_1 j$ in $C_m$ to the canonical form (5.2.9), by lemma 5.2.8. Then each matrix $\mathbf{B}_j$ is divided into $m + 1$ horizontal strips, and we obtain the following form of $\mathbf{R}$:

$$\mathbf{R} \simeq \overbrace{\begin{bmatrix} \mathbf{I} & \mathbf{O} & \cdots & \mathbf{O} \\ \vdots & \vdots & \ddots & \vdots \\ \mathbf{O} & \mathbf{O} & \cdots & \mathbf{I} \\ \mathbf{O} & \mathbf{O} & \cdots & \mathbf{O} \end{bmatrix}}^{C_m} \overbrace{\begin{bmatrix} \mathbf{B}_{11} & \cdots & \mathbf{B}_{1n} \\ \vdots & \ddots & \vdots \\ \mathbf{B}_{m1} & \cdots & \mathbf{B}_{mn} \\ \mathbf{B}_{m+1,1} & \cdots & \mathbf{B}_{m+1,n} \end{bmatrix}}^{Y} = \begin{bmatrix} \mathbf{I} & \mathbf{O} & \cdots & \mathbf{O} & \mathbf{B}'_1 \\ \vdots & \vdots & \ddots & \vdots & \vdots \\ \mathbf{O} & \mathbf{O} & \cdots & \mathbf{I} & \mathbf{B}'_m \\ \mathbf{O} & \mathbf{O} & \cdots & \mathbf{O} & \mathbf{B}'_{m+1} \end{bmatrix}$$

The rows of a horizontal strip $\mathbf{B}'_i$ can be added to the rows of a horizontal strip $\mathbf{B}'_j$ if $j \leq_1 i$ in $C_m$. Therefore, the matrix

$$\mathbf{B} = \begin{bmatrix} \mathbf{B}_{11} & \cdots & \mathbf{B}_{1n} \\ \cdots & \ddots & \cdots \\ \mathbf{B}_{m1} & \cdots & \mathbf{B}_{mn} \\ \mathbf{B}_{m+1,1} & \cdots & \mathbf{B}_{m+1,n} \end{bmatrix}$$

can be considered as a representation of a pair of posets $(Y, C_{m+1})$. $\square$

## 5.5 The Main Construction

Before introducing the main construction, some examples will be treated. They show in which way the representations of a finite poset can be reduced by this construction.

**Example 5.5.1.** Consider the representations of the poset $\mathcal{P}$:

$$\begin{array}{ccc} 1 & 2 & 3 \\ \bullet & \bullet & \bullet \end{array}$$

Let $\mathcal{P}_1$ be the sub-poset of $\mathcal{P}$ with $w(\mathcal{P}_1) = 2$ of the form:

$$
\begin{array}{cc}
1 & 2 \\
\bullet & \bullet
\end{array}
$$

The corresponding matrix representation of $\mathcal{P}$ has the following form

$$
\mathbf{R} = [\ \overbrace{\mathbf{A}_1}^{1} \ \| \ \overbrace{\mathbf{A}_2}^{2} \ \| \ \overbrace{\mathbf{A}_3}^{3} \ ],
$$

where the columns of $\mathbf{A}_i$ are assigned to the elements $i$ of the poset $\mathcal{P}$ for $i = 1, 2, 3$.

By means of elementary row operations in the matrix $\mathbf{R}$ and elementary column operations within each vertical strip $\mathbf{A}_1$ and $\mathbf{A}_2$, the matrix $\mathbf{R}$ can be reduced to the following form:

$$
\mathbf{R} \simeq
\begin{array}{|c|c|c|c|c|c|c|c|}
\hline
\mathbf{I} & \mathbf{0} & \mathbf{0} & \mathbf{0} & \mathbf{I} & \mathbf{0} & \mathbf{A}_{31} & 1 \vee 2 \\
\hline
\mathbf{0} & \mathbf{I} & \mathbf{0} & \mathbf{0} & \mathbf{0} & \mathbf{0} & \mathbf{A}_{32} & 1 \\
\hline
\mathbf{0} & \mathbf{0} & \mathbf{0} & \mathbf{I} & \mathbf{0} & \mathbf{0} & \mathbf{A}_{33} & 2 \\
\hline
\mathbf{0} & \mathbf{0} & \mathbf{0} & \mathbf{0} & \mathbf{0} & \mathbf{0} & \mathbf{A}_{34} & 0 \\
\hline
\end{array}
\qquad (5.5.2)
$$

According to this reduction, the matrix $\mathbf{A}_3$ is divided into four horizontal strips $\mathbf{A}_{3i}$, which correspond the elements of a new poset $\mathcal{A}(\mathcal{P}_1)$ with elements $\{0, 1, 2, 1 \vee 2\}$ as shown in (5.5.2). The ordering relation $\leq^+$ in this poset is given according to possibility for adding horizontal strips of the matrix $\mathbf{R}$ to other horizontal strips of this matrix which do not change the form of the first two vertical strips of $\mathbf{R}$. Since the rows of the last horizontal strip can be added to rows of each other horizontal strip, this means that $0 \leq^+ 1$, $0 \leq^+ 2$ and $0 \leq^+ (1 \vee 2)$. The rows of the second and the third horizontal strips we cannot add each other without spoiling the form of the first two vertical strips of $\mathbf{R}$. Therefore the elements 1 and 2 are not in an ordering relation with each other. The rows of the second horizontal strip can be added only to the rows of the first horizontal strip, since though in this case we spoil the zero matrix at location $(1,2)$ it can be compensated for by column operations in the first vertical strip of the matrix $\mathbf{R}$. Therefore $1 \leq^+ (1 \vee 2)$. Analogously, $2 \leq^+ (1 \vee 2)$.

So the elementary operations between horizontal strips of the third vertical strip of the matrix $\mathbf{R}$ are admissible according to the following partially ordered set (5.5.3).

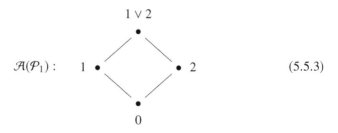

$$
\mathcal{A}(\mathcal{P}_1): \qquad (5.5.3)
$$

Thus the partial reduction (5.5.2) of the original matrix representation of the poset
•       •       • is reduced to the consideration of matrix representations of the
1       2       3
poset (5.5.3) above (which, hopefully, is simpler).

**Example 5.5.4.** Consider a poset $\mathcal{P}$ of the form

and its sub-poset $\mathcal{P}_1$ of the form

Consider a corresponding matrix representation of $\mathcal{P}$:

$$\mathbf{R} = \quad \boxed{\begin{array}{c|c|c|c|c} \mathbf{A}_1 & \mathbf{A}_2 & \mathbf{A}_3 & \mathbf{A}_4 & \mathbf{A}_5 \end{array}}$$

$$\phantom{\mathbf{R} = \quad} 1 \quad\; 2 \quad\; 3 \quad\; 4 \quad\; 5$$

Reducing the poset $\mathcal{P}_1$, i.e. the first four matrices, we obtain the following
equivalent representation:

|   | I |   | I |   |   |   |   | $\mathbf{A}_{51}$ | $1 \vee 2$ |
|---|---|---|---|---|---|---|---|---|---|
|   |   | I |   |   | I |   |   | $\mathbf{A}_{52}$ | $1 \vee 3$ |
|   |   |   | I |   |   |   | I | $\mathbf{A}_{53}$ | $1 \vee 4$ |
|   |   |   |   | I |   |   |   | $\mathbf{A}_{54}$ | $2$ |
| $\mathbf{R} \simeq$ |   |   |   |   |   | I |   | $\mathbf{A}_{55}$ | $3$ |
|   |   |   |   |   |   |   | I | $\mathbf{A}_{56}$ | $4$ |
|   |   |   | I |   |   |   |   | $\mathbf{A}_{57}$ | $1$ |
|   | O |   | O |   | O |   | O | $\mathbf{A}_{58}$ | $0$ |
|   | 1 |   | 2 |   | 3 |   | 4 | 5 |   |

According to this reduction, the matrix $\mathbf{A}_5$ is divided into eight horizontal strips $\mathbf{A}_{5i}$, and the elementary operations between them are admissible according to the following partially ordered set:

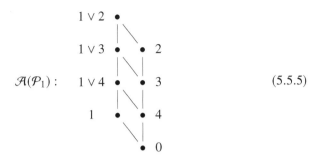

$$\mathcal{A}(\mathcal{P}_1): \qquad\qquad\qquad\qquad\qquad (5.5.5)$$

We now introduce the formal definitions.

Let $\mathcal{P} = \{p_1, p_2, \ldots, p_n\}$ be a poset with partially ordering relation $\leq$. Denote by $\mathcal{A}(\mathcal{P})$ the set of all antichains of $\mathcal{P}$ of width $l \geq 0$. One takes the antichain of width 0 to be the empty set, and we denote this antichain as 0. We identify antichains of width 1 with the elements of $\mathcal{P}$ themselves. We define an order relation $\leq^+$ on $\mathcal{A}(\mathcal{P})$ as follows. If $X, Y \in \mathcal{A}(\mathcal{P})$, then $X \leq^+ Y$ if and only if for any $a \in X$ there exists $b \in Y$ such that $b \leq a$. We also set that $0 <^+ X$ for all $X \neq 0$. Let $X \in \mathcal{A}(\mathcal{P})$. If $X = \{p_i\} \subset \mathcal{P}$, then it is denoted in $\mathcal{A}(\mathcal{P})$ as $p_i$, and if $X = \{p_i, p_j\} \subset \mathcal{P}$, then it is denoted in $\mathcal{A}(\mathcal{P})$ as $p_i \vee p_j$.

Denote by $\mathcal{P}^{op}$ the poset that is dual to $\mathcal{P}$, i.e. the set of elements of $\mathcal{P}^{op}$ is the same as $\mathcal{P}$ and the order relation is $\geq$. Then from the definition of $\mathcal{A}(\mathcal{P})$ it follows that $\mathcal{A}(\mathcal{P}) \supseteq \mathcal{P}^{op}$.

**Remark 5.5.6.** Note that if $\mathcal{P} = \mathcal{P}_1 \sqcup \{a\}$ then $\mathcal{A}(\mathcal{P}_1)$ is a derivative (as defined in [101, section 5.3]) of $\mathcal{P}$ at $a$. In particular, (5.5.3) is a derivative of the poset

$$\begin{matrix} \bullet & \bullet & \bullet \\ 1 & 2 & 3 \end{matrix}$$ at the point 3, and (5.5.5) is a derivative of the poset considered

in Example 5.5.4 at the point 5.

For the poset $\mathcal{P}_1 = \mathcal{A}(\mathcal{P})$, in a similar way one can build $\mathcal{A}(\mathcal{P}_1)$. Then we write $\mathcal{A}^2(\mathcal{P}) = \mathcal{A}(\mathcal{P}_1)$, and in the general case we write

$$\mathcal{A}^n(\mathcal{P}) = \mathcal{A}(\mathcal{A}^{n-1}(\mathcal{P})).$$

From the definitions above it is easy to check the next two statements.

**Proposition 5.5.7.** *If $\mathcal{P}$ is a poset, then $\mathcal{A}(\mathcal{P})$ is a poset with respect to the order relation $\leq^+$ with as least element 0 and the greatest element is the antichain containing all minimal elements of the poset $\mathcal{P}$.*

**Proposition 5.5.8.** *If $\mathcal{Q}$ is a sub-poset of a poset $\mathcal{P}$, then $\mathcal{A}^n(\mathcal{Q}) \subset \mathcal{A}^n(\mathcal{P})$ for all $n$.*

**Example 5.5.9.** Let $\mathcal{P}$ be the chain $C_3$:

Then $\mathcal{A}(\mathcal{P})$ is the chain $C_4$:

$$\bullet \; 1$$
$$\bullet \; 2$$
$$\bullet \; 3$$
$$\bullet \; 0$$

In the general case, if $\mathcal{P}$ is a chain $C_n$, then $\mathcal{A}(\mathcal{P})$ is a chain $C_{n+1}$, i.e.

$$\mathcal{A}(C_n) = C_{n+1}.$$

**Example 5.5.10.** Let $\mathcal{P}$ be an antichain $A_2$:

$$\begin{matrix} 1 & & 2 \\ \bullet & & \bullet \end{matrix}$$

Then $\mathcal{A}(\mathcal{P})$ has the following form:

$$1 \vee 2$$

$$1 \bullet \qquad \bullet \; 2 \qquad\qquad (5.5.11)$$

$$0$$

**Example 5.5.12.** Let $\mathcal{P}$ be the poset of the form:

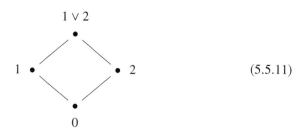

Then $\mathcal{A}(\mathcal{P})$ looks as follows:

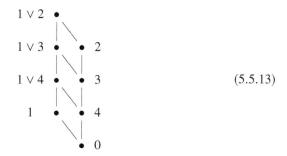

$$(5.5.13)$$

**Example 5.5.14.** Let $\mathcal{P}$ be the poset:

Then the diagram of $\mathcal{A}(\mathcal{P})$ is:

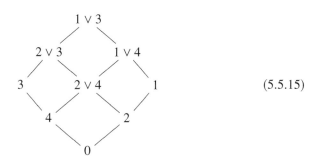

$$(5.5.15)$$

Since the elements $1$, $2 \vee 4$ and $3$ are incomparable, $w(\mathcal{A}(\mathcal{P})) = 3$.

**Theorem 5.5.16.** *Let $\mathcal{P}$ be a poset which is a cardinal sum of two sub-posets $\mathcal{P}_1$ and $\mathcal{P}_2$, where $\mathcal{P}_1 = C_n$ and $\mathcal{P}_2$ is a poset of width 2. Then $\mathcal{P}$ is of finite representation type if and only if the cardinal sum of posets $C_{n-1} \sqcup \mathcal{A}(\mathcal{P}_1)$ is of finite representation type. In particular, if $n = 1$, then $\mathcal{P}$ is of finite representation type if and only if the poset $\mathcal{A}(\mathcal{P}_1)$ is of finite representation type. Moreover, if $\mathcal{A}^s(\mathcal{P})$ is a poset of width 2 for all $s < k$, then $\mathcal{P}$ is of finite representation type if and only if the cardinal sum of posets $C_{n-k} \sqcup \mathcal{A}^k(\mathcal{P}_1)$ is of finite representation type.*

*Proof.* Let a poset $\mathcal{P} = \mathcal{P}_1 \sqcup \mathcal{P}_2 = C_n \sqcup \mathcal{P}_2$ be of finite type. Consider the corresponding matrix representation $\mathbf{R}$ of $\mathcal{P}$:

$$\mathbf{R} = \left[ \ \mathbf{X} \ \| \ \mathbf{Y} \ \right]$$

where the matrix $\mathbf{X}$ corresponds to the chain $C_n$ and the matrix $\mathbf{Y}$ corresponds to a sub-poset $\mathcal{P}_2 = \{\alpha_1, \ldots, \alpha_n\}$. Since $w(\mathcal{P}_2) = 2$, this poset is of finite representation type, by corollary 5.3.6. Moreover, all indecomposable representations of $\mathcal{P}_2$ are elementary, by the same corollary. So the matrix $\mathbf{Y}$ can be reduced to the form of a direct sum of elementary representations. Collect all the representations of the same form in blocks. Thus, we obtain the direct sum of representations of the following form:

I. Zero representation of $\mathcal{P}_2$ with zero matrices with the same number of rows for each $\alpha_i \in \mathcal{P}_2$.

II. Representations of $\mathcal{P}_2$ with one identity matrix in the place $i$ that corresponds to $\alpha_i \in \mathcal{P}_2$ and zero matrices on the other places, and all these matrices have the same number of rows.

III. Representations of $\mathcal{P}_2$ with two identity matrices in the places $i$ and $j$ that correspond to elements $\alpha_i, \alpha_j \in \mathcal{P}_2$ and zero matrices in the other places, and all these matrices have the same number of rows.

We denote the strip of the matrix $\mathbf{X}$ corresponding to the zero representation of the type I by 0; the strip of the matrix $\mathbf{X}$ corresponding to representation of the type II by $i$; the strip of the matrix $\mathbf{X}$ corresponding to the representation of the type III by $i \vee j$. It is easy to see, that the representation of the type III is contained in $\mathbf{R}$ if and only if the elements $\alpha_i$ and $\alpha_j$ are incomparable in the poset $\mathcal{P}_2$. Moreover, the strip 0 can be added to any strip without changing a thing. Any strip $i$ can also be added to the strip $i \vee j$ without changing the representation matrix of the poset $\mathcal{P}_2$, because the new appearing identity matrix in the strip $i \vee j$ can be made zero by means of the identity matrix in this strip.

Therefore, according to this reduction the matrix $\mathbf{X}$ is divided into $|\mathcal{A}(\mathcal{P}_1)|$ vertical strips, and the elementary operations between these strips are admissible in accordance with the order relation in the poset $\mathcal{A}(\mathcal{P}_1)$. Therefore we obtain the problem of the representation of a pair of finite posets $(C_n, \mathcal{A}(\mathcal{P}_1))$. By lemma 5.4.2, this pair of posets is of finite type if and only if the cardinal sum $C_{n-1} \sqcup \mathcal{A}(\mathcal{P}_1)$ is of finite type.

Continuing by induction, we obtain the last statement of the theorem. $\square$

## 5.6 Primitive Posets of Infinite Representation Type

Recall that a finite partially ordered set $\mathcal{P}$ is a **primitive poset** (or an **elementary poset**) if it is a cardinal sum of linearly ordered sets $L_1, \ldots, L_m$, and it is denoted by $\mathcal{P} = L_1 \sqcup \cdots \sqcup L_m$. The following theorem gives a criterium for primitive posets to be of finite representation type.

**Theorem 5.6.1.** *A finite primitive poset $\mathcal{P}$ is of finite representation type over a division ring $K$ if and only if $\mathcal{P}$ does not contain as a full sub-poset any poset from the following list:*

$$(1) \sqcup (1) \sqcup (1) \sqcup (1), \quad (2) \sqcup (2) \sqcup (2), \quad (1) \sqcup (3) \sqcup (3), \quad (1) \sqcup (2) \sqcup (5)$$

We give the proof of this theorem following [30]. This section deals with the necessity part of this theorem, i.e. we will prove that all critical primitive posets from theorem 5.6.1 are of infinite representation type.

**Lemma 5.6.2.** *The poset*

$$\mathcal{P} : \{ \bullet \quad \bullet \quad \bullet \quad \bullet \}$$

*is of infinite representation type.*

*Proof.* For this purpose note that for any $\lambda \in K$ the matrix representation

$$\mathbf{A}^{(n,\lambda)} = \begin{array}{|c|c|c|c|} \hline J(n, \lambda) & \mathbf{I} & \mathbf{I} & \mathbf{O} \\ \hline \mathbf{I} & \mathbf{I} & \mathbf{O} & \mathbf{I} \\ \hline \end{array}$$

with $\mathbf{d}(\mathbf{A}^{(n,\lambda)}) = (2n; n, n, n, n)$ is an indecomposable representation of $\mathcal{P}$, where $\mathbf{I}$ is the identity matrix in $M_n(K)$ and $J(n, \lambda)$ is an $n \times n$-Jordan block. $\square$

**Lemma 5.6.3.** *The poset* $\mathcal{P} = (2) \sqcup (2) \sqcup (2)$:

$$\begin{array}{ccc} \bullet & \bullet & \bullet \\ | & | & | \\ \bullet & \bullet & \bullet \end{array}$$

(5.6.4)

*is of infinite representation type.*

*Proof.* Let $\mathcal{P} = (2) \sqcup (2) \sqcup (2)$. Consider the sub-poset $\mathcal{Q} = (2) \sqcup (2)$ of $\mathcal{P}$ with Hasse diagram

$$\begin{array}{cc} 2 \; \bullet & \bullet \; 4 \\ | & | \\ 1 \; \bullet & \bullet \; 3 \end{array}$$

(5.6.5)

A matrix representation of the poset $\mathcal{P}$ has the following form:

$$\mathbf{R} = \begin{array}{|c|c|c|c|c|c|} \hline \mathbf{A}_1 & \mathbf{A}_2 & \mathbf{A}_3 & \mathbf{A}_4 & \mathbf{A}_5 & \mathbf{A}_6 \\ \hline 1 & 2 & 3 & 4 & 5 & 6 \end{array}$$

and the following operations are admissible in the matrix $\mathbf{R}$:

1. Elementary row operations in the matrix $\mathbf{R}$.

2. Elementary column operations within each matrix $\mathbf{A}_i$ for $i = 1, \ldots, 6$.

3. Addition of columns of matrix $\mathbf{A}_1$ to columns of matrix $\mathbf{A}_2$; addition of columns of matrix $\mathbf{A}_3$ to columns of matrix $\mathbf{A}_4$; addition of columns of matrix $\mathbf{A}_5$ to columns of matrix $\mathbf{A}_6$.

By means of these operations the first four matrices $\mathbf{A}_1$, $\mathbf{A}_2$, $\mathbf{A}_3$ and $\mathbf{A}_4$ can be reduced. As a result the matrix

$$\mathbf{R} = \begin{array}{|c|c|} \hline \mathbf{A}_5 & \mathbf{A}_6 \\ \hline 5 & 6 \end{array}$$

is divided into nine horizontal strips, which are in one-to-one correspondence with the poset $\mathcal{A}(Q)$ of the form (5.5.15) whose width is equal to 3. Thus, our problem is reduced to describe the representations of the pair of posets $(C_2, \mathcal{A}(Q))$. By the Kleiner lemma, this problem is equivalent to describing the representations of the poset that is the cardinal sum $C_1 \sqcup \mathcal{A}(Q)$. Since the width of $\mathcal{A}(Q)$ is equal to 3, this cardinal sum contains a full sub-poset of width 4, whence, by lemma 5.6.2, it is of infinite representation type. $\square$

**Lemma 5.6.6.** *The poset $\mathcal{P} = (1) \sqcup (3) \sqcup (3)$ with Hasse diagram:*

$$
\begin{array}{ccc}
\bullet & \bullet & \\
| & | & \\
\bullet & \bullet & \\
| & | & \\
\bullet & \bullet & \bullet
\end{array}
\tag{5.6.7}
$$

*is of infinite representation type.*

*Proof.* Let $Q = (1) \sqcup (3)$ be the sub-poset of $\mathcal{P}$ with Hasse diagram:

$$
\begin{array}{cc}
\bullet\ 4 & \\
| & \\
\bullet\ 3 & \\
| & \\
1\ \bullet \quad & \bullet\ 2
\end{array}
\tag{5.6.8}
$$

Then $\mathcal{P} = Q \sqcup (3)$, and by theorem 5.5.16, it has the same representation type as the cardinal sum of posets $\mathcal{A}(Q) \sqcup (2)$. Since $\mathcal{A}(Q)$ has the form (5.5.13), it contains a sub-poset with Hasse diagram

$$
\begin{array}{cc}
1 \vee 4\ \bullet & \bullet\ 2 \\
| & | \\
1\ \ \bullet & \bullet\ 3
\end{array}
$$

Therefore, $\mathcal{A}(Q) \sqcup (2)$ contains the sub-poset $(2) \sqcup (2) \sqcup (2)$, whence, by the previous lemma, $\mathcal{P}$ is of infinite representation type. $\square$

**Lemma 5.6.9.** *The poset $\mathcal{P} = (1) \sqcup (2) \sqcup (5)$ is of infinite representation type.*

*Proof.* Let $Q = (1) \sqcup (5)$ be the sub-poset of $\mathcal{P}$ with Hasse diagram:

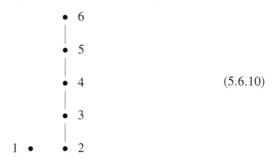

$$
\tag{5.6.10}
$$

Then $\mathcal{P} = Q \sqcup (2)$, and by theorem 5.5.16, it has the same representation type as the cardinal sum of posets $\mathcal{A}(Q) \sqcup (1)$. Since $\mathcal{A}(Q)$ has the following diagram:

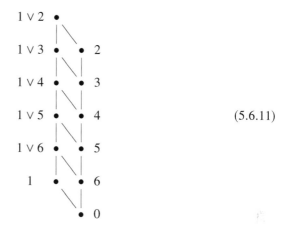

$$(5.6.11)$$

it contains a sub-poset with the Hasse diagram:

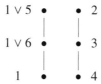

Therefore $\mathcal{A}(Q) \sqcup (1)$ contains a sub-poset $(3) \sqcup (3) \sqcup (1)$, whence, by the previous lemma, $\mathcal{P}$ is of infinite representation type. □

## 5.7 Primitive Posets of Finite Representation Type

In this section we prove the sufficiency part of theorem 5.6.1. Namely, we show that the posets $(1) \sqcup (1) \sqcup (n)$, $(1) \sqcup (2) \sqcup (2)$, $(1) \sqcup (2) \sqcup (3)$ and $(1) \sqcup (2) \sqcup (4)$ are of finite representation type.

**Lemma 5.7.1.** *If the width $w(\mathcal{P})$ of a poset $\mathcal{P}$ is greater than or equal to four, then $\mathcal{P}$ is of infinite representation type.*

*Proof.* Suppose that $w(\mathcal{P}) \geq 4$. Then $\mathcal{P}$ contains a full sub-poset $\mathcal{R}$ consisting of four incomparable elements, which is of infinite representation type, by lemma 5.6.2. □

From this lemma it follows that a poset of finite representation type has width $\leq 3$.

**Lemma 5.7.2.** *The finite poset* $\mathcal{P} = (1) \sqcup (1) \sqcup (n)$ *with Hasse diagram*

$$\begin{array}{c}
\bullet \ \ n+2 \\
| \\
\bullet \ \ n+1 \\
\vdots \\
\bullet \ \ 4 \\
1 \quad 2 \quad | \\
\bullet \ \bullet \ \bullet \ \ 3
\end{array}$$

(5.7.3)

*is of finite representation type.*

*Proof.* Indeed, consider the sub-poset $Q = (1) \sqcup (n)$ with Hasse diagram:

$$\begin{array}{c}
\bullet \ \ n+2 \\
| \\
\bullet \ \ n+1 \\
\vdots \\
\bullet \ \ 4 \\
2 \quad | \\
\bullet \ \bullet \ \ 3
\end{array}$$

Then the Hasse diagram of $\mathcal{A}(Q)$ has the following form:

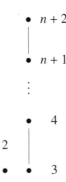

$$2 \vee 3 \ \bullet$$
$$2 \vee 4 \ \bullet \quad \bullet \ 3$$
$$\bullet \ 4$$
$$\vdots$$
$$\vdots$$
$$2 \vee (n+2) \ \bullet \quad \bullet \ n+1$$
$$2 \ \bullet \quad \bullet \ n+2$$
$$\bullet \ 0$$

(5.7.4)

Consequently, the width of $\mathcal{A}(Q)$ equals 2. Hence, by theorem 5.5.16 and corollary 5.3.6, $\mathcal{P} = Q \sqcup (1)$ has a finite representation type. $\square$

**Lemma 5.7.5.** *The finite posets* $\mathcal{P}_1 = (1) \sqcup (2) \sqcup (2)$, $\mathcal{P}_2 = (1) \sqcup (2) \sqcup (3)$, $\mathcal{P}_3 = (1) \sqcup (2) \sqcup (4)$ *are of finite representation type.*

*Proof.* Consider the poset $Q = (1) \sqcup (2)$ with Hasse diagram

$$\begin{array}{ccc} & & \bullet \ \ 3 \\ & & | \\ 1 \ \bullet & & \bullet \ \ 2 \end{array} \qquad\qquad (5.7.6)$$

which is a sub-poset of all three posets $\mathcal{P}_1$, $\mathcal{P}_2$ and $\mathcal{P}_3$:

$$\mathcal{P}_1 = Q \sqcup (2), \quad \mathcal{P}_2 = Q \sqcup (3), \quad \mathcal{P}_3 = Q \sqcup (4).$$

Then the poset $\mathcal{A}(Q)$ has the Hasse diagram

$$ (5.7.7) $$

Therefore, $\mathcal{A}(Q)$ has width 2. Hence, by corollary 5.3.6, it is of finite representation type. The diagram (5.7.7) after renumbering has the following form

Consider the poset $\mathcal{A}^2(Q)$. The new elements of $\mathcal{A}^2(Q)$ are $2 \vee 4$, $3 \vee 4$, $3 \vee 5$ and $0$. So the poset $\mathcal{A}^2(Q)$ has the following diagram:

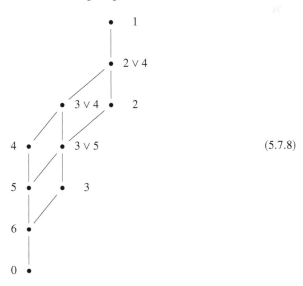

$$ (5.7.8) $$

Therefore $\mathcal{A}^2(Q)$ has width 2, and so, by corollary 5.3.6, it is of finite representation type.

We obtain after renumbering:

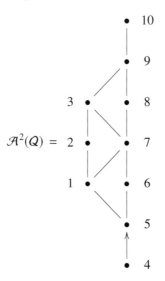

The new elements of $\mathcal{A}^3(Q)$ are $1 \vee 6$, $2 \vee 6$, $2 \vee 7$, $2 \vee 8$, $3 \vee 8$ and $0$. The diagram of the poset $\mathcal{A}^3(Q)$ has the following form:

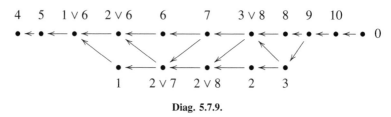

**Diag. 5.7.9.**

Therefore $\mathcal{A}^3(Q)$ has width 2, and so, by corollary 5.3.6, it is of finite representation type.

After renumbering diagram (5.7.9), we obtain:

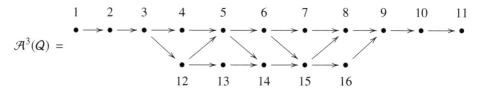

Now consider the poset $\mathcal{A}^4(Q)$. The new elements of $\mathcal{A}^4(Q)$ are $4 \vee 12$, $4 \vee 13$, $5 \vee 13$, $6 \vee 13$, $6 \vee 14$, $7 \vee 13$, $7 \vee 14$, $7 \vee 15$, $7 \vee 16$, $8 \vee 16$ and $0$. Consequently,

$\mathcal{A}^4(Q)$ contains 27 elements and $\mathcal{A}^4(Q)$ has the following diagram:

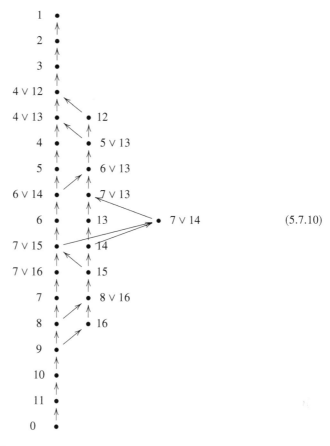

(5.7.10)

Thus, the poset $\mathcal{A}^4(Q)$ has width 3. It will be shown that it is of finite representation type. To this aim we use the trichotomy lemma two times. As the poset $Y$ we take the left chain consisting of 17 elements

$$\{0, 11, 10, 9, 8, 7, 7 \vee 16, 7 \vee 15, 6, 6 \vee 14, 5, 4, 4 \vee 13, 4 \vee 12, 3, 2, 1\}.$$

And as the posets $X$ and $Z$ we take the sub-posets

$$\{16 \prec 8 \vee 16 \prec 15 \prec 14\}, \quad Z = \{13, 7 \vee 14, 7 \vee 13, 6 \vee 13, 5 \vee 13, 12\},$$

respectively. All indecomposable representations have support either in $X \cup Y$ or in $Y \cup Z$. The poset $X \cup Y$ has width equal to 2, whence it is of finite representation type. Thus, it remains only to consider the poset $Y \cup Z$. Consider two sub-posets $X_1 = \{13, 7 \vee 14\}$ and $Z_1 = \{7 \vee 13, 6 \vee 13, 5 \vee 13, 12\}$ with relations $7 \vee 13 \prec 6 \vee 13 \prec 5 \vee 13 \prec 12$ in the poset $Z$. The poset $Y$ is the same as above. It is clear that $Y \cup Z_1$ is a poset of width 2, whence it is of finite type. The poset $X_1 \cup Y$ is of the form $(1) \sqcup (1) \sqcup (17)$ and so it is of finite type, by lemma 5.7.2. Therefore the poset $\mathcal{A}^4(Q)$ is of finite representation type.

    I. Since $\mathcal{P}_1 = Q \sqcup (2)$, and $\mathcal{A}(Q)$, $\mathcal{A}^2(Q)$ are posets of width 2, it follows, by lemma 5.5.16, that $\mathcal{P}_1$ is a poset of finite representation type.

    II. In a similar way, since $\mathcal{P}_2 = Q \sqcup (3)$, and $\mathcal{A}(Q)$, $\mathcal{A}^2(Q)$, $\mathcal{A}^3(Q)$ are posets of width 2, it follows, by lemma 5.5.16, that $\mathcal{P}_2$ is a poset of finite representation type.

    III. In a similar way, since $\mathcal{P}_3 = Q \sqcup (4)$, and $\mathcal{A}^i(Q)$ for $i < 4$ are posets of width 2, and $\mathcal{A}^4(Q)$ is of finite representation type, we obtain, by lemma 5.5.16, that $\mathcal{P}_3$ is a poset of finite representation type. $\square$

Taking into account lemmas 5.7.1, 5.7.2 and 5.7.5 we conclude the sufficiency part of theorem 5.6.1. Thus, theorem 5.6.1 is proved. $\square$

Let $L_1 \sqcup L_2 \sqcup L_3$ be a primitive poset of finite type and let $S_3$ be the symmetric group of degree 3. Then for any $\sigma \in S_3$ the poset $L_{\sigma(1)} \sqcup L_{\sigma(2)} \sqcup L_{\sigma(3)}$ is also of finite representation type.

**Corollary 5.7.11.** *The following primitive posets are of finite representation type over a division ring $K$:*
    a. *Any chain $(n)$.*
    b. *Any cardinal sum $(n) \sqcup (m)$.*
    c. $(1) \sqcup (1) \sqcup (n)$.
    d. $(1) \sqcup (2) \sqcup (2)$.
    e. $(1) \sqcup (2) \sqcup (3)$.
    f. $(1) \sqcup (2) \sqcup (4)$.
    *Conversely, if $\mathcal{P}$ is a primitive poset of finite representation type, then $\mathcal{P}$ is either a poset of the form a, b or a poset of the form $L_{\sigma(1)} \sqcup L_{\sigma(2)} \sqcup L_{\sigma(3)}$, where $L_1 \sqcup L_2 \sqcup L_3$ is a poset of the form c-f.*

**Definition 5.7.12.** A poset $\mathcal{P}$ of width 2 is called a **garland** if for any $x \in \mathcal{P}$ there is at most one element $y \in \mathcal{P}$ such that $x$ and $y$ are incomparable in $\mathcal{P}$.

**Example 5.7.13.** The poset with diagram

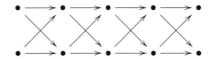

is a garland.

**Corollary 5.7.14.** *Let $\mathcal{P}$ be a poset of width 2 and let $\mathcal{A}^m(\mathcal{P})$ be a poset of width 2 for all $m > 1$. Then $\mathcal{P}$ is a garland.*

*Proof.* Let $Q = (1) \sqcup (2)$. Then, by the proof of lemma 5.7.5, $\mathcal{A}^4(Q)$ has width 3. If $Q$ is a sub-poset of $\mathcal{P}$, then, by proposition 5.5.8, $\mathcal{A}^4(Q) \subseteq \mathcal{A}^4(\mathcal{P})$, whence $w(\mathcal{A}^4(\mathcal{P})) \geq 3$. Therefore, if $\mathcal{P}$ is not a garland, then there exists $n$ such that $w(\mathcal{A}^n(\mathcal{P})) \geq 3$. Thus, we obtain the required statement. $\square$

## 5.8 Notes and References

Representations of partially ordered sets were first introduced by L.A. Nazarova and A.V. Roiter in 1972 [163]. In this paper they gave an algorithm which allows to check whether a given poset is of finite type. Using this algorithm M.M. Kleiner in 1972 characterized posets of finite type [129]. Moreover, M.M. Kleiner classified all the indecomposable $P$-spaces of finite type. He also found that the dimensions of all such indecomposable $P$-spaces are bounded by 6 [130]. In 1975 L.A. Nazarova characterized posets of infinite type (see [167], [166]). These results were independently also obtained by P. Donovan, M.R. Freislich (see [57]). The algorithm of Nazarova-Roiter works only for posets of width at most three. In 1977 A.G. Zavadskij has proposed a new differentiation algorithm for computing representations of posets [232]. This algorithm has been used to give a new proof for the characterization of poset of tame type (see [168]). O. Kerner showed that this algorithm is quite useful also in the case of finite representation type. He has used this algorithm to give a new proof of Kleiner's theorem (see [118]).

The notion of a trichotomy of a finite poset was introduced by P. Gabriel and A.V. Roiter in their book [82]. The trichotomy lemma (lemma 5.3.4) was also proved in this book [82, lemma 5.1].

Note that the main construction given in section 5.5 in some other form was introduced by L.A. Nazarova and A.V. Roiter in [163].

In this book the proof of theorem 5.6.1 follows [30].

Dilworth's decomposition theorem (theorem 5.1.1) was proved by the American mathematician R.P. Dilworth (1914 - 1993) in 1959 [51] and it was named after him. Like a number of other results in combinatorics Dilworth's theorem is equivalent to König's theorem on bipartite graphs. A dual of Dilworth's theorem (Mirsky's theorem) states that the size of the largest chain of a partial order (if finite) equals the smallest number of antichains into which the order may be partitioned [157].

In this chapter we considered the problem of classifying representations of finite posets only in the language of matrix problems and we also showed their connection with space representations of finite posets. Unfortunately we have not touched other interesting approaches for solving this problem.

One of them is module-theoretic. In this setting for any poset $\mathcal{P}$ one introduces a new poset $\mathcal{P}^* = \mathcal{P} \cup \{*\}$ such that $*$ is the unique maximal element in $\mathcal{P}^*$, that is $i < *$ for any point $i \in \mathcal{P}$. By means of the poset $\mathcal{P}^*$ one can construct an incidence $K$-algebra $K\mathcal{P}^*$ and consider the category of finite dimensional right modules over the algebra $K\mathcal{P}^*$. Then there is a close connection between the representations of $\mathcal{P}$ and some subcategory of $\mathrm{mod}_r(K\mathcal{P}^*)$ (see [193] for details).

The original bimodule-theoretic approach for solving matrix problems was proposed by Yu.A. Drozd in [61] and [62]. Having a finite poset $\mathcal{P}$ one can construct an incidence $K$-algebra $A = K(\mathcal{P})$ and consider finitely generated right $A$-modules. If $U$ is finite dimensional $K$-vector space and $V \in \text{mod}_r(A)$ then $U \otimes_K V$ is also $A$-module and there is a bijection between $A$-submodules in $U \otimes_K V$ and representations of the poset $\mathcal{P}$ in the vector space $U$. For details see [61].

The matrix problem associated with a poset $\mathcal{P}$ has also an interesting interpretation in terms of the category of prinjective modules[2] over the incidence algebra $K(\mathcal{P})$. This approach was developed in the work of J.A. de la Peña, D. Simson and S. Kasjan (see e.g. [49], [194], [195], [116], [117]).

Another approach is connected with classifying representations of a certain quiver. In this approach to any poset $\mathcal{P}$ one associates a special quiver $Q(\mathcal{P})$ whose vertices are the points of $\mathcal{P} = \{1, 2, \ldots, n\}$ and an extra point $n + 1$. The set of arrows in $Q(\mathcal{P})$ is defined in the following way:

1. Any point $i$ of $\mathcal{P}$ is connected by arrow with point $n + 1$.
2. The point $i$ of $\mathcal{P}$ is connected by arrow with point $j$ of $\mathcal{P}$ if $j$ covers $i$ in $\mathcal{P}$.

Good information about this approach is presented in [193] and [44].

---

[2]A module $X \in \text{mod}_r A$ is called **prinjective** if it has a projective resolution of the form

$$0 \to P_1 \to P_0 \to X \to 0,$$

where $P_0$ is a projective module and $P_1$ is a semisimple projective module.

# CHAPTER 6

# Representations of Quivers, Species and Finite Dimensional Algebras

An important problem in the theory of representations of finite dimensional algebras (f.d. algebras) is to obtain the full list of different kinds of algebras which are of finite representation type (finite type, or f.r.t.). The first classes of associative f.d. algebras of f.r.t which have been described were the class of algebras with zero square radical and the class of hereditary algebras over algebraically closed fields.

There are different approaches to study the representations of f.d. algebras. One of them is the approach of P. Gabriel [79], which reduces the study of representations of algebras to the study of representations of quivers. Another approach was first considered by L.A. Nazarova and A.V. Roiter [163]. This approach is to solve "matrix problems", that is, reducing some classes of matrices by means of admissible transformations to their simplest form. A third approach is due to M. Auslander and it is connected with the technique of almost split sequences.

The three sections of this chapter can be viewed as an introduction to the theory of representations of quivers and finite dimensional algebras. They present the main notions and some fundamental results of these representations, most of which are given without proof. In section 6.1 we consider the notions of finite quivers and their representations and give the main results of this theory. Section 6.2 is devoted to species (a generalization of quivers) and their representations. In section 6.3 we consider some main notions and results of the representation theory of finite dimensional algebras.

As it turns out the category of representations of finite dimensional algebras is equivalent to the category of representations of a special class of quivers, which are called bound quivers. That is why the quivers play a central role in the theory of finite dimensional associative algebras and their modules.

## 6.1 Finite Quivers and Their Representations

The notion of a quiver of a finite dimensional algebra over an algebraically closed field was first introduced by P. Gabriel in 1972 in connection with classification problems of finite dimensional algebras. This section is concerned with quivers in the sense of P. Gabriel [79].

**Definition 6.1.1.** A **quiver** $Q = (Q_0, Q_1, s, t)$ is a directed graph which consists of sets $Q_0$, $Q_1$ and two mappings $s, t : Q_1 \rightarrow Q_0$. The elements of $Q_0$ are called **vertices** (or **points**), and those of $Q_1$ are called **arrows**. We say that each arrow $\sigma \in Q_1$ starts at the vertex $s(\sigma)$ and ends at the vertex $t(\sigma)$. The vertex $s(\sigma)$ is called the **start** (or **initial**, or **source**) **vertex** and the vertex $t(\sigma)$ is called the **target** (or **end**) **vertex** of $\sigma$.

A **path** $p$ in the quiver $Q$ from the vertex $i$ to the vertex $j$ is a sequence of $r$ arrows $\sigma_1\sigma_2\ldots\sigma_{r-1}\sigma_r$ such that the target vertex of each arrow $\sigma_m$ coincides with the start vertex of the next one $\sigma_{m+1}$ for $1 \leq m \leq r - 1$. Moreover, the vertex $i$ is the start vertex of $\sigma_1$, while the vertex $j$ is the target vertex of $\sigma_r$. The number $r$ of arrows is called the **length of the path** $p$. For such a path $p$ define $s(p) = s(\sigma_1) = i$ and $t(p) = t(\sigma_k) = j$. Such path $p = \sigma_1\sigma_2\ldots\sigma_{r-1}\sigma_r$ can be presented as

$$i = s(\sigma_1) \xrightarrow{\sigma_1} t(\sigma_1) = s(\sigma_2) \xrightarrow{\sigma_2} t(\sigma_2) \longrightarrow \cdots \longrightarrow s(\sigma_r) \xrightarrow{\sigma_r} t(\sigma_r) = j$$

By convention we also include into the set of all paths the trivial path $\varepsilon_i$ of length zero which connects the vertex $i$ with itself without any arrow and we set $s(\varepsilon_i) = t(\varepsilon_i) = i$ for each $i \in Q_0$, and, also, for any arrow $\sigma \in Q_1$ with start at $i$ and end at $j$ we set $\varepsilon_j\sigma = \sigma\varepsilon_i = \sigma$. A path, connecting a vertex of a quiver with itself and of length not equal to zero, is called an **oriented cycle**. An oriented cycle of length equal to one is called a **loop** of a quiver. A quiver which has no oriented cycles is called **acyclic**.

In this chapter only finite quivers will be considered, that is quivers with finite and nonempty sets $Q_0$ and $Q_1$. Without loss of generality the set of vertices $Q_0$ can be taken to $\{1, 2, ..., n\}$.

Some examples of quivers are:

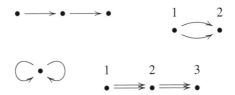

Note that quivers may possess multiply arrows between two points and loops or oriented cycles.

A quiver can be given by its **adjacency** (or **incidence**) **matrix**

$$[Q] = \begin{pmatrix} t_{11} & t_{12} & \cdots & t_{1n} \\ t_{21} & t_{22} & \cdots & t_{2n} \\ \vdots & \vdots & \ddots & \vdots \\ t_{n1} & t_{n2} & \cdots & t_{nn} \end{pmatrix}, \tag{6.1.2}$$

where $t_{ij}$ is the number of arrows from the vertex $i$ to the vertex $j$.

Denote by $\overline{Q}$ the **underlying graph** of $Q$ obtained from $Q$ by deleting the orientation of the arrows. A quiver $Q$ is called **connected** if its underlying graph $\overline{Q}$ is connected.

Two quivers $Q$ and $R$ are called **isomorphic** if there is a bijective correspondence between their vertices and arrows such that starts and ends of corresponding arrows map into one other. It is not difficult to see that $Q \simeq R$ if and only if the adjacency matrix $[Q]$ can be transformed into the adjacency matrix $[R]$ by a simultaneous permutation of rows and columns.

Let $\Gamma$ be a finite undirected graph. This gives a set of vertices $\Gamma_0 = \{1, 2, \ldots, n\}$, a set of edges $\Gamma_1$, and a set of natural numbers $\{t_{ij}\}$, where $t_{ij} = t_{ji}$ is a number of edges $\{i, j\} \in \Gamma_1$ between the vertex $i$ and the vertex $j$. For such a graph there is a quadratic form defined by

$$q(\mathbf{x}) = \sum_{i \in \Gamma_0} x_i^2 - \sum_{\substack{\{i,j\} \in \Gamma_1 \\ i \leq j}} t_{ij} x_i x_j \tag{6.1.3}$$

Given a quadratic form $q(\mathbf{x})$, one can define the corresponding bilinear symmetric form by the formula

$$(\mathbf{x}, \mathbf{y}) = q(\mathbf{x} + \mathbf{y}) - q(\mathbf{x}) - q(\mathbf{y}). \tag{6.1.4}$$

In this case

$$(e_i, e_j) = \begin{cases} 2 - 2t_{ii}, & \text{if } i = j \\ -t_{ij}, & \text{otherwise} \end{cases}$$

where $e_i$ is the $i$-th coordinate vector of the canonical basis of $\mathbf{Z}^n$ and $t_{ij}$ is the number of edges $\{i, j\}$ between $i$ and $j$ in $Q$.

Conversely, given the symmetric bilinear form $(\ ,\ )$, one can recover the quadratic form by the formula

$$q(\mathbf{x}) = \frac{1}{2}(\mathbf{x}, \mathbf{x}). \tag{6.1.5}$$

Recall that a quadratic form $q$ is **positive definite** if $q(\mathbf{x}) > 0$ for all $0 \neq \mathbf{x} \in \mathbf{Z}^n$. A quadratic form $q$ is **positive semi-definite** (or **nonnegative definite**) if $q(\mathbf{x}) \geq 0$ for all $\mathbf{x} \in \mathbf{Z}^n$. The set

$$\{\mathbf{x} \in \mathbf{Z}^n \mid (\mathbf{x}, \mathbf{y}) = 0 \ \text{ for all } \ \mathbf{y} \in \mathbf{Z}^n\}$$

is called the **radical** of $q$ and denoted rad $q$. A vector $\mathbf{x} \in \mathbf{Z}^n$ is called **strict** if no $x_i$ is zero.

The following graphs are called the **Dynkin diagrams**:

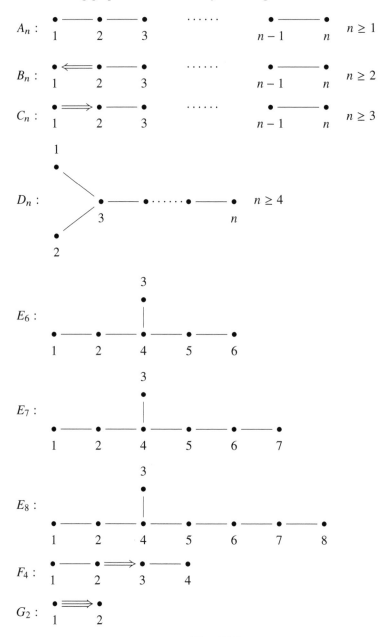

Diag. 6.1.6.

The diagrams $A_n$, $D_n$, $E_6$, $E_7$, $E_8$ are often called **simple Dynkin diagrams**.

The following main statement gives the classification of all graphs without loops and multiply edges from the point of view of their quadratic forms. This theorem is essentially due to Dynkin [67], [68], [69] thought he did not write it up in this way. The full proof of this theorem can be found in [31, chapter IV-VI] (see also [101, theorem 2.5.2]).

**Theorem 6.1.7.** *Let $\Gamma$ be a connected simply laced graph (i.e. a graph without loops and multiply edges), and $q$ its quadratic form. Then the following statements hold:*

1. *$q$ is positive definite if and only if $\Gamma$ is one of the simple Dynkin diagrams $A_n$, $D_n$, $E_6$, $E_7$ or $E_8$ presented in Diag. 6.1.6.*

2. *If $q$ is not positive definite then it is positive semi-definite if and only if $\Gamma$ is a one of the extended Dynkin diagrams[1] $\tilde{A}_n$, $\tilde{D}_n$, $\tilde{E}_6$, $\tilde{E}_7$ or $\tilde{E}_8$; moreover, $\mathrm{rad}(q) = \mathbf{Z}\delta$, where $\delta$ is the vector indicated by the numbers in the graphs presented in Diag. 6.1.8. Note that $\delta$ is strict and $> 0$.*

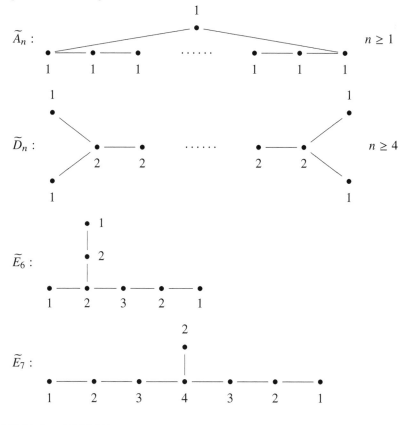

---

[1]The extended Dynkin diagrams are also called **Euclidean diagrams**.

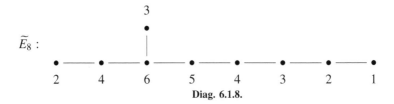

$\widetilde{E}_8$ :

Diag. 6.1.8.

3. *If q is neither positive definite nor positive semi-definite, then there is a vector $\alpha \geq 0$ with $q(\alpha) < 0$ and $(\alpha, e_i) \leq 0$ for all i.*

Suppose that $Q = (Q_0, Q_1, s, t)$ is a finite quiver. For such a quiver $Q$ one defines a (nonsymmetric) bilinear form, which is called the **Euler form** (or **Ringel form**) in the following way:

$$< \mathbf{x}, \mathbf{y} > = \sum_{i \in Q_0} x_i y_i - \sum_{\sigma \in Q_1} x_{s(\sigma)} y_{t(\sigma)}, \qquad (6.1.9)$$

where $\mathbf{x} = (x_1, x_2, \ldots, x_n)$, $\mathbf{y} = (y_1, y_2, \ldots, y_n) \in \mathbf{Z}^n$. The corresponding quadratic form is

$$q(\mathbf{x}) = < \mathbf{x}, \mathbf{x} > = \sum_{i \in Q_0}^{n} x_i^2 - \sum_{\sigma \in Q_1} x_{s(\sigma)} x_{t(\sigma)}. \qquad (6.1.10)$$

This integral quadratic form on $\mathbf{Z}^n$ is called the **Tits quadratic form** of the quiver $Q$.

One also introduces the **symmetric bilinear form** of the quiver $Q$ corresponding to (6.1.9) as follows

$$(\mathbf{x}, \mathbf{y}) = < \mathbf{x}, \mathbf{y} > + < \mathbf{y}, \mathbf{x} > . \qquad (6.1.11)$$

This form is called the **symmetric Euler form** (or **Cartan form**) of the quiver $Q$.

It is easy to see that

$$q(\mathbf{x}) = \frac{1}{2}(\mathbf{x}, \mathbf{x})$$

for any $\mathbf{x} \in \mathbf{Z}^n$ and

$$(e_i, e_j) = \begin{cases} 2 - 2t_{ii}, & \text{if } i = j \\ -t_{ij}, & \text{otherwise} \end{cases}$$

where $t_{ij}$ is the number of arrows in $\sigma \in Q_1$ from $i$ to $j$, and $\{e_1, e_2, \ldots, e_n\}$ is the canonical basis of $\mathbf{Z}^n$. In this case

$$q(\mathbf{x}) = \sum_{i \in Q_0} x_i^2 - \sum_{(i \to j) \in Q_1} t_{ij} x_i x_j \qquad (6.1.12)$$

Note that the Tits form of a quiver $Q$ coincides with the quadratic form of the underlying graph $\overline{Q}$ of $Q$ defined by (6.1.3). Therefore in contrast to the bilinear form (6.1.9) the Tits quadratic form (6.1.10) depends only on the underlying graph $\overline{Q}$ of $Q$, and not on the orientation of the edges.

Taking into account this remark we obtain the following theorem.

**Theorem 6.1.13.** *Let $Q$ be a finite connected and acyclic quiver with quadratic form $q_Q$, and let $\overline{Q}$ be the underlying graph of $Q$.*

1. *$\overline{Q}$ is a Dynkin diagram if and only if the quadratic form $q_Q$ is positive definite.*
2. *$\overline{Q}$ is an extended Dynkin diagram if and only if the quadratic form $q_Q$ is positive semi-definite but not positive definite.*
3. *$\overline{Q}$ is neither a Dynkin diagram nor an extended Dynkin diagram if and only if the quadratic form $q_Q$ is indefinite.*

A quiver $Q$ which underlying graph $\overline{Q}$ is one of the Dynkin diagram is said to be a **quiver of Dynkin type** or a **Dynkin quiver**. A quiver $Q$ which underlying graph $\overline{Q}$ is one of the extended Dynkin diagram is said to be a **quiver of Euclidean type** or an **affine quiver**.

Note that the Tits quadratic form $q(\mathbf{x})$ of a quiver $Q$ can also be written in the matrix form:

$$q(\mathbf{x}) = \frac{1}{2}\mathbf{x}\mathbf{C}_Q\mathbf{x}^T$$

where $\mathbf{C}_Q = (c_{ij})$ is a symmetric matrix of the quiver $Q$ defined by

$$c_{ij} = c_{ji} = \begin{cases} 2 - 2t_{ii}, & \text{if } i = j \\ -t_{ij}, & \text{otherwise} \end{cases} \qquad (6.1.14)$$

The matrix $\mathbf{C}_Q$ is called the **Cartan matrix of the quiver** $Q$.

Recall that an (abstract) **Cartan matrix** is a square matrix $\mathbf{C} = (c_{ij})$ with integer entries satisfying the conditions:

1. $c_{ii} = 2$ for all $i$.
2. $c_{ij} < 0$ for $i \neq j$.
3. $c_{ij} = 0$ if and only if $c_{ji} = 0$.
4. $\mathbf{C}$ is a positive definite matrix.

It is easy to see from (6.1.14) that if a quiver $Q$ contains no loops, i.e. $t_{ii} = 0$ for all $i$, then $\mathbf{C}_Q$ is a symmetric Cartan matrix.

For a quiver $Q = (Q_0, Q_1, s, t)$ and a field $K$ there is the so-called path algebra $KQ$ of $Q$ over $K$.

For a pair $i, j \in Q_0$ let $Q(i, j)$ denote the vector space with basis the set of all paths $p$ with $s(p) = i$ and $t(p) = j$. The obvious composition of paths induces two maps

$$Q(i, p) : Q(i, s(p)) \rightarrow Q(i, t(p)) \qquad (6.1.15)$$

$$Q(p, j) : Q(t(p), j) \rightarrow Q(s(p), j)) \qquad (6.1.16)$$

with $Q(i, p)(\eta) = \eta p$ and $Q(p, i)(\xi) = p\xi$.

**Definition 6.1.17.** The **path algebra** $KQ$ of a quiver $Q$ over a field $K$ is the (free) vector space with as $K$-basis all paths of $Q$. Multiplication in $KQ$ is defined in the obvious way: the product of two paths is given by composition:

$$(\sigma_1 \cdots \sigma_m)(\tau_1 \cdots \tau_n) = \begin{cases} \sigma_1 \cdots \sigma_m \tau_1 \cdots \tau_n & \text{if } t(\sigma_m) = s(\tau_1) \\ 0 & \text{otherwise} \end{cases}$$

In other words, there is a direct decomposition

$$KQ = KQ_0 \oplus KQ_1 \oplus KQ_2 \oplus \cdots \oplus KQ_n \oplus \cdots$$

where $KQ_n$ is a $K$-vector subspace of $KQ$ generated by all paths of length $n$, and multiplication in $KQ$ is defined by composition of paths so that $(KQ_n)(KQ_m) \subseteq KQ_{n+m}$ for all $n, m \geq 0$. Thus, $KQ$ is a $K$-graded algebra.

The path algebra is obviously associative. It is indecomposable if and only if $Q$ is a connected quiver. Moreover, if $Q$ is a disjoint union of two quivers $R$ and $S$, then $KQ = KR \times KS$.

Note that if a quiver $Q$ has infinitely many vertices, then $KQ$ has no an identity element. If $Q$ has infinitely many arrows, then $KQ$ is not finitely generated, and so it is not finite dimensional over $K$. In this chapter $Q_0$ and $Q_1$ are assumed to be finite and $Q_0 = \{1, 2, \ldots, n\}$.

In the path algebra $KQ$ the set of trivial paths $\varepsilon_i$ of length zero forms a set of pairwise orthogonal idempotents, i.e.,

$$\varepsilon_i^2 = \varepsilon_i \quad \text{for all } i \in Q_0$$

$$\varepsilon_i \varepsilon_j = 0 \quad \text{for all } i, j \in Q_0 \text{ such that } i \neq j.$$

Moreover, for each $\sigma \in Q_1$ there hold the relations:

$$\varepsilon_{s(\sigma)} \sigma = \sigma \varepsilon_{t(\sigma)}$$

and $\sum_{i \in Q_0} \varepsilon_i = 1$ is an identity of $KQ$, since $\varepsilon_i \sigma = 0$ unless $i = s(\sigma)$ and $\sigma \varepsilon_j = 0$ unless $j = t(\sigma)$.

The elements $\varepsilon_1, \varepsilon_2, \ldots, \varepsilon_n$ together with the paths of length one, i.e. elements $\sigma \in Q_1$, generate $KQ$ as an algebra. So $KQ$ is a finitely generated algebra.

Since $\{\varepsilon_1, \varepsilon_2, \ldots, \varepsilon_n\}$ is a set of pairwise orthogonal idempotents for $A = KQ$ with sum equal to 1, we have the following decomposition of $A$ into a direct sum:

$$A = \varepsilon_1 A \oplus \varepsilon_2 A \oplus \ldots \oplus \varepsilon_n A$$

of projective right $A$-modules $\varepsilon_i A$. Analogously, each $A\varepsilon_i$ is a projective left $A$-module.

Note that by [101, lemma 2.3.1] each $\varepsilon_i$, for $i \in Q_0$, is a primitive idempotent, and $\varepsilon_i A$ is an indecomposable projective right $A$-module.

**Examples 6.1.18.**

1. Let $Q$ be the quiver with one vertex and two loops:

   Then $KQ$ has two generators $\alpha, \beta$, and a path in $KQ$ is any word in $\alpha, \beta$. Therefore $KQ \simeq K\langle \alpha, \beta \rangle$, the free associative algebra generated by $\alpha, \beta$, which is noncommutative and infinite dimensional over $K$.
   If $Q$ is a quiver with one vertex and $n \geq 2$ loops $\alpha_1, \alpha_2, \ldots, \alpha_n$, then $KQ \simeq K\langle \alpha_1, \alpha_2, \ldots, \alpha_n \rangle$, the free associative algebra generated by $\alpha_1, \alpha_2, \ldots, \alpha_n$, which is also noncommutative and infinite dimensional over $K$.

2. Let $Q$ be the quiver with two vertices and two arrows:

$$\overset{1}{\bullet} \underset{\beta}{\overset{\alpha}{\rightrightarrows}} \overset{2}{\bullet}$$

   i.e. $Q_0 = \{1, 2\}$ and $Q_1 = \{\alpha, \beta\}$. The algebra $KQ$ has a basis $\{\varepsilon_1, \varepsilon_2, \alpha, \beta\}$. This algebra is isomorphic to $A = \begin{pmatrix} K & K \oplus K \\ 0 & K \end{pmatrix}$, which is a four-dimensional algebra over $K$, and called the **Kronecker algebra**.

**Proposition 6.1.19.** *The path algebra $KQ$ is finite dimensional over $K$ if and only if $Q_0$ is finite and $Q$ has no oriented cycles.*

*Proof.* Suppose that $Q$ possesses an oriented cycle. Then we can construct an infinite number of different paths by walking around this cycle $n$ times, for any $n$. Therefore in this case the path algebra $KQ$ is infinite dimensional. Conversely, assume that $Q_0$ is finite and $Q$ has no oriented cycles. In this case all vertices in each nontrivial path $\alpha_1 \alpha_2 \ldots \alpha_n$ must be different. Therefore we have only a finite number of different paths in $Q$, which form a basis of $KQ$. So, $KQ$ is a finite dimensional algebra. $\square$

**Proposition 6.1.20.** *The path algebra $KQ$ of a quiver $Q$ over a field $K$ is hereditary. If $Q$ is a finite quiver without oriented cycles then $KQ$ is a primely triangular ring.*

*Proof.* The first statement is [101, theorem 2.3.4]. Since any finitely dimension algebra is an FDI-ring, the second statement follows from the first one, proposition 6.1.19 and [103, corollary 8.5.7]. $\square$

Write
$$R = R_Q = KQ_1 \oplus KQ_2 \oplus \cdots \oplus KQ_n \oplus \cdots$$

which is the two-sided ideal of $KQ$ generated by the set of all arrows of $Q$. This ideal is called the **arrow ideal** of $KQ$. Since $KQ$ is a $K$-graded algebra, $R^n = \bigoplus_{m \geq n} KQ_m$.

Assume that $Q$ is a finite connected acyclic quiver. Then there is a largest $l \geq 0$ such that $Q$ contains a path of length $l$. It means that any product of $l + 1$ arrows is zero. Therefore $R^{l+1} = 0$, i.e. $R$ is nilpotent. Since $KQ$ is a finite dimensional algebra, $R$ is contained in the Jacobson radical of $KQ$. On the other hand, $KQ/R$ is isomorphic to a product of copies of $K$, i.e. $\text{rad}(KQ/R) = 0$. So $\text{rad}(KQ) \subseteq R$. Thus, if $Q$ is a finite connected acyclic quiver then $R = \text{rad}(KQ)$ is the Jacobson radical of $KQ$.

**Definition 6.1.21.** Let $Q$ be a finite quiver, and $R = R_Q$ is the arrow ideal of $KQ$. A two-sided ideal $I$ of $KQ$ is called **admissible** if there is an integer $m \geq 2$ such that

$$R^m \subseteq I \subseteq R^2$$

If $Q$ is an acyclic quiver then there is an integer $m$ such that $R^m = 0$. Therefore any two-sided ideal $I$ in $KQ$ such that $I \subseteq R^2$ is admissible.

**Definition 6.1.22.** A **relation** in a quiver $Q$ with coefficients in a field $K$ is an element of $KQ$ that is a linear combinations of paths having a common source and a common target, and of length at least 2.

In other words, a relation $\rho$ of $Q$ is an element of $KQ$ which has the following form

$$\rho = \sum_{i=1}^{m} \lambda_i w_i$$

where the $\lambda_i \in K$ are non-zero and the $w_i$ are paths in $Q$ of length at least 2 such that all the $w_i$ have the same source and target, $i = 1, 2, ..., m$.

More generally the term 'relation' is also used for a subspace of $KQ$ that is spanned by such $\rho$.

A **quiver with relations** is a pair $(Q, I)$, where $Q$ is a quiver and $I$ is a two-sided ideal of $KQ$ generated by a finite set $\{\rho_j\}_{j \in J}$ of relations. If $I$ is an admissible ideal generated by a finite set of relations $\{\rho_j\}_{j \in J}$ then the pair $(Q, I)$ is called a **bound quiver** and the quotient algebra $KQ/I$ is called a **bound quiver algebra**.

If $Q$ is a finite quiver and $I$ is an admissible ideal in $KQ$ then the bound quiver algebra $KQ/I$ is finite dimensional. Moreover it is indecomposable if and only if $Q$ is a connected quiver.

**Examples 6.1.23.** Let $Q$ be the quiver with one vertex and two loops:

$$\alpha \, \circlearrowleft \; \bullet \; \circlearrowright \, \beta$$

and the relation $\alpha\beta = \beta\alpha$. Then $KQ$ has two generators $\alpha, \beta$, and a path in $KQ$ is any word in $\alpha, \beta$. Therefore $KQ \simeq K\langle\alpha, \beta\rangle$, the free associative algebra generated by $\alpha, \beta$, which is noncommutative and infinite dimensional over $K$. Then $I = (\alpha\beta - \beta\alpha)$ is a two-sided ideal in $KQ$ and $KQ/I = K\langle\alpha, \beta\rangle/(\alpha\beta - \beta\alpha) \simeq K[\alpha, \beta]$ is just the standard polynomial algebra on 2 generators.

**Definition 6.1.24.** Let $Q = (Q_0, Q_1, s, t)$ be a finite quiver and let $K$ be a field. A **representation** $V = (V_x, V_\sigma)$ of $Q$ over $K$ is a family

$$\{V_x \; : \; x \in Q_0\}$$

of $K$-vector spaces together with a family

$$\{V_\sigma : V_{s(\sigma)} \to V_{t(\sigma)} \; : \; \sigma \in Q_1\}$$

of $K$-linear mappings. It is called **finite dimensional** if all the vector spaces $V_i$ are finite dimensional over $K$.

For any finite dimensional representation $V$ one defines the **dimension vector** of $V$ as the vector

$$\underline{\dim} \, V = (\dim V_i)_{i \in Q_0} \in (\mathbf{N} \cup \{0\})^{Q_0}. \tag{6.1.25}$$

The representations of a quiver $Q$ over a field $K$ form a category denoted by $\mathrm{Rep}_K(Q)$, which objects are representations $V$ and morphisms $f = (f_x) : V \to W$ are defined as a family of linear mappings $f_x : V_x \to W_x$, $x \in Q_0$, such that for each $\sigma \in Q_1$ the diagram

$$
\begin{array}{ccc}
V_{s(\sigma)} & \xrightarrow{\; f_{s(\sigma)} \;} & W_{s(\sigma)} \\
{\scriptstyle V_\sigma} \downarrow & & \downarrow {\scriptstyle W_\sigma} \\
V_{t(\sigma)} & \xrightarrow{\; f_{t(\sigma)} \;} & W_{t(\sigma)}
\end{array}
$$

commutes. If $f_x$ is invertible for every $x \in Q_0$, then $f$ is called an **isomorphism**. We denote the linear space of morphisms from $V$ to $W$ by $\mathrm{Hom}_Q(V, W)$. For two morphisms $f : V \to W$ and $g : W \to U$ one defines the composition of morphisms $gf : V \to U$ in the obvious way as follows $(gf)_i = g_i f_i$.

In this category one has natural (and obvious) definitions of such things as the zero representation, subrepresentations and quotient representations, direct sum of representations, indecomposable and decomposable representations.

Let $V$ and $W$ be representations of $Q$. A representation $W$ is called a **subrepresentation** of $V$ if

1. $W_i$ is a subspace of $V_i$, for every $i \in Q_0$.
2. The restriction of $V_\sigma : V_{s(\sigma)} \to V_{t(\sigma)}$ to $W_{t(\sigma)}$ is equal to $W_\sigma : W_{s(\sigma)} \to W_{t(\sigma)}$.

Similarly one introduces the notion of a quotient representation.

Note that every quiver has a representation $T$ with $T_i = 0$ and $T_\sigma = 0$, for all $i \in Q_0$ and all $\sigma \in Q_1$. This representation is called the **zero representation**. The zero representation and a given representation $V$ itself are called the **trivial subrepresentations** of $V$.

A non-zero representation $V$ is called **irreducible** or **simple** if it has only trivial subrepresentations.

Given a vertex $i \in Q_0$, let $S(i)$ be the representation with

$$S(i)_j = \begin{cases} K, & \text{if } j = i \\ 0, & \text{otherwise} \end{cases} \quad \text{and} \quad S(i)_\sigma = 0$$

for $j \in Q_0$ and $\sigma \in Q_1$. Then $S(i)$ is a simple representation of $Q$ at vertex $i$.

If $Q$ is a finite acyclic quiver then any simple representation $S$ of $Q$ is of the form $S(i)$, i.e. there exists $i \in Q_0$ such that $S \simeq S(i)$.

The **direct sum** $V \oplus W$ of two representations $V$ and $W$ of a quiver $Q$ is defined by

$$(V \oplus W)_i = V_i \oplus W_i$$

for each $i \in Q_0$ and

$$(V \oplus W)_\sigma = V_\sigma \oplus W_\sigma$$

for all $\sigma \in Q_1$.

In other words, $(V \oplus W)_\sigma : V_{s(\sigma)} \oplus W_{s(\sigma)} \to V_{t(\sigma)} \oplus W_{t(\sigma)}$ can be represented by the matrix:

$$\left( \begin{array}{c|c} V_\sigma & \mathbf{0} \\ \hline \mathbf{0} & W_\sigma \end{array} \right)$$

A representation $V$ of $Q$ is called **decomposable** if it is isomorphic to a direct sum of non-trivial representations. Otherwise it is called **indecomposable**.

So the category $\text{Rep}_K(Q)$ contains a (unique) zero element, direct sums of representations, subrepresentations and quotient representations. Thus, $\text{Rep}_K(Q)$ is an Abelian category. We denote by $\text{rep}_K(Q)$ the full subcategory of $\text{Rep}_K(Q)$ consisting of the finitely dimensional representations. Obviously, $\text{rep}_K(Q)$ is also an Abelian category.

**Theorem 6.1.26.** *If $Q = (Q_0, Q_1, s, t)$ is a finite quiver, then the category $\text{Mod}_r(KQ)$ of right $KQ$-modules is equivalent to the category $\text{Rep}_K(Q)$ of representations of $Q$. This equivalence restricts to an equivalence $\text{mod}_r(KQ) \cong \text{rep}_K(Q)$. (Here $\text{mod}_r(KQ)$ is a category of finitely generated right $KQ$-modules.)*

*Proof.* Let $Q$ be a finite quiver with $Q_0 = \{1, 2, \ldots, n\}$, and let $KQ$ be the path algebra of $Q$. Then $KQ$ is generated by $\varepsilon_i$ for each $i \in Q_0$ and by arrows $\sigma \in Q_1$. In this case $1 = \varepsilon_1 + \varepsilon_2 + \cdots + \varepsilon_n$ is a decomposition of the identity of $KQ$ into a sum of pairwise orthogonal idempotents such that $\varepsilon_{s(\sigma)}\sigma = \sigma\varepsilon_{t(\sigma)}$. Let $\overline{V}$ be a right $KQ$-module, then we can define a representation $V = F(\overline{V}) = (V_i, V_\sigma)$ of $Q$ by setting $V_i = \overline{V}\varepsilon_i$ for each vertex $i \in Q_0$, and for each arrow $\sigma \in Q_1$ we define the $K$-linear map $V_\sigma : \overline{V}\varepsilon_{s(\sigma)} \longrightarrow \overline{V}\varepsilon_{t(\sigma)}$ by $V_\sigma(x) = x\sigma$ for every $x \in \overline{V}\varepsilon_{s(\sigma)}$. This is well defined since $\sigma = \varepsilon_{s(\sigma)}\sigma\varepsilon_{t(\sigma)}$ and $\overline{V}$ is a right $KQ$-module, and so $x\sigma = (x\varepsilon_{s(\sigma)})(\varepsilon_{s(\sigma)}\sigma\varepsilon_{t(\sigma)}) \in \overline{V}\varepsilon_{t(\sigma)} = V_{t(\sigma)}$.

Let $f : \overline{V} \longrightarrow \overline{W}$ be a homomorphism of right $A$-modules, and $F(\overline{V}) = (V_i, V_\sigma)$, $F(\overline{W}) = (W_i, W_\sigma)$. Then we define the morphism

$$F(f) : F(\overline{V}) \longrightarrow F(\overline{W})$$

by the follows. For each vertex $i \in Q_0$ and each $x \in V_i = \overline{V}\varepsilon_i$ we set $f(x) = f(x\varepsilon_i) = f(x)\varepsilon_i \in \overline{W}\varepsilon_i = W_i$, since $f$ is a homomorphism of right $KQ$-modules. Therefore for each $i \in Q_0$ we can define a $K$-linear mapping $f_i : V_i \to W_i$ as a restriction $f_i$ of $f$ on $V_i$. Now we can define $F(f) = (f_i)_{i \in Q_0}$. It is easy to show that the diagram

$$
\begin{array}{ccc}
V_{s(\sigma)} & \xrightarrow{\ V_\sigma\ } & V_{t(\sigma)} \\
{\scriptstyle f_{s(\sigma)}}\big\downarrow & & \big\downarrow{\scriptstyle f_{t(\sigma)}} \\
W_{s(\sigma)} & \xrightarrow[\ W_\sigma\ ]{} & W_{t(\sigma)}
\end{array}
$$

is commutative for each $\sigma \in Q_1$, i.e. $F(f)$ is a morphism in $\mathrm{Rep}_K(Q)$. Therefore $F : \mathrm{Mod}_r(KQ) \longrightarrow \mathrm{Rep}_K(Q)$ is a $K$-linear functor.

Conversely, let $V = (V_i, V_\sigma)$ be a representation of $Q$. Then one can consider the vector space $G(V) = \overline{V} = \underset{i \in Q_0}{\oplus} V_i$ and equip it with two families of maps making $\overline{V}$ into a $KQ$-module as follows. For every $i \in Q_0$, $\varepsilon_i$ acts as the projection onto $V_i$. For every $\sigma \in Q_1$, $\sigma|_{V_i} = 0$ if $i \neq s(\sigma)$ and

$$
\sigma|_{V_{s(\sigma)}} = V_\sigma : V_{s(\sigma)} \longrightarrow V_{t(\sigma)}.
$$

Since $KQ$ is generated by all arrows and the $\varepsilon_i$ for all $i \in Q_0$, the action for arbitrary path can only be defined in one way. It is obvious that this indeed defines a right $KQ$-module. Therefore $G(V) = \overline{V} \in \mathrm{Mod}_r(KQ)$.

Let $W = (W_i, W_\sigma)$ be another representation of $Q$, and $g = (g_i)_{i \in Q_0}$ be a morphism from $V$ to $W$, i.e. $g_i : V_i \longrightarrow W_i$ is a $K$-linear mapping such that for each $\sigma \in Q_1$ the diagram

$$
\begin{array}{ccc}
V_{s(\sigma)} & \xrightarrow{\ V_\sigma\ } & V_{t(\sigma)} \\
{\scriptstyle g_{s(\sigma)}}\big\downarrow & & \big\downarrow{\scriptstyle g_{t(\sigma)}} \\
W_{s(\sigma)} & \xrightarrow[\ W_\sigma\ ]{} & W_{t(\sigma)}
\end{array}
$$

is commutative. Therefore we can define

$$
G(g) = \bigoplus_{i \in Q_0} g_i : G(V) = \bigoplus_{i \in Q_0} V_i \longrightarrow G(W) = \bigoplus_{i \in Q_0} W_i
$$

which is a homomorphism of $KQ$-modules. Indeed, assume that $p \in KQ$ with $s(p) = i$ and $t(p) = j$. Then for any $x \in V_i$ we have

$$
G(g)(xp) = g_i(xp) = g_j V_p(x) = W_p g_i(x) = g_i(x)p = G(g)(x)p.
$$

So $G(g)$ is a homomorphism of $KQ$-modules, and

$$
G : \mathrm{Rep}_K(Q) \longrightarrow \mathrm{Mod}_r(KQ)
$$

is a $K$-linear functor.

Now it is easy to verify that $F$ and $G$ give equivalence of functors

$$FG \simeq \mathbf{1}_{\text{Rep}_K(Q)} \quad \text{and} \quad GF \simeq \mathbf{1}_{\text{Mod}_r(KQ)}.$$

Both functors can be restricted on categories $\text{rep}_K(Q)$ and $\text{mod}_r(KQ)$ and give equivalence on these categories. □

Thus, $\text{rep}_K(Q)$ is an Abelian category which is equivalent to the category $\text{mod}_r(KQ)$ of all finitely generated right $KQ$-modules. Therefore $\text{rep}_K(Q)$ is a Krull-Schmidt category[2]. So the problem of classifying all finite dimensional representations of a finite quiver $Q$ over a field $K$ is reduced to classifying the indecomposable finite dimensional representations of $Q$.

**Examples 6.1.27.**
   1. Let $Q$ be the following quiver:

$$1 \qquad 2$$
$$\bullet \xrightarrow{\ \sigma\ } \bullet$$

Then the path algebra $KQ$ has a basis $\{\varepsilon_1, \varepsilon_2, \sigma\}$ and is isomorphic to the upper triangular matrix algebra $\begin{bmatrix} K & K \\ 0 & K \end{bmatrix}$.

The representations of $Q$ are classified (up to isomorphism) by a linear transformation $V_\sigma$ which can be put in the normal form. So we can choose the bases in $V_1$ and $V_2$ such that the corresponding matrix of transformation $V_\sigma$ is given by a block matrix $\begin{bmatrix} \mathbf{I}_r & 0 \\ 0 & 0 \end{bmatrix}$, where $\mathbf{I}_r$ is the $r \times r$ identity matrix with $r$ the rank of $V_\sigma$. Therefore the quiver $Q$ has 3 indecomposable finite dimensional representations over a field $K$:
   $E_1$: $V_1 = K$, $V_2 = 0$, $V_\sigma = 0$;
   $E_2$: $V_1 = 0$, $V_2 = K$, $V_\sigma = 0$;
   $E_3$: $V_1 = K$, $V_2 = K$, $V_\sigma = \text{id}$,
which correspond to the indecomposable modules: $(K, 0)$, $(0, K)$ and $(K, K)$.
   2. Let $Q$ be the quiver with one vertex and one loop:

 $\alpha$

Then the path algebra $KQ$ has the basis $\{\varepsilon, \alpha, \alpha^2, \dots, \alpha^n, \dots\}$. Therefore $KQ \simeq K[x]$, the polynomial algebra in one variable $x$. Obviously, this algebra is finitely generated but it is not finite dimensional.

---

[2]Recall that an additive category $\mathfrak{C}$ is said to be a **Krull-Schmidt category** provided the endomorphism ring $\text{End}(X)$ of any indecomposable object $X$ of $\mathfrak{C}$ is a local ring. In a Krull-Schmidt category any object is isomorphic to a finite direct sum of indecomposable objects, and such a decomposition is unique up to isomorphism and permutation of summands.

The representations of $Q$ are classified up to isomorphism by endomorphisms of $K$-vector spaces. From standard linear algebra all indecomposable f.d. endomorphisms are described by single Jordan blocks $J_{n,\lambda}$. Each such a Jordan block

$$J_{n,\lambda} = \begin{pmatrix} \lambda & 1 & 0 & \cdots & 0 & 0 \\ 0 & \lambda & 1 & \cdots & 0 & 0 \\ 0 & 0 & \lambda & \cdots & 0 & 0 \\ \vdots & \vdots & \vdots & \ddots & \vdots & \vdots \\ 0 & 0 & 0 & \cdots & \lambda & 1 \\ 0 & 0 & 0 & \cdots & 0 & \lambda \end{pmatrix}$$

is characterized by two numbers: the integer $n$ is the size of the block, the real number $\lambda$ is the eigenvalue of the block.

As it turns out theorem 6.1.26 can be considered as a corollary of more general theorem. To this end we introduce the notion of representation of a bound quiver.

Let $Q$ be a finite quiver and $V = (V_i, V_\sigma)$ be a representation of $Q$. For any nontrivial path $p = \sigma_1\sigma_2\ldots\sigma_n$ from the vertex $i$ to the vertex $j$ in $Q$ we define a $K$-linear map from $V_i$ to $V_j$ as follows:

$$V_p = V_{\sigma_n} V_{\sigma_{n-1}} \cdots V_{\sigma_2} V_{\sigma_1}$$

which is said to be the **evaluation** of $V$ on the path $p$.

Suppose $\rho = \sum_{i=1}^m \lambda_i p_i$ is a linear combination of paths of $Q$. Then one can define

$$V_\rho = \sum_{i=1}^m \lambda_i V_{p_i}$$

**Definition 6.1.28.** Let $(Q, I)$ be a bound quiver, where $I$ is an admissible ideal of the path algebra $KQ$. A representation $V = (V_i, V_\sigma)$ of $Q$ is said to be **bound by** $I$ if $V_\rho = 0$ for any relation $\rho \in I$.

Denote by $\mathrm{Rep}_K(Q, I)$ (resp. $\mathrm{rep}_K(Q, I)$) the full subcategory of $\mathrm{Rep}_K(Q)$ (resp. $\mathrm{rep}_K(Q)$) consisting of the representations of $Q$ bound by $I$.

Then there is the following theorem.

**Theorem 6.1.29.** *Let $A = KQ/I$, where $Q$ is a finite connected quiver and $I$ is an admissible ideal of $KQ$. Then the category $\mathrm{Mod}_r(A)$ of right $A$-modules is equivalent to the category $\mathrm{Rep}_K(Q, I)$ of representations of $(Q, I)$. This equivalence restricts to an equivalence $\mathrm{mod}_r A \cong \mathrm{rep}_K(Q, I)$.*

**Remark 6.1.30.** Theorem 6.1.26 follows immediately from this theorem, since $I = 0$ for any connected acyclic quiver.

In what follows in this chapter we will consider only finite quivers and their f.d. representations.

**Definition 6.1.31.** A finite quiver $Q$ is said to be of **finite representation type** (or **finite type**) if there are only a finite number of indecomposable representations of $Q$ (up to isomorphism). Otherwise it is said to be of **infinite type**. A quiver $Q$ is said to be of **tame representation type** if there are infinitely many isomorphism classes of indecomposable representations but these classes can be parametrized by a finite set of integers together with a polynomial irreducible over $K$; a quiver $Q$ is said to be of **wild representation type** if for every finite dimensional algebra $E$ over $K$ there are infinitely many pairwise non-isomorphic representations of $Q$ which have $E$ as their endomorphism algebra.

These three classes of quivers are clearly exclusive. They are, as it turns out, also exhaustive.

For a graph $\Gamma$ which is one of the Dynkin type or Euclidean type a **root** is an integral vector $\alpha \in \mathbf{Z}^n$, where $n$ is the number of vertices in $\Gamma$, such that $q(\alpha) \leq 1$, where $q$ is the quadratic form of the graph $\Gamma$. A root $\alpha$ is called **real** if $q(\alpha) = 1$ and it is **imaginary** if $q(\alpha) = 0$. Each root $\alpha$ can be written in the form $\sum_i k_i \mathbf{e}_i$, where $(\mathbf{e}_i)$ is the standard basis of $\mathbf{Z}^n$. A root $\alpha$ is called **positive** if $\alpha$ is not a zero vector and all $k_i \geq 0$. As it turns out all roots of Dynkin diagrams are real, and all imaginary roots of Euclidean diagrams are of the form $r\delta$ for $r \in \mathbf{Z}$, where $\delta$ is the vector indicated by the numbers in the Euclidean diagrams presented in Diag. 6.1.8.

The main result in the theory of quiver representations is the famous Gabriel theorem classifying finite quivers of finite representation type. It turns out that such quivers are closely connected with Dynkin diagrams.

**Theorem 6.1.32.** (P. Gabriel, [79], I.N. Bernstein, I.M. Gelfand and V.A. Ponomarev [28]). *Let $Q$ be a finite connected and acyclic quiver, and let $K$ be a field. Then the following statements hold*:

1. *The quiver $Q$ is of finite representation type over the field $K$ if and only if $Q$ is a Dynkin quiver.*

2. *The number of isomorphism classes of indecomposable representations of $Q$ is finite if and only if the corresponding quadratic form $q_Q$ is positive definite. In this case, the map*

$$V \mapsto \underline{\dim} V$$

   *sets up a one-to-one correspondence between the isomorphism classes of indecomposable representations of $Q$ and positive roots of the Tits quadratic form of $Q$. For a Dynkin quiver $Q$, the dimension vectors of indecomposable representations do not depend on the orientation of the arrows in $Q$.*

3. *The number of isomorphism classes of indecomposable representations of the quiver $Q$ equals $\frac{1}{2}n(n + 1)$, $n(n - 1)$, 36, 63 and 120, if $\overline{Q}$ is one of the forms $A_n$, $D_n$, $E_6$, $E_7$ and $E_8$, respectively.*

This theorem for the case of an algebraically closed field was first proved by P. Gabrial [79]. Another proof of this theorem (statements 1 and 2) for an arbitrary field, using reflection functors and Coxeter functors, has been given by I.N. Bernstein, I.M. Gelfand and V.A. Ponomarev [28] (for the proof of this theorem see also [102, Theorem 1.19.1]).

Representations of quivers in the tame case were studied by L.A. Nazarova [164] and P. Donovan, M.R. Freislich [57], who proved independently the following classical theorem under the assumption that the base field $K$ is algebraically closed. In the general case, if $K$ is an arbitrary field, the result was established by V. Dlab and C.M. Ringel in [54]. The following theorem is similar to the first part of the Gabriel theorem.

**Theorem 6.1.33.** (L.A. Nazarova [164], P. Donovan, M.R. Freislich [57]). *Let $Q$ be a finite connected and acyclic quiver. Then $Q$ is of tame type if and only if $Q$ is a quiver of Euclidean type.*

Let $Q = (Q_0, Q_1)$ be a finite quiver. For each $i \in Q_0$ one defines a **projective representation** $P(i)$ and an **injective representation** $I(i)$ by the following way.

$$P(i)_j = K[Q(i,j)] \quad \text{and} \quad P(i)_\sigma = K[Q(i,\sigma)]$$

$$I(i)_j = \operatorname{Hom}_K(K[Q(j,i)], K) \quad \text{and} \quad I(i)_\sigma = \operatorname{Hom}_K(K[Q(\sigma,i)], K)$$

for $j \in Q_0$ and $\sigma \in Q_1$.

Here $K[Q(i,j)]$ is the $K$-vector space with basis $Q(i,j)$, which is the set of all paths from $i$ to $j$, and $K[Q(i,\sigma)]$ and $K[Q(\sigma,i)]$ are the linear maps $K[Q(i,s(\sigma))] \longrightarrow K[Q(i,t(\sigma))]$ and $K[Q(t(\sigma),i)] \longrightarrow K[Q(s(\sigma),i)]$ of suitable compositions of a path with $\sigma$, described by (6.1.15) and (6.1.16).

If $Q$ is a finite acyclic quiver then $P(i)$ and $I(i)$ are finite dimensional representations with $\operatorname{End}_K(P(i)) \simeq P(i)_i = K[Q(i,i)] = K$ and $\operatorname{End}_K(I(i)) \simeq K$.

The representations $P(i)$ and $I(i)$ are projective and injective objects in $\operatorname{Rep}_K(Q)$ in the sense that for each $V, W \in \operatorname{Rep}_K(Q)$ and every epimorphism $V \longrightarrow W$ with $V, W \in \operatorname{Rep}_K(Q)$ the induced map $\operatorname{Hom}(P(i), V) \longrightarrow \operatorname{Hom}(P(i), W)$ is surjective, and for every monomorphism $V \longrightarrow W$ the induced map $\operatorname{Hom}(W, I(i)) \longrightarrow \operatorname{Hom}(V, I(i))$ is injective.

Analogously the representations $P(i)$ and $I(i)$ are projective and injective objects in $\operatorname{rep}_K(Q)$.

In the proof of the Gabriel theorem for Dynkin quivers I.N. Bernstein, I.M. Gelfand and V.A. Ponomarev used Coxeter functors[3] $C^- : \operatorname{Rep}_K(Q) \longrightarrow \operatorname{Rep}_K(Q)$ and $C^+ : \operatorname{Rep}_K(Q) \longrightarrow \operatorname{Rep}_K(Q)$. These functors also play an important role in studying more general quivers.

Let $Q$ be a connected and acyclic quiver. As it turns out projective and injective representations are closely connected with Coxeter functors.

---

[3]See e.g. [102, Section 1.18].

**Proposition 6.1.34.** (V. Dlab, C.M. Ringel [52, Proposition 2.4]). *Let V be an indecomposable representation of a quiver Q without oriented cycles. Then*

1. $C^+V = 0$ *if and only if* $V \simeq P(i)$ *for some* $i \in Q_0$ *if and only if V is a projective representation.*
2. $C^-V = 0$ *if and only if* $V \simeq I(i)$ *for some* $i \in Q_0$ *if and only if V is an injective representation.*

For $r \in \mathbf{Z}$ we write:

$$C^r = \begin{cases} (C^+)^r & \text{if } r > 0 \\ id & \text{if } r = 0 \\ (C^+)^r & \text{if } r < 0 \end{cases}$$

Taking into account proposition 6.1.34 we introduce three classes of indecomposable representations of a quiver $Q$ without oriented cycles.

**Definition 6.1.35.** Let $V$ be an indecomposable representation of a quiver $Q = (Q_0, Q_1)$.

1. $V$ is **preprojective** if $V \simeq C^r P(i)$ for some $i \in Q_0$ and some $r \leq 0$.
2. $V$ is **preinjective** if $V \simeq C^r I(i)$ for some $i \in Q_0$ and some $r \geq 0$.
3. $V$ is **regular** if $C^r V \neq 0$ for all $r \in \mathbf{Z}$.

As it turns out there are no other indecomposable representations for acyclic quivers. Indeed, from proposition 6.1.34 it follows that $V$ is a preprojective representation if and only if $C^r V = 0$ for some $r > 0$, and $V$ is a preinjective representation if and only if $C^r V = 0$ for some $r < 0$.

For the case of a quiver of Euclidian type these indecomposable representations can be characterized by the numerical invariant, called the defect.

If $Q$ is a quiver of Euclidian type with quadratic form $q(\mathbf{x})$ then by theorem 6.1.7 there is the unique positive and strict vector $\delta$ such that $\text{rad}(q) = \mathbf{Z}\delta$.

**Definition 6.1.36.** Let $Q$ be a quiver of Euclidian type, then the **defect** of a vector $\mathbf{x} \in \mathbf{Z}^n$ is

$$\partial \mathbf{x} = < \delta, \mathbf{x} > = - < \mathbf{x}, \delta > \tag{6.1.37}$$

where $<, >$ is a symmetric bilinear form defined by (6.1.9).

If $V$ is a representation of a quiver $Q$ which underlying graph is one of the Euclidian diagram with dimension vector $\underline{\dim}V$, then the **defect** of the representation $V$ is $\partial(V) = \partial(\underline{\dim}V)$.

**Proposition 6.1.38.** (V. Dlab, C.M. Ringel [54], see also [134, Proposition 5.2.1]). *Let Q be a quiver of Euclidian type, and V an indecomposable representation of Q. Then*

1. *V is preprojective if and only if* $\partial V < 0$.

2. *V is preinjective if and only if $\partial V > 0$.*
3. *V is regular if and only if $\partial V = 0$.*

The following theorem improves the previous result of the Gabriel theorem in the part 2.

**Theorem 6.1.39.** (V. Dlab, C.M. Ringel [54, Theorem]). *Let $Q = (Q_0, Q_1)$ be a quiver of Euclidean type. Then the mapping $V \mapsto \underline{\dim}V$ induces a bijection between the isomorphism classes of finite dimensional indecomposable preprojective or preinjective representations of $Q$ and the positive roots of quadratic form of $Q$ with non-zero defect. The indecomposable preprojective and preinjective representations form $2|Q_0|$ countably infinite series $C^{-r}P(i)$ and $C^r I(i)$, for $i \in Q_0$ and $r > 0$, of pairwise non-isomorphic representations.*

We now define some interesting class of algebras which are preprojective algebras of quivers. They play an important role in mathematics and were first introduced by I.M. Gelfand and V.A. Ponomarev [85].

Let $Q = (Q_0, Q_1, s, t)$ be a finite quiver. First associate with $Q$ the quiver $\widehat{Q}$ having the same vertices as $Q$ and for every arrow $i \xrightarrow{\alpha} j$ in $Q$ we add a new arrow $i \xleftarrow{\alpha^*} j$ in $\widehat{Q}$. So each arrow $i \xrightarrow{\alpha} j$ in $Q$ is replaced by a pair of arrows $i \underset{\alpha^*}{\overset{\alpha}{\rightleftarrows}} j$ in $\widehat{Q}$. If $Q$ has loops, then, by construction, the number of loops of $\widehat{Q}$ is twice the number of loops of $Q$.

Let $K\widehat{Q}$ be the path algebra associated to $\widehat{Q}$. Consider the element $r = \sum_{\alpha \in Q_1} (\alpha^*\alpha - \alpha\alpha^*)$ in $K\widehat{Q}$ and the ideal $(r)$ generated by $r$. The algebra $\Pi(Q) = K\widehat{Q}/(r)$ is called the **preprojective algebra of the quiver** $Q$. It is easy to see that $\Pi(Q)$ does not depend on the orientation on $Q$.

Note that a preprojective algebra $\Pi(Q)$ is not necessarily finite dimensional over $K$.

**Theorem 6.1.40** (I. Reiten, [174, Theorem 2.2]). *Let $Q$ be a finite quiver without loops, and let $\Pi(Q)$ be the associated preprojective algebra over a field $K$. Then*
1. *$\Pi(Q)$ is a finite dimensional algebra (or an Artinian ring) if and only if the underlying graph $\overline{Q}$ is a Dynkin diagram.*
2. *$\Pi(Q)$ is a Noetherian ring if and only if the underlying graph $\overline{Q}$ is a Dynkin or an extended Dynkin diagram.*

## 6.2 Species, Valued Quivers and Valued Graphs

In this section we present some preliminary information about valued quivers and species which can be considered as a generalization of quivers considered in the previous section. They play a key role in the theory of representations of associative algebras and rings. Historically species were first introduced by P. Gabriel in 1973.

Later V. Dlab and C.M. Ringel found their close connections with valued graphs and valued quivers.

We begin this section with an introduction to valued graphs which can be considered as some generalizations of ordinary graphs.

Let $\Gamma = (\Gamma_0, \Gamma_1)$ be a finite graph (without orientation) with set of vertices $\Gamma_0 = \{1, 2, \ldots, n\}$ and a finite set of edges $\Gamma_1$. A (relative) **valuation** $d$ of $\Gamma$ is a mapping $d : \Gamma_0 \times \Gamma_0 \to \mathbf{N} \times \mathbf{N}$, where $d(i, j) = (d_{ij}, d_{ji})$ and the $d_{ij}$ are required to satisfy the following conditions:

   (i) If $i \neq j$ then $d_{ij} \neq 0$ if and only if $\{i, j\} \in \Gamma_1$.
   (ii) There exists natural numbers $f_i \in \mathbf{N}$ such that $d_{ij} f_j = d_{ji} f_i$.
   (iii) $d_{ii} = 0$ for all $i \in \Gamma_0$.

A pair $(\Gamma, d)$ is called a (relative) **valued graph**[4].

If $d_{ij} > 1$ or $d_{ji} > 1$ then we depict this as

$$\bullet \overset{(d_{ij}, d_{ji})}{\underline{\qquad\qquad}} \bullet$$
$$i \qquad\qquad j$$

If $d_{ij} = d_{ji} = 1$ we simply write

$$\bullet \underline{\qquad\qquad} \bullet$$
$$i \qquad\qquad j$$

Note that condition (ii) says that the matrix $\mathbf{D} = [d_{ij}]$, is symmetrizable[5]. Indeed, if $\mathbf{F} = \mathrm{diag}(f_1, f_2, \ldots, f_n)$ is the diagonal matrix defined by the $f_i$ then $\mathbf{DF}$ is a symmetric matrix.

If $(\Gamma, d)$ is a (relative) valued graph and $\mathbf{x}, \mathbf{y} \in \mathbf{Z}^n$ then a (not necessarily symmetric) bilinear form of $\Gamma$ is defined as follows[6]:

$$B(\mathbf{x}, \mathbf{y}) = \sum_{i \in \Gamma_0} f_i x_i y_i - \frac{1}{2} \sum_{i,j \in \Gamma_1} d_{ij} f_j x_i y_j, \tag{6.2.1}$$

where $\mathbf{x} = (x_1, x_2, \ldots, x_n), \mathbf{y} = (y_1, y_2, \ldots, y_n) \in \mathbf{Z}^n$.

The corresponding quadratic form is

$$q(\mathbf{x}) = \sum_{i \in \Gamma_0} f_i x_i^2 - \sum_{\{i,j\} \in \Gamma_1} d_{ij} f_j x_i x_j. \tag{6.2.2}$$

---

[4]We introduce bellow one more valuation of a graph, which will be called an absolute valuation following J. Lemay [142], and a corresponding graph as an absolute valued graph.

[5]An integer matrix $\mathbf{C}$ is said to be **symmetrizable** if there exists a diagonal matrix $\mathbf{F}$ with positive entries in its diagonal such that $\mathbf{CF}$ is symmetric.

[6]Here it is customary to take the $f_i$ minimal, i.e. so that their greatest common divisor is 1.

These forms were independently introduced by V. Dlab, C.M. Ringel [176], [52] and S.A. Ovsienko, A.V. Roiter [172]. Therefore the quadratic form (6.2.2) is often called the **Dlab-Ringel-Ovsienko-Roiter quadratic form** of $(\Gamma, d)$.

The following theorem gives a characterization of all valued graphs from the point of view their quadratic forms:

**Theorem 6.2.3.** (V. Dlab, C.M. Ringel [52, Proposition 1.2]). *Suppose* $(\Gamma, d)$ *is a connected (relative) valued graph with quadratic form* $q$. *Then*

1. $(\Gamma, d)$ *is a Dynkin diagram if and only if its quadratic form* $q$ *is positive definite, i.e. a (relative) valued graph is one of the forms*:

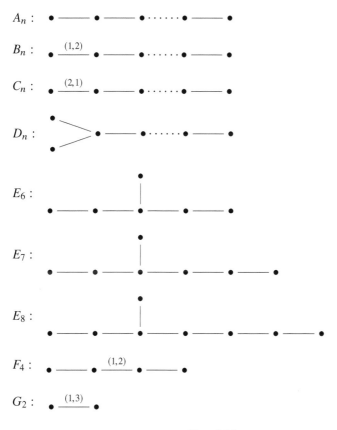

Diag. 6.2.4.

2. $(\Gamma, d)$ *is an extended Dynkin diagram if and only if* $q$ *is positive semi-definite, i.e. a (relative) valued graph is one of the forms* $\widetilde{A}_n$, $\widetilde{D}_n$, $\widetilde{E}_6$, $\widetilde{E}_7$, $\widetilde{E}_8$, *presented in Diag. 6.1.8 and*:

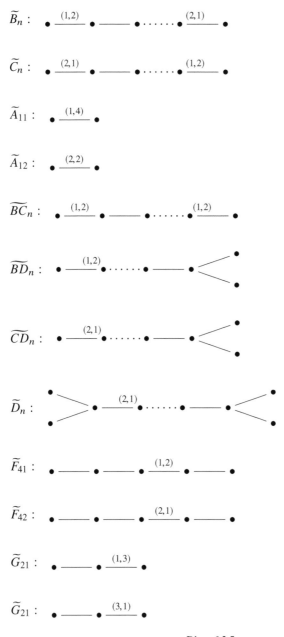

Diag. 6.2.5.

There is a natural bijection between (relative) valued graphs and symmetrizable generalized Cartan matrices. Recall that a **generalized Cartan matrix** is a square matrix $\mathbf{C} = (c_{ij})$ with integer entries satisfying the conditions:

1. $c_{ij} \leq 0$ for all $i \neq j$.
2. $c_{ij} = 0$ if and only if $c_{ji} = 0$.
3. $c_{ii} = 2$ for all $i$.

To a valued graph with edge labels $(d_{ij}, d_{ji})$ associate the generalized Cartan matrix $\mathbf{C} = (c_{ij})$ with $c_{ii} = 2$, $c_{ij} = -d_{ij}$ if there is an edge between $i$ and $j$, and $c_{ij} = 0$ otherwise. This matrix is symmetrizable because, by the definition of the notion of a (relative) valued graph, there are natural numbers $f_i$ such that $d_{ij} f_j = d_{ji} f_i$.

On the other hand, if $\mathbf{C} = (c_{ij})$ is a symmetrizable generalized Cartan matrices, then a corresponding diagram $\Gamma(\mathbf{C}) = (\Gamma_0, \Gamma_1)$ of $\mathbf{C}$ can be constructed. It has as a set of vertices $\Gamma_0 = \{1, 2, \ldots, n\}$ and a $\{i, j\} \in \Gamma_1$ if and only if $c_{ij} \neq 0$. A valuation on $\Gamma(\mathbf{C})$ is given by a set of numbers $d_{ij}$ with $d_{ii} = 0$ and $d_{ij} = -c_{ij}$ for $i \neq j$. Since $\mathbf{C}$ is a symmetrizable matrix there are positive integers $f_1, f_2, \ldots, f_n$ such that $f_i c_{ij} = f_j c_{ji}$ which shows that the diagram $\Gamma(\mathbf{C})$ thus obtained is a (relative) valued graph.

In this correspondence connectedness of the valued graph turns out to be the same as indecomposability of the generalized Cartan matrix.

Then the quadratic form (6.2.2) can be written in the following form:

$$q(\mathbf{x}) = \frac{1}{2} \mathbf{x} \mathbf{C} \mathbf{D} \mathbf{x}^T,$$

where $\mathbf{C} = (c_{ij})$ with

$$c_{ij} = \begin{cases} 2, & \text{if } i = j \\ -d_{ij}, & \text{otherwise} \end{cases}$$

is the symmetrizable generalized Cartan matrix of $(\Gamma, d)$, and $\mathbf{D} = \mathrm{diag}(f_1, f_2, \ldots, f_n)$ with positive integers $f_i$.

V. Dlab and C.M. Ringel [52] introduced the notions of the Weyl group, roots, positive and negative roots for (relative) valued graphs in a similar way as for quivers.

Given a (relative) valued graph $(\Gamma, d)$ denote by $\mathbf{Q}^\Gamma$ the vector space of all $\mathbf{x} = (x_i)_{i \in \Gamma_0}$ over the rational numbers. For each $k \in \Gamma_0$ denote by $\mathbf{e}_k \in \mathbf{Q}^\Gamma$ the vector with $x_k = 1$ and $x_i = 0$ otherwise. Also for each $k \in \Gamma_0$ define the **reflection** $s_k : \mathbf{Q}^\Gamma \to \mathbf{Q}^\Gamma$ by $s_k x = y$, where

$$y_k = -x_k + \sum_{i \in \Gamma_0} d_{ik} x_i$$

and $y_i = x_i$ for $i \neq k$.

The **Weyl group** $W = W_\Gamma$ of a (relative) valued graph $(\Gamma, d)$ is the group of all transformations of $\mathbf{Q}^\Gamma$ generated by reflections $s_k$, $k \in \Gamma_0$. A vector $\mathbf{x} \in \mathbf{Q}^\Gamma$ is called

**stable**[7] if $w\mathbf{x} = \mathbf{x}$ for all $w \in W$. A vector $\mathbf{x} \in \mathbf{Q}^\Gamma$ is called a **root** of $(\Gamma, d)$ if there exists $w \in W$ and $k \in \Gamma_0$ such that $\mathbf{x} = w\mathbf{e}_k$. The vector $e_k$ is called a **simple root**. A root $\mathbf{x}$ is called **positive** (or **negative**) if $x_i \geq 0$ (or $x_i \leq 0$) for all $i \in \Gamma_0$.

An **orientation** $\Omega$ of a (relative) valued graph $(\Gamma, d)$ consists of two function $s : \Gamma_1 \longrightarrow \Gamma_0$ and $t : \Gamma_1 \longrightarrow \Gamma_0$ given by $\sigma \mapsto s(\sigma)$ and $\sigma \mapsto t(\sigma)$, called the starting vertex and the ending vertex for any $\sigma \in \Gamma_1$. So that we prescribe for each edge $\{i, j\} \in (\Gamma, d)$ an ordering indicated by an oriented edge, that is

$$\text{either} \quad \overset{(d_{ij}, d_{ji})}{\underset{i \qquad j}{\bullet \longrightarrow \bullet}} \quad \text{or} \quad \overset{(d_{ij}, d_{ji})}{\underset{i \qquad j}{\bullet \longleftarrow \bullet}}$$

The triple $(\Gamma, d, \Omega)$ is called a (relative) **valued quiver**. In a similar way as for an ordinary quiver in any valued quiver we define a path connecting a vertex $i \in \Gamma_0$ with a vertex $j \in \Gamma_0$ as a sequence of arrows $\sigma_1, \sigma_2, \ldots \sigma_n$ such that $s(\sigma_1) = i$, $t(\sigma_n) = j$ and $t(\sigma_k) = s(\sigma_{k+1})$ for $1 < k < n$. In this case $n$ is called the **length** of the path. A path of length $> 0$ which connects some vertex $i \in \Gamma_0$ with itself is called an **oriented cycle** of a (relative) valued quiver $(\Gamma, d, \Omega)$. If $(\Gamma, d, \Omega)$ does not have oriented cycles, then $(\Gamma, d, \Omega)$ is called **acyclic**.

Given an orientation $\Omega$ and a vertex $k \in \Gamma_0$, define a new orientation $s_k\Omega$ of $(\Gamma, d)$ by reversing the direction of arrows along the all edges containing $k$. A vertex $k \in \Gamma_0$ is said to be a **sink** (or a **source**) with respect to $\Omega$ if $i \rightarrow k$ (or $k \rightarrow i$) for all neighbors $i \in \Gamma_0$ of $k$.

An orientation $\Omega$ of $(\Gamma, d)$ is said to be **admissible** if there is an ordering $k_1, k_2, \ldots, k_n$ of $\Gamma_0$ such that each vertex $k_t$ is a sink with respect to the orientation $s_{k_{t-1}} \ldots s_{k_2} s_{k_1} \Omega$ for all $1 \leq t \leq n$; such an ordering is called an **admissible ordering** for $\Omega$.

It is easy to see that an orientation $\Omega$ of a (relative) valued graph $(\Gamma, d)$ is admissible if and only if there is no circuit with orientation

$$i_1 \rightarrow i_2 \rightarrow i_3 \rightarrow \cdots \rightarrow i_{t-1} \rightarrow i_t = i_1$$

Therefore an orientation $\Omega$ of a (relative) valued graph $(\Gamma, d)$ is admissible if and only if the (relative) valued quiver $(\Gamma, d, \Omega)$ does not contain oriented cycles, i.e. it is acylic. In particular, every orientation of a tree is admissible.

Along with (relative) valued quiver as defined above there is another kind of valued quivers which are called (absolute) valued quivers following J. Lemay [142]. An (absolute) **valuation** of an acyclic quiver $\Gamma = (\Gamma_0, \Gamma_1, s, t)$ is given by two mappings $g : \Gamma_0 \longrightarrow \mathbf{N}$ given by $g(i) = g_i$ for all $i \in \Gamma_0$ and $m : \Gamma_1 \longrightarrow \mathbf{N}$, given by $m(\sigma) = m_\sigma$ for all $\sigma \in \Gamma_1$ such that $m_\sigma$ is a common multiple of $g_{s(\sigma)}$ and $g_{t(\sigma)}$. If $\sigma$ is an edge which connects the vertices $i$ and $j$ we write $m_\sigma = m_{ij}$. The triple $(\Gamma, g, m)$ is called an (absolute) **valued quiver**.

---

[7]Later, in the statement of theorem 6.2.19, the notion of a least stable positive integral vector is used. Stable means invariant under the Weyl group. So this is an appropriate place to define this notion.

An ordinary acyclic quiver $\Gamma = (\Gamma_0, \Gamma_1, s, t)$ can be considered as an (absolute) valued quiver with trivial valuation, i.e. $g_i = 1$ for all $i \in \Gamma_0$ and $m_\sigma = 1$ for all $\sigma \in \Gamma_1$.

Let us consider the connection between these two kinds of valued quivers, both of which are assumed to be acyclic. An (absolute) valued quiver $(\Gamma, g, m)$ can be considered as a (relative) valued quiver $(Q, d, \Omega)$ in the following way:

1. $(Q, \Omega) = \Gamma$
2. The valuation is given by

$$d_{s(\sigma),t(\sigma)} = \frac{m_\sigma}{f_{t(\sigma)}}, \quad \text{and} \quad d_{t(\sigma),s(\sigma)} = \frac{m_\sigma}{f_{s(\sigma)}} \tag{6.2.6}$$

for every $\sigma \in (Q, \Omega)$.

Conversely, if $(Q, d, \Omega)$ is a (relative) valued quiver then it may be considered as an (absolute) valued quiver $(\Gamma, g, m)$ by setting:

1. $\Gamma = (Q, \Omega)$.
2. $g_i = f_i$ for all $i \in \Gamma_0 = Q_0$.
3. $m_\sigma = d_{s(\sigma),t(\sigma)} f_{t(\sigma)} = d_{t(\sigma),s(\sigma)} f_{s(\sigma)}$.

If we restrict the sets of valued quivers assuming that $\gcd(g_i)_{i \in \Gamma_0} = 1$ and $\gcd(f_i)_{i \in Q_0} = 1$ then the sets of valued quivers $(\Gamma, g, m)$ and $(Q, d, \Omega)$ will be the same.

Like for (relative) valued quiver we can define the Euler form of an (absolute) valued acyclic quiver $(\Gamma, g, m)$ which in this case is given by

$$\langle \mathbf{x}, \mathbf{y} \rangle = \sum_{i \in \Gamma_0} g_i x_i y_i - \sum_{\sigma \in \Gamma_1} m_\sigma x_{t(\sigma)} y_{s(\sigma)}, \tag{6.2.7}$$

The corresponding quadratic form is

$$q(\mathbf{x}) = \sum_{i \in \Gamma_0} g_i x_i^2 - \sum_{\sigma \in \Gamma_1} m_\sigma x_{t(\sigma)} x_{s(\sigma)}. \tag{6.2.8}$$

Note that the quadratic form (6.2.8) does not depend on the orientation of the quiver $\Gamma$ and it is coincides with the Tits form 6.1.12 if the valuation $(g, m)$ of the valued quiver $(\Gamma, g, m)$ is trivial.

Now we consider the concepts of species which were first introduced by P. Gabriel [80] and their connection with valued graphs and valued quivers.

**Definition 6.2.9.** (P. Gabriel [80]). A **species** $\mathfrak{L} = (F_i, {}_iM_j)_{i,j \in I}$ is a finite family $(F_i)_{i \in I}$ of division rings together with a family $({}_iM_j)_{i,j \in I}$ of $(F_i, F_j)$-bimodules.

We say that $(F_i, {}_iM_j)_{i,j \in I}$ is a $K$-**species** if all $F_i$ are finite dimensional and central over a common commutative subfield $K$ which acts centrally on ${}_iM_j$, i.e. $\lambda m = m\lambda$

for all $\lambda \in K$ and all $m \in {}_iM_j$. We also assume that each bimodule ${}_iM_j$ is a finite dimensional vector space over $K$. It is a $K$-**quiver** if moreover $F_i = K$ for each $i$.

From any species $\mathfrak{L} = (F_i, {}_iM_j)_{i,j\in I}$ we derive an oriented graph $Q_{\mathfrak{L}}$ with arrows $i\bullet \longrightarrow \bullet j$ if and only if ${}_iM_j \neq 0$. We say that a species $\mathfrak{L}$ is **connected** if $Q_{\mathfrak{L}}$ is a connected graph. Analogously, a species $\mathfrak{L}$ is **acyclic** if and only if $Q_{\mathfrak{L}}$ is an acyclic graph. In particular, this means that if ${}_iM_j \neq 0$, then ${}_jM_i = 0$. A species $\mathfrak{L}$ is without loops if and only if $Q_{\mathfrak{L}}$ is without loops, which means that ${}_iM_i = 0$.

If $\mathfrak{L} = (F_i, {}_iM_j)_{i,j\in I}$ is a $K$-species, then we can associate with it an (absolute) valued quiver $(\Gamma_{\mathfrak{L}}, f, m)$ in the following way:
1. The set of vertices is the finite set $I$.
2. $f_i = \dim_K F_i$ and $m_{ij} = \dim_K ({}_iM_j)$.

The corresponding quadratic form of this valued quiver $(\Gamma_{\mathfrak{L}}, f, m)$ is defined on the rational vector space $\mathbf{Q}^{(I)}$ as follows:

$$q(\mathbf{x}) = \sum_{i \in I} f_i x_i^2 - \sum_{i,j \in I} m_{ij} x_i x_j. \qquad (6.2.10)$$

Note that the quadratic form (6.2.10) is defined up to a positive rational multiple.

**Definition 6.2.11.** A **representation** $(V_i, {}_j\varphi_i)$ of a species $\mathfrak{L} = (F_i, {}_iM_j)_{i,j\in I}$ (or an $\mathfrak{L}$-**representation**) is a family of right $F_i$-modules $V_i$ and $F_j$-linear mappings:

$${}_j\varphi_i : V_i \otimes_{F_i} {}_iM_j \longrightarrow V_j$$

for each $i, j \in I$. Such a representation is called **finite dimensional** provided all the $V_i$ are finite dimensional $F_i$-vector spaces.

Let $V = (V_i, {}_j\varphi_i)$ and $W = (W_i, {}_j\psi_i)$ be two $\mathfrak{L}$-representations. An $\mathfrak{L}$-morphism $\Psi : V \to W$ is a set of $F_i$-linear maps $\alpha_i : V_i \to W_i$ such that the following diagram commutes

$$
\begin{array}{ccc}
V_i \otimes_{F_i} {}_iM_j & \xrightarrow{\ {}_j\varphi_i\ } & V_j \\
{\scriptstyle \alpha_i \otimes 1}\downarrow & & \downarrow{\scriptstyle \alpha_j} \\
W_i \otimes_{F_i} {}_iM_j & \xrightarrow{\ {}_j\psi_i\ } & W_j
\end{array}
\qquad (6.2.12)
$$

Two representations $(V_i, {}_j\varphi_i)$ and $(W_i, {}_j\psi_i)$ are called **equivalent** if there is a set of isomorphisms $\alpha_i$ from the $F_i$-module $V_i$ to the $F_i$-module $W_i$ such that the diagram (6.2.12) is commutative for each $i \in I$.

A representation $(V_i, {}_j\varphi_i)$ is called **indecomposable**, if there are no non-zero sets of subspaces $(V_i')$ and $(V_i'')$ such that $V_i = V_i' \oplus V_i''$ and ${}_j\varphi_i = {}_j\varphi_i' \oplus {}_j\varphi_i''$, where

$${}_j\varphi_i' : V_i' \otimes_{F_i} {}_iM_j \longrightarrow V_j'$$
$${}_j\varphi_i'' : V_i'' \otimes_{F_i} {}_iM_j \longrightarrow V_j''.$$

One defines the direct sum of two $\mathfrak{L}$-representations in the obvious way.

Denote by Rep $\mathfrak{L}$ the category of all $\mathfrak{L}$-representations, and by rep $\mathfrak{L}$ the category of finite dimensional $\mathfrak{L}$-representations, whose objects are $\mathfrak{L}$-representations and whose morphisms are as defined above.

With any species $\mathfrak{L} = (F_i, {}_iM_j)_{i,j\in I}$ one can associate a tensor algebra in the following way. Let $B = \prod_{i\in I} F_i$, and let $M = \underset{i,j\in I}{\oplus} \ {}_iM_j$. Then $B$ is a ring and $M$ naturally becomes a $(B, B)$-bimodule. The **tensor algebra** of the $(B, B)$-bimodule $M$ is the graded ring

$$\mathfrak{T}(\mathfrak{L}) = \mathfrak{T}_B(M) = \bigoplus_{n=0}^{\infty} M^{\otimes n} \tag{6.2.13}$$

where $M^{\otimes 0} = B$, $M^{\otimes n} = M^{\otimes(n-1)} \otimes_B M$ for $n > 0$ with component-wise addition and the multiplication induced by taking tensor products.

If $\mathfrak{L}$ is a $K$-species then $\mathfrak{T}(\mathfrak{L})$ is a finite dimensional $K$-algebra.

**Theorem 6.2.14.** (V.Dlab, C.M. Ringel [53, Proposition 10.1]). *Let $\mathfrak{L}$ be a $K$-species. Then the category* Rep $\mathfrak{L}$ *of all representations of $\mathfrak{L}$ and the category* $\mathrm{Mod}_r \ \mathfrak{T}(\mathfrak{L})$ *of all right $\mathfrak{T}(\mathfrak{L})$-modules are equivalent.*

Given a species $\mathfrak{L} = (F_i, {}_iM_j)_{i,j\in I}$ with $I = \{1, 2, \dots, n\}$ consider the mapping

$$\underline{\dim} : \mathrm{Rep}\ \mathfrak{L} \longrightarrow \mathbf{Q}^{(I)}$$

defined by $\underline{\dim}V = (d_1, d_2, \dots, d_n)$ where $d_i = \dim(V_i)_{F_i}$ is the dimension of $V_i$ as a vector space over $F_i$. The vector $\underline{\dim}V$ is called the **dimension vector** of the representation $V$. Set $d_0 = \sum_{i\in I} \dim(V_i)_{F_i}$. The representation $V = (V_i, {}_j\varphi_i)$ is finite dimensional if $d_0 < \infty$.

**Definition 6.2.15.** ([53].) A species $\mathfrak{L} = (F_i, {}_iM_j)_{i,j\in I}$ is called of **finite type**, if the number of indecomposable non-isomorphic finite dimensional representations is finite.

A species $\mathfrak{L} = (F_i, {}_iM_j)_{i,j\in I}$ is said to be of **strongly unbounded type** if it possesses the following three properties:

(i) $\mathfrak{L}$ has indecomposable objects of arbitrary large finite dimension.

(ii) If $\mathfrak{L}$ contains a finite dimensional object with an infinite endomorphism ring, then there is an infinite number of (finite) dimensions $d$ such that, for each $d$, the species $\mathfrak{L}$ has infinitely many (non-isomorphic) indecomposable objects of dimension $d$.

(iii) $\mathfrak{L}$ has indecomposable objects of infinite dimension.

Following V. Dlab and C.M. Ringel [52] to a species $\mathfrak{L} = (F_i, {}_iM_j)_{i,j\in I}$ one can associate a **diagram** in the following way.

1. The set of vertices is the finite set $I$.
2. The vertex $i$ is connected with the vertex $j$ by $t_{ij}$ edges where

$$t_{ij} = \dim_{K_i}({}_iM_j) \times \dim({}_iM_j)_{K_j} + \dim_{K_k}({}_jM_i) \times \dim({}_jM_i)_{K_i}.$$

If ${}_iM_j \neq 0$ and $\dim_{K_i}({}_iM_j) > \dim({}_iM_j)_{K_j}$, then we denote this by the following arrow:

$$i \Longrightarrow j$$

Note that by convention and historical continuity the multiple edges on Dynkin graphs are depicted in a special case and their number does not always equal the number of edges in the diagrams. For more see in Appendix to section 2.5 in [101].

In the case when all $F_i = F$, where $F$ is a fixed division ring, and ${}_F({}_iM_j)_F = ({}_F F_F)^{t_{ij}}$, P. Gabriel has characterized $K$-species of finite type (see [79]). This result was extended by V. Dlab and C.M. Ringel [53] to the case where $\mathfrak{L}$ is an arbitrary $K$-species.

**Theorem 6.2.16.** (V. Dlab, C.M. Ringel [53, Theorem B, Theorem E]).
   a. *A $K$-species is of finite type if and only if its diagram is a finite disjoint union of Dynkin diagrams $A_n$, $B_n$, $C_n$, $D_n$, $E_6$, $E_7$, $E_8$, $F_4$, $G_2$ as depicted in Diag. 6.1.6.*
   b. *A $K$-species is either of finite or of strongly unbounded type.*

Given a Dynkin diagram it is easy to construct a corresponding $K$-species. P. Gabriel in [80] has shown that the numbers of indecomposable representations of the $K$-species of type $A_n$, $D_n$, $E_6$, $E_7$ and $E_8$ are, respectively, $\frac{1}{2}n(n+1)$, $n(n-1)$, 36, 63 and 120. V. Dlab and C.M. Ringel in [53] have shown that there are $n^2$ indecomposable representations of the $K$-species of type $B_n$ or $C_n$, whereas the numbers of indecomposable representations of the $K$-species of type $F_4$ and $G_2$ are 24 and 6, respectively. Thus, the number of indecomposable representations of the $K$-species of a given type coincides with the number of positive roots of the corresponding quadratic form.

With any (relative) valued graph one can associate its so-called modulations in the form of some species of a special form.

Given a (relative) valued quiver $(\Gamma, d, \Omega)$ define a special species $\mathfrak{L} = (F_i, M_\sigma)_{i \in \Gamma_0, \sigma \in \Gamma_1}$, called a **modulation** of $(\Gamma, d, \Omega)$, in the following way.
   1. To any vertex $i \in \Gamma_0$ corresponds a division ring $F_i$ and to any arrow $\sigma \in \Gamma_1$ corresponds $M_\sigma \neq 0$; (If $s(\sigma) = i$ and $t(\sigma) = j$ then we write $M_\sigma = {}_iM_j$).
   2. For any $\sigma \in \Gamma_1$ with $s(\sigma) = i$ and $t(\sigma) = j$ there is an $F_j$-$F_i$-bimodule isomorphism

$$\mathrm{Hom}_{F_i}({}_iM_j, F_i) \cong \mathrm{Hom}_{F_j}({}_iM_j, F_j)$$

3. For any $\sigma \in \Gamma_1$ with $s(\sigma) = i$ and $t(\sigma) = j$ there hold $\dim_{F_j}({}_iM_j) = d_{ij}^\sigma$ and $\dim_{F_i}({}_iM_j) = d_{ji}^\sigma$.

In the same way as for an arbitrary species one can define the notion of a representation of $(\mathfrak{L}, \Omega)$. The representations of $(\mathfrak{L}, \Omega)$ form an Abelian category which is denoted $\mathrm{Rep}(\mathfrak{L}, \Omega)$, and $\mathrm{rep}(\mathfrak{L}, \Omega)$ denotes the category of finite dimensional representations of $(\mathfrak{L}, \Omega)$.

An Abelian category is said to be of **finite type** if there are up to isomorphism only finitely many indecomposable objects.

**Remark 6.2.17.** Note that for arbitrary additive category there is the notion of finite representation type which was introduced by M. Auslander in [14]. Let $\mathcal{B}$ be an additive category closed under direct summands. Then $\mathcal{B}$ is said to be of **finite representation type** if there is an object $M$ in $\mathcal{B}$ such that every object in $\mathcal{B}$ is a direct summands of $M^{\oplus n}$ for some $n$. Such an object is called a **representation generator** for $\mathcal{B}$.

The notions of finite type and finitely representation type does not coincide for arbitrary Abelian category. But if $\mathcal{B}$ is a Krull-Schmidt category[8], then these notions are the same.

**Definition 6.2.18.** An indecomposable representation $V$ of a species (or a quiver) is of **discrete dimension type** if it is the unique indecomposable representation (up to isomorphism) with graded dimension vector $\underline{\dim}V$. Otherwise, it is of **continuous dimension type**.

The indecomposable representations of discrete dimension type are determined by their dimension vectors: these are precisely the positive roots of the corresponding quadratic form $q(\mathbf{x})$. The indecomposable representations of continuous dimension type are determined by their continuous dimension vectors which are in this case the positive integral vectors in the radical space $\mathrm{rad}(q)$, and thus they are the positive multiplies of a fixed dimension vector.

V. Dlab and C.M. Ringel proved the following theorem:

**Theorem 6.2.19** (V. Dlab, C.M. Ringel, [54], [52].) *Let $(\mathfrak{L}, \Omega)$ be a modulation of a (relative) valued acyclic quiver $(\Gamma, d, \Omega)$.*

1. *Then $\mathrm{rep}(\mathfrak{L}, \Omega)$ is of finite type if and only if $(\Gamma, d)$ is a Dynkin diagram, i.e. a valued graph of one of the forms $A_n$, $B_n$, $C_n$, $D_n$, $E_6$, $E_7$, $E_8$, $F_4$, $G_2$ presented in Diag. 6.1.6. Moreover, the mapping $\underline{\dim} : \mathrm{rep}(\mathfrak{L}, \Omega) \to \mathbf{Q}^\Gamma$ provides a one-to-one correspondence between all indecomposable representations in $\mathrm{rep}(\mathfrak{L}, \Omega)$ and all positive roots of $(\Gamma, d)$.*

2. *If $(\Gamma, d)$ is an extended Dynkin diagram, i.e. a valued graph of one of the forms $\widetilde{A}_{11}$, $\widetilde{A}_{12}$, $\widetilde{A}_n$, $\widetilde{B}_n$, $\widetilde{C}_n$, $\widetilde{BC}_n$, $\widetilde{CD}_n$, $\widetilde{BD}_n$, $\widetilde{D}_n$, $\widetilde{E}_6$, $\widetilde{E}_7$, $\widetilde{E}_8$, $\widetilde{F}_{41}$, $\widetilde{F}_{42}$,*

---

[8]Recall that a category $\mathcal{B}$ is a **Krull-Schmidt category** if every object of it decomposes into a finite direct sum of objects having local endomorphism rings.

$\widetilde{G}_{21}$, $\widetilde{G}_{22}$, *presented in Diag.* 6.1.8 *and Diag.* 6.2.5. *then the mapping* $\underline{\dim}$ : rep$(\mathfrak{L}, \Omega) \rightarrow \mathbf{Q}^\Gamma$ *provides a one-to-one correspondence between all indecomposable representations in* rep$(\mathfrak{L}, \Omega)$ *of non-zero defect and all positive roots of* $(\Gamma, d)$ *of non-zero defect. Moreover, the category* rep$(\mathfrak{L}, \Omega)$ *has two kinds of indecomposable representations: those of discrete dimension types and those of continuous dimension types. The continuous dimension types are the positive integral multiplies of the least stable positive integral vector of* $\mathbf{Q}^\Gamma$. *Moreover, the indecomposable representations of continuous dimension types can be derived from the indecomposable representation of continuous dimension type of a suitable modulation of the graphs* $\widetilde{A}_{11}$ *or* $\widetilde{A}_{12}$.

Some other generalization of this theorem was obtained by P. Dowbor, C.M. Ringel and D. Simson in [59]. They introduced a new kind of valued quivers and described species of finite representation type whose corresponding quiver edges are labeled by the lengths of the dimension sequences of the bimodules belonging to these edges. To this end they introduced dimension sequences which are valuations for new valued quivers and graphs.

**Definition 6.2.20.** A finite sequence $(d_1, d_2, \ldots, d_m)$ of length $m \geq 2$ of non-negative integers is called a **dimension sequence** if there exist two sequences $(x_0, x_1, \ldots, x_m, x_{m+1})$ and $(y_0, y_1, \ldots, y_m, y_{m+1})$ of integers such that the following properties hold:
1. $x_0 = y_1 = x_m = y_{m+1} = 0$, $y_0 = x_{m+1} = -1$, $x_1 = y_m = 1$.
2. $x_i > 0$, $y_i > 0$, for all $i = 1, 2, \ldots, m$.
3. $d_i x_i = x_{i-1} + x_{i+1}$, $d_i y_y = y_{i-1} + y_{i+1}$ for all $i = 1, 2, \ldots, m$.

**Definition 6.2.21.** The set of vectors $\{v_1, v_2, \ldots, v_m\}$, where $v_i = \begin{pmatrix} x_i \\ y_i \end{pmatrix}$, is called the **branch system** of a dimension sequence $(d_1, d_2, \ldots, d_m)$.

It can be shown that all finite dimension sequences are obtained as follows:
1. $(0, 0)$ is a dimension sequence.
2. If $(d_1, d_2, \ldots, d_m)$ is a dimension sequence then so is
$(d_1, d_2, \ldots, d_{i-1}d_i + 1, 1, d_{i+1} + 1, d_{i+2}, \ldots d_m)$.

Examples of dimension sequences are: (0,0), (1,1,1), (1,2,1,2), (1,2,2,1,3), (1,2,2,2,2,1,5).

Let $\mathfrak{D}$ be the set of all dimension sequences.

**Definition 6.2.22.** Let $_F M_G$ be a bimodule over two skew fields $F$ and $G$. Then $M^L = {}_G(\mathrm{Hom}_F(M, F))_F$ is called the **left dual** of $M$ and $M^R = {}_G(\mathrm{Hom}_G(M, G))_F$ is called the **right dual** of $M$.

If $_F M_G$ is finite dimensional over $F$ then $M \simeq M^{LR}$ as bimodules. Dually, if $_F M_G$ is finite dimensional over $G$ then $M \simeq M^{RL}$.

**Definition 6.2.23.** Let $_F M_G$ be a bimodule over two skew fields $F$ and $G$. Suppose that there exists $n$ such that $M^{nR} \simeq M$ as bimodules and

$$(\dim_G M, \dim_F M^R, \dim_G M^{RR}, \ldots, \dim_K M^{(n-1)R}) \tag{6.2.24}$$

(here $K = F$ if $n$ is even and $K = G$ otherwise) is a dimension sequence. Then the sequence (6.2.24) is called the **dimension sequence** of the bimodule $_F M_G$.

Note that this notion was introduced by P. Dowbor, C.M. Ringel, D. Simson when considering the Artinian ring

$$A = \begin{pmatrix} F & _F M_G \\ 0 & G \end{pmatrix}$$

of finite representation type. They proved in [59] that this ring is of finite representation type if and only if there is $n$ such that $M^{nR} \simeq M$ and the sequence (6.2.24) is a dimension sequence.

From any species $\mathfrak{L} = (F_i, {}_i M_j)_{i,j \in I}$ we derive a finite quiver $\Gamma_{\mathfrak{L}}$ with set of vertices $I$ and arrows $i\bullet \longrightarrow \bullet j$ if and only if ${}_i M_j \neq 0$. Let $\mathfrak{D}_{i,j}$ be the set of all dimension sequences of a bimodule ${}_i M_j$, and $\mathfrak{D} = \bigcup_{i,j} \mathfrak{D}_{i,j}$. We define a valuation $d$ on $\Gamma_{\mathfrak{L}}$ as a function $d : \Gamma_{\mathfrak{L}} \times \Gamma_{\mathfrak{L}} \to \mathfrak{D}$ with $d(i, j) \in \mathfrak{D}_{i,j} \setminus \{(0,0)\}$ if there an arrow $i\bullet \longrightarrow \bullet j$ in $\Gamma_{\mathfrak{L}}$ and $d(i, j) = (0, 0)$ otherwise. The thus obtained valued quiver $(\Gamma_{\mathfrak{L}}, d)$ is called the **valued quiver** of the species $\mathfrak{L}$. Along with this valued quiver we will consider the valued graph $(\Gamma_{\mathfrak{L}}, |d|)$ which is the underlying graph of $\Gamma_{\mathfrak{L}}$ and a valuation $|d| : \Gamma_{\mathfrak{L}} \times \Gamma_{\mathfrak{L}} \to \mathbf{N} \cup \{0\}$ is defined as $|d|(i, j) = |d(i, j)|$ the length of a dimension sequence $d(i, j)$. $(\Gamma_{\mathfrak{L}}, |d|)$ is called the **valued graph** of the species $\mathfrak{L}$.

**Theorem 6.2.25.** (P. Dowbor, C.M. Ringel, D. Simson [59, Theorem 1], S. Oppermann [171, Theorem 8.10]). *A connected species* $\mathfrak{L} = (F_i, {}_i M_j)_{i,j \in I}$ *is of finite representation type if and only if its corresponding valued graph* $(\Gamma_{\mathfrak{L}}, |d|)$, *where an edge is labeled by the length of the dimension sequence of the bimodule belonging to the edge, is a Coxeter diagram:*

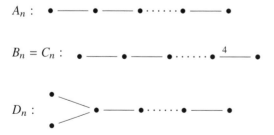

$E_6$ :

$E_7$ :

$E_8$ :

$F_4$ :   $\bullet \longrightarrow \bullet \overset{4}{\longrightarrow} \bullet \longrightarrow \bullet$

$G_2$ :   $\bullet \overset{6}{\longrightarrow} \bullet$

$H_3$ :   $\bullet \longrightarrow \bullet \overset{5}{\longrightarrow} \bullet$

$H_4$ :   $\bullet \longrightarrow \bullet \longrightarrow \bullet \overset{5}{\longrightarrow} \bullet$

$I_n$ :   $\bullet \overset{n}{\longrightarrow} \bullet$ ,   $n = 5, n \geq 7$.

**Diag. 6.2.26.**

*Here instead of* $\bullet \overset{3}{\longrightarrow} \bullet$ *in a Coxeter diagram we write* $\bullet \longrightarrow \bullet$ .

**Remark 6.2.27.** This theorem was stated in [59], where it is remarked that the proof is "rather technical" and details are left out. A full proof of this theorem is given by S. Oppermann in [171].

## 6.3 Finite Dimensional Algebras and Their Representations

There is a close connection between representations of quivers, species and representations of finite dimensional associative algebras.

Let $K$ be a base field, and let $A$ be an associative finite dimensional $K$-algebra with $1 \neq 0$.

**Definition 6.3.1.** A **representation** of a $K$-algebra $A$ is an algebra homomorphism $T : A \rightarrow \mathrm{End}_K(V)$, where $V$ is a $K$-vector space.

In other words, to define a representation $T$ is to assign to every element $a \in A$ a linear operator $T(a)$ in such a way that

1. $T(a + b) = T(a) + T(b)$.
2. $T(\alpha a) = \alpha T(a)$.
3. $T(ab) = T(a)T(b)$.
4. $T(1) = id$ (the identity operator)

for arbitrary $a, b \in A$, $\alpha \in K$.

The action of the operators $T(a)$ on $V$ will be written on the right, i.e. $T(a) : V \rightarrow V$, $v \mapsto vT(a)$.

If the vector space $V$ is finite dimensional over $K$, then its dimension is called the **dimension** (or **degree**) of the representation $T$.

Obviously, the image of a representation $T$ forms a subalgebra in $\mathrm{End}_K(V)$. If $T$ is a monomorphism, then this subalgebra is isomorphic to the algebra $A$. In this case the representation $T$ is called **faithful**.

Let $T : A \rightarrow \mathrm{End}_K(V)$ and $S : A \rightarrow \mathrm{End}_K(W)$ be two representations of a $K$-algebra $A$. A **morphism** from the representation $T$ to the representation $S$ is a $K$-linear map $\varphi : V \rightarrow W$ such that the diagram

$$
\begin{array}{ccc}
V & \xrightarrow{\varphi} & W \\
\downarrow{\scriptstyle T(x)} & & \downarrow{\scriptstyle S(x)} \\
V & \xrightarrow{\varphi} & W
\end{array}
$$

is commutative for all $x \in A$, that is, $\varphi T(x) = S(x)\varphi$, for all $x \in A$.

If $\varphi$ is an invertible morphism, then it is called an **isomorphism** of representations. Two representations $T$ and $S$ are called **isomorphic** if there is an isomorphism $\varphi$ from the representation $T$ to the representation $S$, and in this case we have

$$S(x) = \varphi T(x)\varphi^{-1}. \tag{6.3.2}$$

There is also a close connection between the representations of an algebra $A$ and its modules. For any representation of $A$ one can construct a right module over this algebra, and, vice versa, for any right module one can construct a representation of $A$. It is easy to show that the category of all representations of $A$ is equivalent to the category of all right $A$-modules. The formulas are rather obvious. If $T$ is a representation on $V$, then $V$ becomes a right $A$-module by $va = vT(a)$ and the other way around.

A representation of $A$ is said to be **simple** (or **irreducible**) provided it is non-zero and the only proper subrepresentation is the zero representation.

A representation $T$ of a $K$-algebra $A$ is said to be **indecomposable** if its corresponding right $A$-module is indecomposable. In other words, a representation $T$ is indecomposable if it cannot be written as a direct sum of non-zero representations. Otherwise it is called **decomposable**. A simple module is obviously indecomposable. But an arbitrary indecomposable module may have proper submodules.

A representation $T$ of a $K$-algebra $A$ is said to be **completely reducible** if it is a direct sum of irreducible representations.

Taking into account that any f.d. algebra is an Artinian and Noetherian ring, we obtain that any finitely generated module over a f.d. algebra can be decomposed into a direct sum of indecomposable modules by [103, theorem 9.2.3]. By Fitting lemma for a f.d. algebra[9] a module over a finite dimensional algebra is indecomposable if and only if its endomorphism algebra is local. So that from the Krull-Remak-Schmidt theorem[10] we obtain that any finite dimensional module over a f.d. algebra can be uniquely written (up to isomorphism) in the form of a direct sum of indecomposable modules. This means that for many questions one can restrict attention to the consideration of indecomposable modules.

With any finite dimensional algebra $A$ over a field $K$ one can associate the so-called **Gabriel quiver** $Q(A)$ in the following way. We can restrict ourselves to basic algebras[11], since for every f.d. algebra $A$ there is a (unique up to isomorphism) basic algebra $B$ that is Morita equivalent to $A$. This follows from [100, Theorem 10.7.2 10.7.4]. Indeed, $A_A \simeq P_1^{n_1} \oplus \cdots \oplus P_k^{n_k}$, where $P_i$ are indecomposable projective right $A$-modules and $P_i \neq P_j$ for $i \neq j$. Then $P = P_1 \oplus \cdots \oplus P_k$ is a progenerator for $\mathrm{Mod}_r A$ and so $A$ is Morita equivalent to a ring $\mathrm{End}_A(P) = B$ which is a basic algebra.

Let $1 = e_1 + e_2 + \cdots + e_n$ be a decomposition of the identity of $A$ into a sum of pairwise orthogonal primitive idempotents, and let $P_1, P_2, \ldots, P_n$ be the pairwise non-isomorphic principal (i.e. projective indecomposable) right $A$-modules. Here $P_i = e_i A$ because $A$ is basic. Denote by $R_i = P_i R$ the Jacobson radical of $P_i$. Note that, by [103, theorem 1.8.3], there is a one-to-one correspondence between simple $A$-modules $S_i$ and principal $A$-modules which is given by the following

---

[9]See e.g. [100, Proposition 10.1.5].

[10]See e.g. theorem 1.9.5 for the case of semiperfect rings.

[11]An algebra $A$ with a complete set $\{e_1, e_2, \ldots, e_n\}$ of pairwise orthogonal primitive idempotents is called **basic** if $e_i A \not\cong e_j A$ for all $i \neq j$, which is equivalent to saying that $A/R$ is isomorphic to a product of division algebras, where $R$ is the Jacobson radical of $A$.

correspondences $P_i \mapsto P_i/R_i = S_i$ and $S_i \mapsto P(S_i)$, where $P(S_i)$ is a projective cover of $S_i$. To each simple module $S_i$ (or principal module $P_i$, which amount to the same) assign a vertex $i$ in the plane, and join the vertex $i$ with the vertex $j$ by $t_{ij}$ arrows, where

$$t_{ij} = \dim_K \operatorname{Ext}^1(S_i, S_j) \tag{6.3.3}$$

The so constructed oriented graph is called the $\operatorname{Ext}^1$-**quiver** (or the Gabriel quiver) of the algebra $A$, and denoted by $Q(A)$.

There is another equivalent definition of the Gabriel quiver of a basic algebra $A$. Write $V_i = P_iR/P_iR^2 = R_i/R_i^2$. Since $V_i$ is a semisimple module and each simple module over a basic finitely dimensional algebra $A$ is one-dimensional, $V_i \cong \bigoplus_{k=1}^{s} S_k^{m_{ik}}$. Therefore there is an isomorphism $P(R_i) \cong \bigoplus_{k=1}^{s} P_k^{m_{ik}}$ where $P(R_i)$ is a projective cover of $R_i$. Taking into account this isomorphism we obtain a projective resolution of $S_i$ in the form

$$\cdots \to \bigoplus_{k=1}^{s} P_k^{m_{ik}} \to P_i \to S_i \to 0,$$

which is a composition of two exact sequences:

$$0 \to R_i \to P_i \to S_i \to 0$$

$$0 \to \operatorname{Ker}(g_i) \to P(R_i) \xrightarrow{g_i} R_i \to 0.$$

So that

$$\operatorname{Ext}^1_A(S_i, S_j) \cong \operatorname{Hom}_A(\bigoplus_{k=1}^{s} P_k^{m_{ik}}, S_j).$$

Therefore

$$t_{ij} = \dim_K \operatorname{Ext}^1_A(S_i, S_j) = \dim_K \operatorname{Hom}_A(\bigoplus_{k=1}^{s} P_k^{m_{ik}}, S_j)$$

$$= \dim_K \bigoplus_{k=1}^{s} \operatorname{Hom}_A(P_k^{m_{ik}}, S_j)$$

$$= \bigoplus_{k=1}^{s} m_{ik} \dim_K (P_k, S_j)$$

$$= \bigoplus_{k=1}^{s} m_{ik} \dim_K S_j e_k$$

$$= m_{ij},$$

since $S_i$ is a one-dimensional $K$-vector space. Thus, the number of arrows from the vertex $i$ to the vertex $j$ in the Gabriel quiver of a basic finite dimensional algebra $A$ is equal to the number of principal modules $P_j$ in the projective cover $P(R_i) = P(P_iR)$.

Note that in the case of a path algebra $A = KQ$ of some quiver $Q$ without oriented cycles, the $\text{Ext}^1$-quiver allows us to recover the original quiver in an invariant way.

**Examples 6.3.4.**

1. Let $A = \begin{pmatrix} K & K \\ 0 & K \end{pmatrix}$ with Jacobson radical $R = \begin{pmatrix} 0 & K \\ 0 & 0 \end{pmatrix}$. Then $1 = e_1 + e_2$, where $e_1 = \begin{pmatrix} 1 & 0 \\ 0 & 0 \end{pmatrix}$ and $e_2 = \begin{pmatrix} 0 & 0 \\ 0 & 1 \end{pmatrix}$. $A$ has two principal modules $P_1 = e_1 A = \begin{pmatrix} K & K \\ 0 & 0 \end{pmatrix}$, $P_2 = e_2 A = \begin{pmatrix} 0 & 0 \\ 0 & K \end{pmatrix}$, and two simple modules $S_1 = P_1/P_1 R = \begin{pmatrix} 0 & K \\ 0 & 0 \end{pmatrix}$ and $S_2 = P_2/P_2 R = P_2$. We have $R_1 = P_1 R \cong S_2$, $R_1^2 = 0$, and $R_2 = P_2 R = 0$. So $V_1 \cong S_2$ and $V_2 = 0$. Therefore the Gabriel quiver is:

$$
\overset{1}{\bullet} \longrightarrow \overset{2}{\bullet}
\tag{6.3.5}
$$

2. Let $A = K[x, y]/(x, y)^2$. Then $A$ is a local ring with Jacobson radical $R = (x, y)/(x, y)^2$. $A$ has a unique simple module $S$. Since $\dim_K A = 3$, $\dim_K R = 2$, $\dim_K S = 1$, $R \cong S \oplus S$. Therefore the Gabriel quiver of $A$ has the following form:

$$
\overset{\curvearrowleft\,\bullet\,\curvearrowright}{}
\tag{6.3.6}
$$

As was shown in section 6.1 for any finite quiver $Q$ without oriented cycles there is the path algebra $KQ$ which is finite dimensional over the field $K$. The representations of arbitrary finite dimensional algebras can be described in terms of bound quivers. We will show more precisely that the category of representations of the algebra $A$ is equivalent to the category of representations of the algebra $KQ/I$ for some quiver $Q$ and some two-sided admissible[12] ideal $I$ of $KQ$. See also theorem 6.1.29.

**Lemma 6.3.7.** *Let $Q$ be a finite connected quiver, $I$ an admissible ideal of $KQ$, and $A = KQ/I$. Then $Q$ is the Gabriel quiver of $A$.*

*Proof.* Suppose that $I$ is an admissible ideal of $KQ$, i.e. there is an integer $m$ such that $J^m \subseteq I \subseteq J^2$. Let $A = KQ/I$ with Gabriel quiver $Q_A$. Assume that $Q_0 = \{1, 2, \ldots, n\}$, and $e_1, e_2, \ldots, e_n$ is a complete set of primitive pairwise orthogonal idempotents of $KQ$. Then it is easy to show that $\{e_i + I : i \in Q_0\}$ is a complete set of primitive pairwise primitive idempotents of $KQ/I$. Hence the sets of points of the quivers $Q$ and $Q_A$ are the same. Let $R_Q$ be the arrow ideal of $KQ$. Then $R = \text{rad } A = R_Q/I$, and so $R^2 = (R_Q/I)^2$. Therefore $R/R^2 = (R_Q/I)/(R_Q/I)^2 \cong R_Q/(R_Q)^2$. This means that the arrows from the vertex $i$ to the vertex $j$ in $Q$ are in

---

[12]Recall that a two-sided ideal $I$ of an algebra $A$ is called **admissible** if $J^n \subseteq I \subseteq J^2$ for some $n$, where $J$ is the Jacobson radical of $A$.

bijective correspondence with the vectors in a basis of the vector space $e_i(R/R^2)e_j$, i.e. with the set of arrows from $i$ to $j$ in $Q_A$. $\square$

**Proposition 6.3.8.** *Let A be a indecomposable finite dimensional basic algebra over a field K with Gabriel quiver Q. Then there exists an admissible ideal I of KQ such that $A \cong KQ/I$. So the category $\text{Rep}_K A$ of representations of A is equivalent to the category of representations of the algebra $KQ/I$ for the Gabriel quiver Q of A and some two-sided admissible ideal I of KQ.*

*Proof.* By the previous lemma, a finite dimensional algebra $A$ can be written as a bound quiver algebra $KQ/I$, where $Q$ is the Gabriel quiver of $A$ and $I$ is an admissible ideal of the path algebra $KQ$. Let $J$ be the ideal of algebra $KQ$ generated by all arrows of the quiver $Q$. Obviously, $J = \text{rad}A$ is the Jacobson radical of $A$. Since $A$ is finite dimensional, $J$ is nilpotent. The ideal $I$ contains a power of $J$, so that $J^n \subseteq I \subseteq J^2$. Hence the path algebra $KQ/I$ is finite dimensional. The category $\text{Rep}A$ is equivalent to the category $\text{Rep}(Q, I)$ as defined in the obvious way. (See also [10, theorem 3.7].) $\square$

One of the main problems in the theory of representations is to get information about the structure of its indecomposable modules. All algebras are divided into different types of representation classes.

**Definition 6.3.9.** A $K$-algebra $A$ is said to be of **finite representation type** (or **finite type**) if $A$ has only a finite number of non-isomorphic finite dimensional indecomposable representations up to isomorphism. Otherwise $A$ is said to be of **infinite representation type**.

All $K$-algebras of infinite type are further divided into **algebras of wild representation type** and **algebras of tame representation type**.

**Definition 6.3.10.** An algebra $A$ is said to be of **tame representation type** (or a **tame algebra**) if it is of infinite type but all families of indecomposable representations are 1-parametric. In other words, for any $r$ there are $(A, K[x])$-bimodules $M_1, \ldots, M_n$ (where the natural number $n$ may depend on $r$), which are finitely generated and free over $K[x]$ such that any indecomposable $A$-module of dimension $r$ is isomorphic to some $A$-module of the form $M_i \otimes K[x]/(x - \lambda)$.

An algebra $A$ is said to be of **wild representation type** (or a **wild algebra**) if there is an $(A, K\langle x, y\rangle)$-bimodule $M$ which is finitely generated and free over $K\langle x, y\rangle$ and such that the functor $M \otimes_{K\langle x,y\rangle} *$ sends non-isomorphic finite dimensional $K\langle x, y\rangle$-modules to non-isomorphic finite dimensional $A$-modules. In this case the category of all finite dimensional $A$-modules includes the classification problem for pairs of square matrices up to simultaneous equivalence.

The first important result in the theory of representations of finite dimensional algebras was obtained in 1968 by A.V. Roiter [183], who proved the following remarkable theorem (which is the first Brauer-Thrall conjecture):

**Theorem 6.3.11.** (A.V. Roiter, [183], see also [101, Theorem 3.5.1].) *A finite dimensional algebra is either of finite representation type or there are indecomposable modules of arbitrary large finite dimension.*

An additive category $\mathfrak{U}$ is said to be a **dimension category** if there exists a mapping dim : $\mathfrak{U} \to \mathbf{N} \cup \infty$ satisfying the condition

$$\dim(X \oplus Y) = \dim(X) + \dim(Y).$$

The representations of $K$-species, and of finite dimensional algebras are examples of dimension categories. The notion of a category to be of finite type is defined as expected.

A dimension category $\mathfrak{U}$ is said to be of **strongly unbounded type** if it possesses the following properties:
  I. $\mathfrak{U}$ has indecomposable objects of an arbitrarily large finite dimension.
  II. If $\mathfrak{U}$ contains a finite dimensional object with an infinite endomorphism ring, then there is an infinite number of finite dimensions $d$ such that $\mathfrak{U}$ has infinitely many non-isomorphic indecomposable objects of dimension $d$, for each $d$.
  III. $\mathfrak{U}$ has indecomposable objects of infinite dimension.

The well known Brauer-Thrall conjecture states that any $K$-algebra is either of finite type or strongly unbounded type. This conjecture is said to be the first Brauer-Thrall conjecture when strongly unbounded type is understood only as property I. So A.V. Roiter in the theorem above proved property I for the category of modules over $K$-algebra. V. Dlab and C.M. Ringel extended this result in [53]:

**Theorem 6.3.12.** (V. Dlab, C.M. Ringel, [53].)
  1. *A $K$-species is either of finite representation type or of strongly unbounded type.*
  2. *A finite dimensional $K$-algebra $A$ which is hereditary or which satisfies $(\operatorname{rad} A)^2 = 0$ is either of finite representation type or of strongly unbounded type.*

As was conjectured by P. Donovan & M. Freislich and established by Yu.A. Drozd [63] there is a trichotomy between three different representation classes: finite, tame and wild representation type in the case of an algebraically closed field $K$.

**Theorem 6.3.13.** (Yu.A. Drozd, [63].) *Let $A$ be a finite dimensional algebra over an algebraically closed field. Then $A$ is of finite, tame or wild representation type.*

A full description of algebras of finite or tame type and their representations has been obtained only for some particular classes of algebras, for example, for hereditary algebras, algebras in which the square of the Jacobson radical equals zero, and self-injective algebras[13].

---

[13]Recall that an algebra $A$ is **self-injective** if the right and left regular modules are injective. See e.g. [101, section 4.12].

**Theorem 6.3.14.** (P. Gabriel, [79].) *Suppose that A is a finite dimensional basic algebra over a field K. Let $Q = Q(A)$ be its $\text{Ext}^1$-quiver. Then there is a surjective algebra morphism $\pi : KQ \rightarrow A$ such that the $\text{Ker}\,\pi$ is contained in the ideal of all paths of length at least two. If A is hereditary, then $KQ \simeq A$ (as K-algebras).*

From this theorem and theorem 6.1.23 the next theorem immediately follows:

**Theorem 6.3.15** (P. Gabriel, [79].) *Let A be a hereditary finite dimensional basic algebra over an algebraically closed field K. Then A is of finite type if and only if its $\text{Ext}^1$-quiver is a finite disjoint union of simple Dynkin diagrams of the form $A_n, D_n, E_6, E_7, E_8$.*

The description of representations of f.d. algebras often uses bipartite quivers.

**Definition 6.3.16.** A quiver $Q = (Q_0, Q_1, s, t)$ is **bipartite** (or **separated**) if there are two nonempty sets $X_1 \neq \emptyset$, $X_2 \neq \emptyset$ such that $Q_0 = X_1 \cup X_2$, $X_1 \cap X_2 = \emptyset$, and for any $\sigma \in Q_1$, $s(\sigma) \in X_1$ and $t(\sigma) \in X_2$, i.e. any vertex in $Q_0$ is either a start vertex or an end vertex.

This definition is equivalent to the following: a quiver $Q = (Q_0, Q_1, s, t)$ is bipartite if and only if it contains neither loops nor oriented paths of the following form:

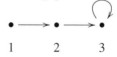 (6.3.17)

For any finite quiver $Q = (Q_0, Q_1, s, t)$ one can construct a (canonical) associated bipartite quiver $Q^b = (Q_0^b, Q_1^b, s_1, t_1)$ in the following way. Let $Q_0 = \{1, 2, ..., s\}$, $Q_1 = \{\sigma_1, \sigma_2, ..., \sigma_k\}$. Then $Q_0^b = \{1, 2, ..., s, b(1), b(2), ..., b(s)\}$ and $Q_1^b = \{\tau_1, \tau_2, ..., \tau_k\}$, such that for any $\sigma_j \in Q_1$ we have $s(\tau_j) = s(\sigma_j)$ and $t(\tau_j) = b(t(\sigma_j))$. In other words, in the quiver $Q^b$ from the vertex $i$ to the vertex $b(j)$ go $t_{ij}$ arrows if and only if in the quiver $Q$ from the vertex $i$ to the vertex $j$ there are $t_{ij}$ arrows. As before, denote by $\overline{Q}$ the undirected graph which is obtained from $Q$ by deleting the orientation of all arrows.

**Example 6.3.18.** Let $Q$ be the following quiver:

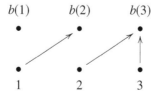

Then the corresponding bipartite quiver $Q^b$ has the following form:

$b(1)$      $b(2)$      $b(3)$

     1      2      3

If $A$ is a finite dimensional algebra over an algebraically closed field $K$ with zero square radical and with associated quiver $Q$, then an explicit connection between the category $\mathrm{mod}_r(A)$ and the category $\mathrm{rep}\,Q^b(A)$ was established by P. Gabriel [79]. He proved the following theorem:

**Theorem 6.3.19.** (P. Gabriel, [79].) *Let $A$ be a finite dimensional algebra over an algebraically closed field $K$ with zero square radical and quiver $Q$. Then $A$ is of finite type if and only if the quiver $\Gamma = Q^b(A)$ is of finite type, i.e. the underlying undirected graph $\overline{\Gamma}$ is a finite disjoint union of simple Dynkin diagrams of the form $A_n, D_n, E_6, E_7, E_8$.*

Theorems 6.3.15 and 6.3.19 have been generalized to the case of finite dimensional algebras over arbitrary fields or, more generally, to the case of certain class of Artinian rings using the results on representations of valued graphs and species created by P. Gabriel, V. Dlab and C.M. Ringel.

To any finitely dimensional algebra (and more generally some classes of Artinian rings) $A$ one can associate a species $\mathfrak{M}$ and a valued graph $\Gamma_A = (\Gamma, d)_A$ for which this species will be modification as follows.

We can assume without loss of generality that $A$ is a basic ring with Jacobson radical $R$. Then

$$A/R = \prod_{i=1}^{n} F_i$$

and

$$R/R^2 = \prod_{1 \le i, j \le n} {}_i M_j$$

with uniquely determined division rings $F_i$ and $F_i$-$F_j$-bimodules ${}_i M_j$. Thus one obtain a species $\mathfrak{L} = (F_i, {}_i M_j)_{1 \le i, j \le n}$.

Each such a species one can associate the tensor algebra $\mathfrak{T}(\mathfrak{L}) = \oplus_{n=0}^{\infty} M^{\otimes n}$ as was shown above in section 6.2. If $A$ is a finite dimensional $K$-algebra then $\mathfrak{T}(\mathfrak{L}) = \oplus_{n=0}^{\infty} M^{\otimes n}$ is also a $K$-algebra.

With this species one can associate a (relative) valued quiver $\Gamma_A = (\Gamma, d)$ in the following way.

Let $\Gamma_0 = \{1, 2, \ldots, n\}$ be the set of vertices of the (to be constructed) relative valued quiver $\Gamma_A$. Define the set of edges of $\Gamma_A$ as follows. An arrow $\sigma \in \Gamma_1$ with $s(\sigma) = i$ and $t(\sigma) = j$ if and only if ${}_i M_j \ne 0$. We define a (relative) valuation of $\Gamma$ by setting each $\sigma \in \Gamma_1$ with $s(\sigma) = i$ and $t(\sigma) = j$ two integer numbers $(d_{ij}^{\sigma}, d_{ji}^{\sigma})$, where $d_{ij}^{\sigma} = \dim({}_i M_j)_{F_j}$ and $d_{ji}^{\sigma} = \dim_{F_i}({}_i M_j)$, i.e. we have a valued arrow in $\Gamma_A$:

$$\begin{array}{ccc} & (d_{ij}^{\sigma}, d_{ji}^{\sigma}) & \\ \bullet & \longrightarrow & \bullet \\ i & & j \end{array}$$

Now if $\mathfrak{M}$ is an acyclic species, in particular ${}_i M_j = 0$ or ${}_j M_i = 0$ for each pair $\{i, j\}$, and the graph $\Gamma_A$ is a relative valued graph then $\mathfrak{L} = (F_i, {}_i M_j)_{1 \le i, j \le n}$ is a modulation of this quiver provided that there exist numbers $f_i$ with $d_{ij}f_j = d_{ji}f_i$ and that the dualization conditions

$$\operatorname{Hom}_{F_i}({}_iM_j, F_i) \cong \operatorname{Hom}_{F_j}({}_iM_j, F_j)$$

hold. In particular, it takes place in the case that $A$ is a finitely generated algebra over a central field $K$. In this case the numbers $f_i$ can be interpreted as the indices $[F_i : K]$.

Conversely, given a relative valued quiver $(\Gamma, d)$ and a modulation $\mathfrak{L} = (F_i, M_\sigma)_{i\in\Gamma_0, \sigma\in\Gamma_1}$ of this quiver one can define the tensor algebra in the following way. Let $B = \prod_{i\in\Gamma_0} F_i$, and let $M = \bigoplus_{\sigma\in\Gamma_1} M_\sigma$. Then $B$ is a ring and $M$ naturally becomes a $(B, B)$-bimodule. The **tensor algebra** of the $(B, B)$-bimodule $M$ is the graded ring

$$\mathfrak{T}(\mathfrak{L}) = \mathfrak{T}_B(M) = \bigoplus_{n=0}^{\infty} M^{\otimes n}$$

where $M^{\otimes 0} = B$, $M^{\otimes n} = M^{\otimes(n-1)} \otimes_B M$ for $n > 0$ with component-wise addition and the multiplication induced by taking tensor products.

If $\mathfrak{L}$ is a $K$-species then $\mathfrak{T}(\mathfrak{L})$ is a $K$-algebra. If $(\Gamma, d, \Omega)$ is a valued quiver with trivial valuation $d$, i.e. $(\Gamma, d)$ is an ordinary quiver, and if $F_i = K$ for all $i \in \Gamma_0$, then the tensor algebra $\mathfrak{T}(\mathfrak{L}) = K\Gamma$ is the path algebra of the quiver $\Gamma$, i.e. the tensor algebra $\mathfrak{T}(\mathfrak{L})$ of a species $\mathfrak{L}$ is a generalization of the path algebra. If $(\Gamma, d)$ is an acyclic quiver then the Jacobson radical $\operatorname{rad}(\mathfrak{T}(\mathfrak{L})) = \oplus_{n=1}^{\infty} M^{\otimes n}$. And in this case $\mathfrak{T}(\mathfrak{L})$ is an Artinian hereditary ring. So in this case theorem 6.2.14 implies:

**Theorem 6.3.20.** (P. Gabriel, V. Dlab, C.M. Ringel [53, Theorem C].) *A finite dimensional $K$-algebra $A$ is a hereditary algebra of finite type if and only if $A$ is Morita equivalent to a tensor algebra $\mathfrak{T}(\mathfrak{L})$, where $\mathfrak{L}$ is a $K$-species of finite type.*

Taking into account the results of theorem 6.2.16 there is the following theorem which is a generalization of the Gabriel theorem to the case of an arbitrary field:

**Theorem 6.3.21.** (V. Dlab, C.M. Ringel [54]). *Let $A$ be a finite dimensional hereditary $K$-algebra with a valued graph $(\Gamma, d)_A$. Then $A$ is of finite type if and only if $\Gamma_A$ is a finite disjoint union of Dynkin diagrams of the form $A_n$, $B_n$, $C_n$, $D_n$, $E_6$, $E_7$, $E_8$, $F_4$, $G_2$.*

Let $C$ be a commutative Artinian ring. A ring $A$ is called an **Artin algebra** if $A$ is finitely generated as $C$-module. It is clear that $A$ is an Artinian ring.

A species $\mathfrak{L} = (F_i, {}_iM_j)_{i,j\in I}$ is called an **Artin species** if there is a commutative ring $C$ such that $F_i$ are finitely generated $C$-modules, $C$ acts centrally on all $F_i$ and ${}_iM_j$, and every $C$-module ${}_iM_j$ has finite length.

A species $\mathfrak{L} = (F_i, {}_iM_j)_{i,j\in I}$ is called **quasi-Artin** if $\dim({}_iM_j)_{F_j}$ and $\dim_{F_i}({}_iM_j)$ are finite and $F_i$ is finitely generated over its center for all $i, j \in I$.

**Theorem 6.3.22.** (P. Dowbor, D. Simson [58, Theorem 1.1]). *Let $\mathfrak{L} = (F_i, {}_iM_j)_{i,j\in I}$ be a quasi-Artin species and suppose that its relative valued quiver $(\Gamma, d)$ is acyclic.*

*Then $\mathfrak{L}$ is of finite type if and only if $(\Gamma, d)$ is a finite disjoint union of Dynkin diagrams of the form $A_n$, $B_n$, $C_n$, $D_n$, $n \geq 1$, $E_6$, $E_7$, $E_8$, $F_4$, $G_2$.*

As a corollary of this theorem P. Dowbor and D. Simson obtained the following result.

**Corollary 6.3.23.** (P. Dowbor, D. Simson [58, Corollary 1.4]). *Let A be a hereditary Artinian basic ring with Jacobson radical R. Suppose that $A/R$ is an Artin algebra. Then A is of finite representation type if and only if the species $\mathfrak{L}$ of A is of finite representation type.*

In a similar way, a complete description was obtained by V. Dlab and C.M. Ringel for the category of all modules over a finite dimensional algebra $A$ over a field $K$ with $R^2 = 0$, where $R = \text{rad } A$, using the so-called separated diagram of $A$.

Let $\mathfrak{L} = (F_i, {}_iM_j)_{i,j \in I}$ be a species corresponding to a finite dimensional algebra $A$ with $R^2 = 0$ where $R$ is the Jacobson radical of $A$. Then as was shown in section 6.2 to this species associate the diagram $\Gamma$. We now construct so called **separated diagram** $\Gamma^*$ of $A$ as follows. Let $\Gamma_0 = I = \{1, 2, \ldots, n\}$. Then $\Gamma_0^* = I \times \{0, 1\}$ is the set of vertices, and all edges of $\Gamma$ connect only the vertices of the form $(i, 0)$ and $(j, 1)$. Note that there are no vertices neither between $(i, 0)$ and $(j, 0)$, nor between $(i, 1)$ and $(j, 1)$. The number of edges between vertices $(i, 0)$ and $(j, 1)$ in $\Gamma^*$ is the same as in the diagram $\Gamma$ between vertices $i$ and $j$, i.e. there are $t_{ij} = \dim_{K_i}({}_iM_j) \times \dim({}_iM_j)_{K_j}$ edges between $(i, 0)$ and $(j, 1)$. And in the same way we denote by the arrow the multiple edge $\Longrightarrow$ provided $\dim_{K_i}({}_iM_j) < \dim({}_iM_j)_{K_j}$, which connects $(i, 0)$ with $(j, 1)$ and contains $t_{ij}$ edges.

**Theorem 6.3.24.** (P. Gabriel [80], V. Dlab, C.M. Ringel [53].) *Let A be a finite dimensional basic K-algebra A over an arbitrary field K with $(\text{rad}A)^2 = 0$. Then A is of finite type if and only if the separated diagram of its K-species is a finite disjoint union of Dynkin diagrams of the form $A_n$, $B_n$, $C_n$, $D_n$, $E_6$, $E_7$, $E_8$, $F_4$, $G_2$.*

**Remark 6.3.25.** This theorem was proved by P. Gabriel [80] in the case when the associated $K$-species $\mathcal{L}_A = (F_i, {}_iM_j)_{i,j \in I}$ has the property that all $F_i$ are equal to a fixed skew field $F$ and $_F({}_iM_j)_F = ({}_FF_F)^{n_{ij}}$ for some natural number $n_{ij}$. Also P. Gabriel has shown that the structure of a $K$-algebra $A$ of finite type with $(\text{rad}A)^2 = 0$ can be recovered from the known results in the case when $K$ is a perfect field [79], [80].

In the case $A$ is an Artinian basic ring with $(\text{rad}A)^2 = 0$ instead of separated diagram we can construct its **separated valued graph** $(\Gamma^*, d)$ using the relative valued graph $(\Gamma, d)$ constructed above. The set of vertices of a new graph $\Gamma_0^* = I \times \{0, 1\}$, where $I = \Gamma_0$. An edge $\sigma^* \in \Gamma_1^*$ connects the vertex $(i, 0)$ with the vertex $(j, 1)$ precisely when there is an edge $\sigma \in \Gamma_1$ connecting the vertex $i$ with the vertex $j$ in $\Gamma$, i.e. when ${}_iM_j \neq 0$. In this case the valuation of an edge $\sigma$ is transferred to the valuation of the edge $\sigma^*$, i.e. we have an edge:

$$\underset{(i,0) \qquad (j,1)}{\bullet \overset{(d_{ij}^{\sigma}, d_{ji}^{\sigma})}{\rule{3cm}{0.4pt}} \bullet}$$

**Theorem 6.3.26.** (P. Dowbor, D. Simson [58, Corollary 1.5]). *Let A be an Artinian basic ring with Jacobson radical R such that $R^2 = 0$. If $A/R$ is an Artin algebra then A is of finite representation type if and only if the separated valued graph $(\Gamma^*, d)$ of A is a disjoint union of Dynkin diagrams.*

The description of hereditary finite dimensional algebras of tame type and wild type was obtained by V. Dlab and C.M. Ringel in [54], [176], [177].

**Theorem 6.3.27.** (V. Dlab, C.M. Ringel [54], [176], [177]). *Let A be a finite dimensional algebra over an algebraically closed field K with Gabriel quiver $Q = Q(A)$. If A is of tame representation type then the underlying undirected graph $\overline{\Gamma}$ of the separated quiver $\Gamma = Q^b$ is a union of extended Dynkin diagrams of types $\tilde{A}_n$, $\tilde{D}_n$, $\tilde{E}_6$, $\tilde{E}_7$, $\tilde{E}_8$.*

## 6.4 Notes and References

The classification of finite dimensional algebras of f.r.t. with zero square radical was first obtained independently by P. Gabriel [79] by means of quivers and S.A. Kruglak [137] by means of matrix problems.

The category of absolute valued quivers $\mathcal{D}_{abs}$ and the category of relative valued quivers $\mathcal{D}_{rel}$ were considered by J. Lemay in [142] where he, in particular, showed that these categories are not isomorphic even under restrictions to the subcategories of objects having greatest common divisor 1.

# CHAPTER 7

# Artinian Rings of Finite Representation Type

Finite dimensional algebras of finite representation type were considered in section 6.3. For right Artinian rings one can also introduce the notion of a ring of finite representation type. As has been shown by D. Eisenbud and P. Griffith [71] this notion is left-right symmetric. They proved this fact using the duality theory of Auslander and Bridger. This result is treated in section 7.1.

For finite dimensional algebras along with the notion of finite representation type there is also the notion of bounded representation type. Recall that a finite dimensional algebra $A$ is said to be of **bounded representation type** if there is a limit on the length of the indecomposable finite dimensional $A$-modules. The first Brauer-Thrall conjecture says that these notions are the same in the case of a finite dimensional algebra $A$ (as was proved by A.V. Roiter [101, theorem 3.5.1]) and in the case of Artin algebras (as was proved by M. Auslander [15]).

For right Artinian rings one can also introduce the notion of a ring of bounded representation type. And, as proved by M. Auslander in [15] and [20], these notions, finite and bounded representation type, coincide for right Artinian rings. This famous result will be dealt with in section 7.2.

In section 7.3 it will be shown that any finite dimensional algebra of finite representation type is a semidistributive ring.

The structure of Artinian semidistributive hereditary rings of finite representation type is given in section 7.4.

Throughout this chapter $A$ will be a right Artinian ring and all modules will be finitely generated right $A$-modules.

## 7.1 Eisenbud-Griffith Theorem

**Definition 7.1.1.** A right Artinian ring $A$ is of **right finite representation type** if $A$ has up to isomorphism only a finite number of indecomposable finitely generated right $A$-modules.

The important result that for right Artinian rings the notion of right finite representation type coincides with the notion of left finite representation type was obtained by D. Eisenbud and P. Griffith in [71]. And so in this case one can simply talk about rings of finite representation type. This section is devoted to a proof of this theorem.

By the length of a $A$-module $M$ we will mean the length of a composition series of $M$ (if one exists). Recall that a ring $A$ is said to be **semiprimary** if the Jacobson radical $R$ of $A$ is nilpotent and $A/R$ is semisimple. Any semiprimary ring is obviously a semiperfect ring.

**Lemma 7.1.2.** (D. Eisenbud, P. Griffith [71, Lemma 1.1]). *Let $A$ be a semiprimary ring, and $P$ a principal right $A$-module. Suppose that $T$ is a submodule of $P$ and that $T$ has a finite length. Then* $\operatorname{End}_A(P/T)$ *is a local ring.*

*Proof.* If $T = 0$ then $\operatorname{End}_A(P/T) = \operatorname{End}_A(P)$ is a local ring by [100, theorem 10.3.8].

Suppose $T \neq 0$. Let $R = \operatorname{rad} A$, then $PR$ is the unique maximal submodule of $P$, by [103, theorem 1.8.3]. So one can assume that $T \subset PR$. Since $P = eA$ for some local idempotent $e \in A$, the set

$$Y = \{\varphi \in \operatorname{End}_A(P/T) \ : \ \operatorname{Im}(\varphi) \subseteq PR/T\}$$

is an ideal in $\operatorname{End}_A(P/T)$. We now show that $Y$ is the unique maximal ideal in $\operatorname{End}_A(P/T)$.

Let $\varphi \in \operatorname{End}_A(P/T)$ and $\operatorname{Im}(\varphi) \not\subseteq PR/T$. Since $PR$ is the unique maximal submodule of $P$, $\varphi$ is an epimorphism. Let $\pi : P \to P/T$ be the canonical projection. Then both $\pi$ and $\varphi\pi$ are epimorphisms onto $P/T$. So, by Schanuel's lemma (see [100, lemma 12.3.3])

$$T \oplus P \simeq \operatorname{Ker}(\varphi\pi) \oplus P.$$

Since $\operatorname{End}_A(P)$ is local, $T \simeq \operatorname{Ker}(\varphi\pi)$, by the cancelation property [103, proposition 7.3.5]. Thus, $\operatorname{Ker}(\varphi\pi)$ has the same finite length as $T$. But $T \subseteq \operatorname{Ker}(\varphi\pi)$ as a submodule of $P$, hence $T = \operatorname{Ker}(\varphi\pi)$. Thus $\varphi$ is a monomorphism, and so $\varphi$ is an isomorphism. The lemma is proved. $\square$

**Theorem 7.1.3.** (D. Eisenbud, P. Griffith [71, Theorem 1.2]). *A left Artinian ring which has a finite number of non-isomorphic finitely generated indecomposable left $A$-modules is right Artinian.*

*Proof.* Since $A$ is a left Artinian ring, it is semiperfect. Therefore, by [103, proposition 1.9.6], every finitely generated projective right $A$-module is a finite direct sum of principal right $A$-modules. Suppose that $A$ is not right Artinian, then there exists a principal right $A$-module $X$ and an infinite strictly increasing sequence of submodules

$$S_1 \subsetneqq S_2 \subsetneqq \ldots \subsetneqq X$$

such that each $S_i$ has finite length. Since $X$ has a local endomorphism ring, from Schanuel's lemma and the cancelation property it follows that $X/S_i \neq X/S_j$ for $i \neq j$.

Let $U_1, \ldots, U_n$ be representatives of the finitely generated indecomposable non-projective left $A$-modules. Let Tr be the Auslander-Bridger translation, whose properties were studied in section 4.3. For each $i$ one can write

$$\mathrm{Tr}(X/S_i) = V_i \oplus P_i \qquad (7.1.4)$$

where $V_i = \bigoplus\limits_{j \in J(i)} U_j$ is a direct sum of certain $U_j$'s, and $P_i$ is a projective $A$-module.

Taking into account the properties of Tr (lemma 4.3.4 and theorem 4.3.7), and applying the translation Tr to (7.1.4) there results

$$X/S_i \oplus P \simeq \bigoplus\limits_{j \in J(i)} \mathrm{Tr}(U_j) \oplus Q, \qquad (7.1.4)$$

where $P$ and $Q$ are certain finitely generated projective modules. By lemma 7.1.2, the left hand side of (7.1.4) is a direct sum of strongly indecomposable modules. Therefore, by [103, propositions 7.1.30, 7.3.5], any such module has the exchange property and the cancelation property. Then, using [103, proposition 7.3.5] repeatedly, one sees that there is an index $j = j(i)$ such that $U_j$ is a summand of $V_i$, and $X/S_i$ is a summand of $\mathrm{Tr}(U_j)$, and so

$$X/S_i \oplus P' \simeq \mathrm{Tr}(U_{j(i)}) \oplus Q', \qquad (7.1.5)$$

where $P'$ and $Q'$ are some suitable projective modules.

Since there are finitely many $U_j$'s and infinitely many $S_i$'s, there are indices $i, i'$ such that $i \neq i'$ but $j(i) = j(i')$. For such a pair of indices we have

$$X/S_i \oplus P'' \simeq X/S_{i'} \oplus Q'', \qquad (7.1.6)$$

where $P''$ and $Q''$ are projective. Both sides of (7.1.6) are sums of strongly indecomposable modules, so by the Krull-Remak-Schmidt-Azumaya theorem [103, theorem 7.2.7], $X/S_i \simeq X/S_{i'}$, a contradiction. Hence $A$ is a right Artinian ring. $\square$

**Theorem 7.1.7.** (D. Eisenbud, P. Griffith [71, Theorem 1.2]). *Suppose $A$ is a left Artinian ring of left finite representation type. Then it also is a right Artinian ring of right finite representation type. The same statement hold when "left" is replaced by "right" and vice versa, i.e. a right Artinian ring of right finite representation type is a left Artinian ring of left finite representation type.*

*Proof.* Since, by the previous theorem, $A$ is a right and left Artinian ring, it is easy to see that for any finitely generated right or left $A$-module $M$, the modules $M$ and $\mathrm{Tr}(M)$ have the same number of non-projective indecomposable summands. Also, two finitely generated $A$-modules without projective summands are isomorphic if and only if they are stably isomorphic. So $A$ has the same number of finitely generated indecomposable modules on the right as on the left. $\square$

Taking into account the symmetry of theorem 7.1.7, one can introduce the following definition.

**Definition 7.1.8.** An Artinian ring $A$ is said to be of **finite representation type** if $A$ has up to isomorphism only a finite number of indecomposable finitely generated $A$-modules.

Let $A$ be a ring, and let $P$ be a finitely generated projective $A$-module which can be decomposed into a direct sum of $n$ indecomposable modules. The endomorphism ring $B = \mathrm{End}_A(P)$ of the module $P$ is called a **minor** of order $n$ of the ring $A$.

Many properties of a ring are reflected by its minors. By [100, theorem 10.3.8], a ring $A$ is semiperfect if and only if the identity of $A$ decomposes into a sum of finite number of pairwise orthogonal local idempotents, i.e. any minor of the first order of this ring is local.

From [100, theorem 3.6.1] it immediately follows that minors of (right) Noetherian, (right) Artinian rings are (right) Noetherian, (right) Artinian, respectively. By [100, corollary 12.3.2], the ring is serial if and only if all its minors of order three are serial.

Let $e$ be a non-zero idempotent of a ring $A$, and let $M$ be a right $eAe$-module. Consider $\widetilde{M} = M \otimes_{eAe} eA$. Obviously, $\widetilde{M}$ has a right $A$-module structure.

**Lemma 7.1.9.** *The modules $M$ and $\widetilde{M}$ are indecomposable simultaneously.*

*Proof.* For any nonzero $e^2 = e \in A$ and a right $A$-module $M$ we have

$$\mathrm{Hom}_A(eA, M) = Me.$$

By the Adjoint isomorphism [103, proposition 1.3.1]:

$$\mathrm{Hom}_A(\widetilde{M}, \widetilde{M}) = \mathrm{Hom}_A(M \otimes_{eAe} eA, \widetilde{M}) \simeq \mathrm{Hom}_{eAe}(M, \mathrm{Hom}_A(eA, \widetilde{M})).$$

By what was said above,

$$\mathrm{Hom}_A(eA, \widetilde{M}) = \widetilde{M}e = (M \otimes_{eAe} eA)e \simeq M.$$

Therefore,

$$\mathrm{Hom}_A(\widetilde{M}, \widetilde{M}) \simeq \mathrm{Hom}_{eAe}(M, M).$$

Consequently, $M$ and $\widetilde{M}$ are indecomposable simultaneously. $\square$

**Proposition 7.1.10.** *Let $A$ be an Artinian ring, and let $e^2 = e$ be a non-zero idempotent of $A$. If $A$ is of finite representation type then $eAe$ is of finite representation type as well.*

*Proof.* Let $P = eA$ and $B = eAe$. If $N$ is an indecomposable $B$-module then $\widetilde{N} = N \otimes_B P$ is an indecomposable $A$-module, by lemma 7.1.9. Therefore if $B$ has an infinite number of indecomposable representations then $A$ has an infinite number of representations as well. $\square$

Recall that the **opposite ring** $A^{op}$ of a ring $A$ is $A$ as Abelian additive group and multiplication $*$ in $A^{op}$ is given by $a * b = ba$, where the latter denotes the usual product in $A$. Obviously, $(A^{op})^{op} \simeq A$. If $A$ is a commutative ring then $A^{op} = A$.

Note that any right $A$-module $M$ can be considered as a left $A^{op}$-module $M^{op}$. More exactly, the Abelian group of $M^{op}$ is $M$ but the left scalar multiplies $\circ$ : $A \times M \to M$ is defined by $a \circ m = ma$ for all $a \in A$ and all $m \in M$.

From Eisenbud-Griffith's theorem 7.1.3 we obtain the following important fact.

**Proposition 7.1.11.** *A is an Artinian ring of finite representation type if and only if $A^{op}$ is an Artinian ring of finite representation type.*

## 7.2 Auslander's Theorem for Right Artinian Rings

Recall that a $T$-algebra $A$ over an Artinian commutative ring $T$ is called an **Artin algebra**, if $A$ is f.g. as $T$-module.

The first Brauer-Thrall conjecture asserts that an Artin algebra of infinite representation type has indecomposable submodules of arbitrary large finite length. The proof of this conjecture for finite dimensional algebras over arbitrary fields, which was obtained by A.V. Roiter in [183], was given in [100, section 3.5]. C.M. Ringel improved the assertion of the first Brauer-Thrall conjecture for Artin algebras by proving the following theorem.

**Theorem 7.2.1.** (C.M. Ringel [179].) *Let A be an Artin algebra. An A-module M is either the direct sum of copies of a finite number of indecomposable modules of finite length, or M contains indecomposable submodules of arbitrarily large finite length.*

The first Brauer-Thrall conjecture in a more general categorical form for one-sided Artinian rings was proved by M. Auslander in [15]. In this section a proof of the Auslander theorem will be given for right Artinian rings following K. Yamagata [226]. K. Yamagata not only gave a module-theoretical simple proof of this theorem, he also showed in his paper how to construct all indecomposable modules from simple modules over an Artinian ring of finite representation type. Namely, in this case every indecomposable $A$-module appears as a direct summand of

(a) The radical of a projective indecomposable $A$-module,
    or
(b) The middle term of an almost split sequence, which is obtained from a simple $A$-module by successive almost split extensions.

Throughout in this section we assume that $A$ is a right Artinian ring and that all modules are finitely generated right $A$-modules.

In section 4.4 there was considered the notion of almost split sequences and the main properties of them. In particular, in the category of finitely generated modules over an Artin algebra for any finitely generated indecomposable module which is not projective there exists an almost split sequence, by proposition 4.4.22. The main properties and existence of almost split sequences for modules over semiperfect rings were discussed in section 4.7. By theorem 4.7.8, for each finitely presented non-projective module with local endomorphism ring there exists an almost split sequence.

By definition, each almost split sequence can not be ended by projective module. The following definition introduces a generalization of almost split sequences which eliminates this drawback.

**Definition 7.2.2.** A homomorphism $g : M \rightarrow N$ of two modules $M, N$ is said to be an **almost split extension** over $N$ provided that

    (a) If $N$ is projective, then $g$ is left almost split homomorphism[1] and $M$ is the unique maximal submodule of $N$.

    (b) If $N$ is non-projective, then $g$ is a right almost split homomorphism and $\mathrm{Ker}\,g$ is indecomposable.

There is an equivalent definition of the notion of an almost split extension in the case of finitely generated modules over right Artinian rings, which is given by the following proposition.

**Proposition 7.2.3.** *Let $A$ be a right Artinian ring with Jacobson radical $R$, and let $M, N$ be finitely generated right $A$-modules. Then the following statements are equivalent*:

    1. *A homomorphism $g : M \rightarrow N$ is almost split extension over $N$.*

    2. *A homomorphism $g : M \rightarrow N$ satisfies the following conditions*:

        (a) *If $N$ is projective then $g$ is injective and $\mathrm{Im}\,g = NR$.*

        (b) *If $N$ is non-projective then the sequence*

$$0 \rightarrow \mathrm{Ker}\,g \longrightarrow M \xrightarrow{g} N \rightarrow 0 \qquad\qquad (7.2.4)$$

    *is almost split.*

*Proof.*

    $1 \Longrightarrow 2$.

    (a) Suppose that $N$ is projective. Then $g$ is an injection and $M$ is the unique maximal submodule of $N$. Since $N$ is a projective module over a right Artinian ring which is semiperfect, the unique maximal submodule of $N$ is equal to $NR$. So $\mathrm{Im}\,g = NR$.

    (b) Suppose that $N$ is non-projective. Then $g$ is a right almost split homomorphism and $\mathrm{Ker}\,g$ is indecomposable. Since $A$ is a right Artinian ring and $\mathrm{Ker}\,g$ is a f.g. indecomposable module, it has a composition series, and so its endomorphism ring is local. Then by theorem 4.4.14 the exact sequence (7.2.4) is almost split.

---

[1]See definition 4.4.7 and properties of almost split homomorphisms in section 4.4.

$2 \Longrightarrow 1$.

(a) Suppose that $N$ is projective. Then $g$ is injective and $\mathrm{Im}\, g = NR$, which is a unique maximal submodule in $N$. Moreover, it is clear that the inclusion $g : NR \to N$ is right almost split.

(b) Suppose that $N$ is non-projective. Then the sequence (7.2.4) is almost split which means that $g$ is an epimorphism and $\mathrm{Ker}\, g$ is indecomposable by lemma 4.4.8. $\square$

**Lemma 7.2.5.** (K. Yamagata [226, Lemma 2]). *Let $A$ be a right Artinian ring with Jacobson radical $R$ and all $A$-modules are finitely generated. Let $g : M \to N$ be an almost split extension over an indecomposable right $A$-module $N$. Assume that $M = \bigoplus\limits_{i \in I} M_i$ is a direct sum decomposition into indecomposable submodules of $M$, and $v_i : M_i \to M$ is a canonical injection for $i \in I$. Then for every $i \in I$ the homomorphism $g v_i$ is not an isomorphism.*

*Proof.*

1. Suppose that $N$ is a projective module. Then by proposition 7.2.3 $g$ is injective and $\mathrm{Im}\, g = NR$. Suppose that $\varphi_i = g v_i : M_i \to N$ is an isomorphism for all $i \in I$. Then there exists $\varphi_i^{-1} = (g v_i)^{-1} : N \to M_i$. Write $f_i = v_i \varphi_i^{-1}$. Then $g f_i = (g v_i) \varphi_i^{-1} = \varphi_i \varphi_i^{-1} = 1_{M_i}$. Setting $f = \sum_{i \in I} f_i$ we obtain $g f = \sum_{i \in I} g f_i = 1_M$. Therefore $g$ is surjective. Since $g$ is injective, $g$ is an isomorphism. So that $N \simeq NR$, which implies that $N = 0 = M$, since $R$ is nilpotent for a right Artinian ring.

2. Suppose that $N$ is non-projective. Then by proposition 7.2.3 there exists an almost split sequence (7.2.4). Therefore $g$ is an epimorphism which is not split. Suppose that $\varphi_i = g v_i : M_i \to N$ is an isomorphism for all $i \in I$. Then there exists $\varphi_i^{-1} = (g v_i)^{-1} : N \to M_i$. Let $\pi_i : M \to M_i$ be a natural projection. Then $\pi_i = \varphi_i^{-1} g$ for all $i \in I$. Consider $f = \sum_{i \in I} (v_i \varphi_i^{-1}) : N \to M$. Then

$$f g = \sum_{i \in I} (v_i \varphi_i^{-1}) g = \sum_{i \in I} (v_i \pi_i) = 1_M$$

which implies that $g$ is a split epimorphism. A contradiction. $\square$

**Definition 7.2.6.** A family $\{M_i \; : \; i \in I\}$ of finitely generated indecomposable modules is called **Noetherian** if for any sequence of non-isomorphisms

$$M_{i_1} \xrightarrow{f_{i_1}} M_{i_2} \xrightarrow{f_{i_2}} M_{i_3} \longrightarrow \cdots$$

there is an integer $n$ such that $f_{i_n} \cdots f_{i_2} f_{i_1} = 0$ (here we do not assume that $M_{i_j} \neq M_{i_k}$ for $j \neq k$).

This family is called **co-Noetherian** if for any sequence of non-isomorphisms

$$\cdots \longrightarrow M_{i_3} \xrightarrow{f_{i_2}} M_{i_2} \xrightarrow{f_{i_1}} M_{i_1}$$

there is an integer $n$ such that $f_{i_1} f_{i_2} \cdots f_{i_n} = 0$ (here we do not assume that $M_{i_j} \neq M_{i_k}$ for $j \neq k$).

If the family of all finitely generated indecomposable right $A$-modules is Noetherian (resp. co-Noetherian), one says that the ring $A$ satisfies the **Noetherian** (resp. **co-Noetherian**) **condition** for finitely generated indecomposable right $A$-modules.

In the following two lemmas one uses the notation

$$\theta(j, i) = f_j f_{j-1} \ldots f_{i+1} f_i$$

for $j \geq i$. The proofs of these lemmas as lemmas 11 and 12 in [99].

**Lemma 7.2.7.** ([99, Lemma 11]). *Let* $\{M_i : i \in I\}$ *be a family of indecomposable modules such that* $l(M_i) = n < \infty$ *for all* $i \in I$. *Then* $l(\theta(2^m, 1)(M_1)) \leq n - m - 1$ *for any $m$ and for any non-isomorphism* $f_i : M_i \to M_{i+1}$. *Here* $l(\theta(2^m, 1)(M)) \leq 0$ *is interpreted as* $\theta(2^m, 1)(M) = 0$.

*Proof.* Since $l(M_i) = n$ for all $i$, each $f_i$ is neither a monomorphism nor an epimorphism. This lemma is proved by induction on $m$.

1. If $m = 0$, then $l(f_1(M_1)) \leq n - 1 = n - 0 - 1$.
2. Assume that $l(\theta(2^m, 1)(M_1)) \leq n - m - 1$ and $n - m - 1 \neq 0$.
3. It is clear that $l(\theta(2^{m+1}, 1)(M_1)) \leq n - m - 1$. If $l(\theta(2^{m+1}, 1)(M_1)) = n - m - 1$, then $l(\theta(2^m, 1)(M_1)) = n - m - 1$. Hence $\theta(2^{m+1}, 2^m + 1)\theta(2^m, 1)(M_1)$ is monomorphic. Furthermore, from assumption 2 we have $l(\theta(2^{m+1}, 2^m + 1)(M_{2^m+1})) \leq n - m - 1$. Since $l(\theta(2^{m+1}, 2^m + 1)(M_{2^m+1})) \geq l(\theta(2^{m+1}, 1)(M_1)) = n - m - 1$, $\theta(2^{m+1}, 2^m + 1)(M_{2^m+1}) = \theta(2^{m+1}, 1)(M_1)$. Hence $M_{2^m+1} = \theta(2^m, 1)(M_1) \oplus \mathrm{Ker}\theta(2^{m+1}, 2^m + 1)$. The sum is direct because $\theta(2^{m+1}, 2^m + 1)$ is monomorphic on $\theta(2^m, 1)(M_1)$ as noted above. Hence $\theta(2^m, 1)(M_1)) \neq 0$, $\mathrm{Ker}\theta(2^{m+1}, 2^m + 1) \neq 0$, which is a contradiction. Therefore $l(\theta(2^{m+1}, 1)(M_1)) \leq n - m - 2$. $\square$

**Comment.** Note that $\theta(2^{m+1}, 2^m + 1)$ is also the composition of $2^m$ $f$'s. That is the reason for working with $2^m$ instead of simply $m$.

**Lemma 7.2.8.** ([99, Lemma 12]). *Let* $\{M_i : i \in I\}$ *be a set of indecomposable modules such that there exists some integer $m$, such that* $l(M_i) \leq m$ *for all $i \in I$. Then there exists an integer $n$ such that* $f_n \cdots f_2 f_1 = 0$ *for any family of non-isomorphisms* $f_j : M_j \to M_{j+1}$.

*Proof.* It is clear from the assumption that for any $i$ at least one $f_{i+j}$ among the $f_{i+k}$, $k = 0, 1, \ldots, m$ is not a monomorphism. Let $f_{i_1}, f_{i_2}, \ldots$ be not monomorphisms, then $g_1 = \theta(i_2 - 1, i_1)$, $g_2 = \theta(i_3 - 1, i_2)$, $\ldots$ are not monomorphisms. Consider the family $\{M_{i_j}\}$. Since $l(M_{i_j}) < m$, there exist some $r \leq m$ and an infinite sub-family $\{M_{k_i}\}$ such that $l(M_k) = r$ for all $k$. Put $h_1 = g_{k_2-1} g_{k_2-2} \cdots g_{k_1}$, $h_2 = g_{k_3-1} g_{k_3-2} \cdots g_{k_2}$, $\ldots$ and apply lemma 7.2.7 to the system of non-isomorphisms $\{h_i\}$. It follows that there is a fixed number $n$ such that $\theta(n, 1) = 0$. It is clear that one can find this $n$ independently of the choice of $M_i$ and $f_i$. $\square$

In what follows, $[M]$ will be denote the isomorphism class of a given module $M$.

For an indecomposable module $M$, define a family $\mathbf{E}_n(M)$ $(n \geq 0)$ of a finite number of isomorphism classes of indecomposable modules as follows:

(i) $\mathbf{E}_0(M) = \{[M]\}$;

(ii) $[X] \in \mathbf{E}_{n+1}(M)$ if and only if $X$ is a direct summand of an almost split extension over some module whose isomorphism class belongs to $\mathbf{E}_n(M)$.

**Theorem 7.2.9.** (M. Auslander [15], K. Yamagata [226]). *Let A be a right Artinian ring and* $\{[S_i] \ : \ 1 \leq i \leq n\}$ *the family of all isomorphism classes of simple right A-modules. Assume that there is an almost split extension over any finitely generated indecomposable right A-module. Then the following are equivalent.*

1. *A is of finite representation type.*
2. *A satisfies the co-Noetherian condition for finitely generated indecomposable right A-modules.*
3. *There is an integer m such that*

$$\mathbf{E}_{m+1}(S) \subset \bigcup_{\substack{1 \leq i \leq n \\ 0 \leq j \leq m}} \mathbf{E}_j(S_i)$$

*for every simple right A-module S.*

*Further, in this case* $\displaystyle\bigcup_{\substack{1 \leq i \leq n \\ 0 \leq j \leq m}} \mathbf{E}_j(S_i)$ *is the family of all isomorphism classes of finitely generated indecomposable right A-modules.*

*Proof.*

$1 \Longrightarrow 2$ follows from lemma 7.2.8.

$2 \Longrightarrow 3$. For each $S_i$ define a finite family $\mathbf{H}_r(S_i)$ of non-isomorphisms $f_r :$ $M_r \rightarrow M_{r-1}$ with $[M_r] \in \mathbf{E}_r(S_i)$ and $M_{r-1} \in \mathbf{E}_{r-1}(S_i)$ as follows: if $M_r$ is a direct summand of an almost split extension $N_r$ over $M_{r-1}$ with its defining homomorphism $\sigma_r : N_r \rightarrow M_{r-1}$, then we put $f_r = \sigma_r v_r$, where $v_r$ is a canonical injection of $M_r$ to $N_r$, otherwise we put $f_r = 0$. Here $M_0$ denotes the given $S_i$, and in either case, $f_r$ is not an isomorphism, by lemma 7.2.5.

Let $A_{i,n}$ be the finite family $\{f_1 \cdots f_n \ : \ f_j \in \mathbf{H}_j(S_i), \ f_1 \cdots f_n \neq 0\}$ and $F_i$ a family of functions $\{\theta_{i,n}\}$, where $\theta_{i,n}$ is the function of $A_{i,n}$ to the power set of $A_{i,n+1}$ giving by $\theta_{i,n}(a_n) = \{a_n f_{n+1} \ : \ f_{n+1} \in \mathbf{H}_{n+1}(S_i), \ a_n f_{n+1} \neq 0\}$ for each $a_n \in A_{i,n}$. Then, applying the König graph theorem [101, lemma 4.7.2], there results an integer $m_i$ for each $S_i$ such that $f_1 \cdots f_{m_i} = 0$ for all $f_j \in \mathbf{H}_j(S_i)$ $(1 \leq j \leq m_i)$. Put $m = \max\{m_i \ : \ 1 \leq i \leq n\}$.

Assume that there exists a module $M$ such that

$$[M] \notin \bigcup_{\substack{1 \leq i \leq n \\ 0 \leq j \leq m}} \mathbf{E}_j(S_i) = \mathbf{E}.$$

Since all isomorphism classes of simple modules are contained in $\mathbf{E}$, $M$ is not simple. Hence there is a non-zero homomorphism $g : M \to S$ such that it is not a split epimorphism and $S$ is a non-projective simple module. By the definition 7.2.2, there are homomorphisms $g_1' : M \to N_1$ and $f_1' : N_1 \to S$ such that $g = f_1' g_1'$, where $f_1'$ is an almost split extension over $S$. This means that there are homomorphisms $g_1 : M \to M_1$ and $f_1 : M_1 \to S$ such that $[M_1] \in \mathbf{E}_1(S)$, $f_1 g_1 \neq 0$ and $f_1 \in \mathbf{H}_1(S)$, by lemma 7.2.5. Consider $g_1$ instead of $g$ in the above. Then, by the same method, there exist $g_2 : M \to M_2$ and $f_2 : M_2 \to M_1$ such that $[M_2] \in \mathbf{E}_2(S)$, $f_1 f_2 g_2 \neq 0$ and $f_2 \in \mathbf{H}_2(S)$. Repeating this process we get homomorphisms $f_i \in \mathbf{H}_i(S)$ ($1 \leq i \leq m$) such that $f_1 \cdots f_{m-1} f_m \neq 0$, which contradicts the choice of $m$. This completes the proof $2 \Rightarrow 3$.

$3 \Rightarrow 1$. What needs to be proved is that the process of taking repeated almost split extensions exhausts the isomorphism classes of indecomposables. Given that $3 \Rightarrow 1$ is immediate.

Assume that there is an indecomposable module $M$ such that

$$[M] \notin \bigcup_{\substack{1 \leq i \leq n \\ 0 \leq j \leq m}} \mathbf{E}_j(S_i).$$

Then $M$ is not simple and therefore there is a non-projective simple module $S$ which is a proper homomorphic image of $M$. By the same reasoning as in the proof $2 \Rightarrow 3$, we have non-isomorphisms $f_i : M_i \to M_{i-1}$ such that $f_i \in \mathbf{H}_i(S)$ and $f_1 \cdots f_{m+1} \neq 0$, where $M_0$ is $S$. Furthermore, by the definition of an almost split extension and the assumption on $M$, we also find a homomorphism $f_{m+2} : M_{m+2} \to M_{m+1}$ such that $f_1 \cdots f_{m+1} f_{m+2} \neq 0$ and $[M_{m+2}] \in \mathbf{E}_{m+2}(S)$. Here one can choose a non-isomorphism $f_{m+2}$, by means of lemma 7.2.5. Repeating this process we obtain a sequence of non-isomorphisms

$$\cdots \to M_{m+2} \xrightarrow{f_{m+2}} M_{m+1} \xrightarrow{f_{m+1}} M_m \to \cdots \to M_2 \xrightarrow{f_2} M_1 \xrightarrow{f_1} S$$

such that each

$$[M_k] \in \mathbf{E} = \bigcup_{\substack{1 \leq i \leq n \\ 0 \leq j \leq m}} \mathbf{E}_j(S_i)$$

and $f_1 \cdots f_{k-1} f_k \neq 0$ for any $k > 0$. This, however, contradicts lemma 7.2.8, because $\mathbf{E}$ is a finite set and so there is an integer $t$ such that $l(M_k) \leq t$ for all $k > 0$. $\square$

**Remark 7.2.10.** The equivalence $1 \iff 2$ in theorem 7.2.9 was proved by M. Auslander in [15], and the other equivalences were proved by K. Yamagata in [226]. The proof is as in [226].

**Lemma 7.2.11.** *Let a right Artinian ring A satisfy the Noetherian condition*[2] *for indecomposable modules, and let $\mathcal{E}$: $0 \to L \to N \to M \to 0$ be a non-split exact sequence, where M and L are indecomposable. Then there are an indecomposable module $L_1$ and a homomorphism $f : L \to L_1$ such that*[3]

1. $[f\mathcal{E}] \neq [O]$ *in* $\mathcal{E}xt^1_A(M, L_1)$.
2. *If* $[gf\mathcal{E}] \neq [O]$ *for a homomorphism* $g : L_1 \to X_1$, *then g is a split monomorphism.*

Note that here $O$ stands for a split exact sequence. Elsewhere (in Chapter 4, Chapter 8 e.g.) it stands for a discrete valuation ring.

*Proof.* Suppose that for any homomorphism $f : L \to L'$ such that $[f\mathcal{E}] \neq [O]$ and $L'$ is indecomposable, there is a non-isomorphism $g : L' \to L''$ such that $[gf\mathcal{E}] \neq [O]$ and $L''$ is indecomposable. Then for the identity $1_L$ there is a non-isomorphism $f_2 : L_2 \to L_3$ with $[f_2 f_1 \mathcal{E}] \neq [O]$ (where $f_1 = 1_L$ and $L = L_1 = L_2$), in particular $f_2 f_1 \neq 0$. In this way, there results a sequence of non-isomorphisms

$$L_1 \xrightarrow{f_1} L_2 \xrightarrow{f_2} L_3 \to \cdots$$

such that $f_n \cdots f_2 f_1 \neq 0$ for any $n \geq 1$, where all $L$'s are indecomposable. This contradicts the Noetherian condition. Hence there is a homomorphism $f : L \to L_1$ such that

   a. $L_1$ is indecomposable.
   b. $[f\mathcal{E}] \neq [O]$.
   c. $[gf\mathcal{E}] = [O]$ for every non-isomorphism $g : L_1 \to X$ with $X$ indecomposable.

Let's now show that $f$ satisfies property 2. Let $g : L_1 \to X$ be a non-zero homomorphism such that $[gf\mathcal{E}] \neq [O]$. For a decomposition into indecomposables $X = \bigoplus_{i=1}^m X_i$, let $v_i : X_i \to X$ and $\sigma_i : X \to X_i$ be the canonical injections and projections, respectively, and set $g_i = v_i \sigma_i g$. Then $g = \sum_{i=1}^m g_i$ and $[O] \neq [gf\mathcal{E}] = \sum_{i=1}^m [g_i f\mathcal{E}]$. Hence there is some $g_i$, say $g_1$, with $[g_1 f\mathcal{E}] \neq [O]$. Then it is easy to show that $[g_1 f\mathcal{E}] \neq [O]$ implies $[(\sigma_1 g)f\mathcal{E}] \neq [O]$. Therefore, by result c, $\sigma_1 g$ must be an isomorphism and so $g$ is a split monomorphism. $\square$

**Lemma 7.2.12.** *Assume that a right Artinian ring A satisfies the Noetherian condition for indecomposable modules. Then for any indecomposable A-module M there is an almost split extension over M.*

---

[2]See definition 7.2.6.

[3]Recall that $[\mathcal{E}]$ denotes the equivalent class of a short exact sequence $\mathcal{E}$ in a group of all equivalent classes $\mathcal{E}xt^1_A(M, L_1)$, and $[O]$ denotes the zero element in this group. In section 3.4 it was shown that the group $\mathcal{E}xt^1_A(M, L_1)$ can be identified with the group $Ext^1_A(M, L_1)$.

*Proof.* If $M$ is projective, it has the unique maximal submodule $MR$, where $R$ is the Jacobson radical of $A$. Hence it is clear that the inclusion $f : MR \to M$ is almost split. So $f$ is an almost split extension by definition 7.2.2.

Assume that $M$ is not projective. Then there is an indecomposable module $L'$ such that $\mathcal{E}xt_A^1(M, L') \neq 0$. Let $\mathcal{E}' \in \mathcal{E}xt_A^1(M, L')$ be an exact non-split sequence $\mathcal{E}' : 0 \to L' \to N \to M \to 0$. By lemma 7.2.11, there is a homomorphism $a : L' \to L$ which satisfies conditions 1 and 2 of this lemma. For $a\mathcal{E}'$ write $\mathcal{E} : 0 \to L \xrightarrow{u} N' \xrightarrow{v} M \to 0$. Let $f : X \to M$ be a non-split epimorphism. Then one can show that there is a homomorphism $g : X \to N$ with $f = vg$. For this, consider the commutative diagram:

$$
\begin{array}{ccccccccc}
\mathcal{E}: & 0 & \longrightarrow & L & \xrightarrow{\;u\;} & N & \xrightarrow{\;v\;} & M & \longrightarrow 0 \\
& & & \downarrow{\scriptstyle s} & & \downarrow{\scriptstyle (1_N,0)} & & \downarrow{\scriptstyle 1_M} & \\
s\mathcal{E}: & 0 & \longrightarrow & \operatorname{Ker}\varphi & \longrightarrow & N \oplus X & \xrightarrow[\varphi=(v,f)]{} & M & \longrightarrow 0
\end{array}
$$

where $(1_N, 0) : N \to N \oplus X$ and $\varphi = (v, f) : N \oplus X \to M$ (here we represent homomorphisms by matrices according to the decompositions of the given modules). If we would be able to show that $[s\mathcal{E}] \neq [O]$, then $s$ must be a split monomorphism by lemma 7.2.11, say $t : \operatorname{Ker}\varphi \to L$ and $ts = 1_L$. Since $[\mathcal{E}] = [1_L\mathcal{E}] = [ts\mathcal{E}]$, there exists a homomorphism $(w, g) : N \oplus X \to N$ such that $(v, f) = v(w, g)$, and so $f = vg$. Thus, it remains only to show that $[s\mathcal{E}] \neq [O]$. Suppose the contrary. Then there is a $\psi = (v', f') : M \to N \oplus X$ with $1_M = \phi\psi$. Since the endomorphism ring of $M$ is local and $1_M = vv' + ff'$, either $vv'$ or $ff'$ must be an automorphism. On the other hand, $vv'$ is not an automorphism, because $v : N \to M$ is not split. Moreover, since $f$ is not a split epimorphism by assumption, $ff'$ is not an automorphism, either. The obtained contradiction concludes the proof of the lemma. $\square$

**Theorem 7.2.13.** (A. Roiter [183], M. Auslander [15].) *For a right Artinian ring $A$ the following are equivalent.*

1. *$A$ is of finite representation type.*
2. *$A$ satisfies the Noetherian and co-Noetherian conditions for finitely generated indecomposable right $A$-modules.*

*In particular, if there is an integer $n$ such that $l(M) \leq n$ for all finitely generated indecomposable right $A$-modules $M$, then there is only a finite number of finitely generated indecomposable right $A$-modules.*

*Proof.*

$1 \implies 2$. Let $A$ be a right Artinian ring of finite representation type. Then from lemma 7.2.8 it follows that $A$ satisfies the Noetherian condition. Therefore for any indecomposable $A$-module $M$ there is an almost split extension over $M$ by lemma 7.2.12, i.e. $A$ satisfies the conditions of theorem 7.2.9. So the statement follows from this theorem.

$2 \Longrightarrow 1$. Since $A$ satisfies the Noetherian condition, from lemma 7.2.12 it follows that for any indecomposable $A$-module $M$ there is an almost split extension over $M$. Therefore the statement follows from theorem 7.2.9. $\square$

**Remark 7.2.14.** We now point out some of the basic duality for main morphisms considered above.

Recall that the **opposite category** $C^{op}$ of $C$ is defined as follows: the objects of $C^{op}$ are the same as $C$, the morphism sets in $C^{op}$ are defined as

$$\operatorname{Hom}_{C^{op}}(X, Y) = \operatorname{Hom}_{C^{op}}(Y, X)$$

and the composition of morphisms:

$$\operatorname{Hom}_{C^{op}}(Y, Z) \times \operatorname{Hom}_{C^{op}}(X, Y) \to \operatorname{Hom}_{C^{op}}(X, Z)$$

are given by $(g, f) \mapsto g \circ^{op} f = f \circ g$.

Therefore a morphism $f : X \to Y$ in $C$ is a monomorphism if and only if $f : Y \to X$ is an epimorphism in $C^{op}$. A morphism $1_X$ is the identity in $C$ if and only if it is the identity morphism in $C^{op}$.

This gives the possibility to obtain the dual statements for opposite category[4]. For example, if we have some commutative diagram $D$ in a category $C$ then reversing all arrows in this diagram we obtain a commutative diagram in the opposite category $C^{op}$ which is called a **dual diagram** of $D$. In particular, the projective objects in the category $C$ correspond to injective objects in the opposite category $C^{op}$ and vice versa. The dual to any exact sequence in $C$ is also an exact sequence in $C^{op}$.

Using definitions 4.4.1, 4.4.6 and 4.4.7 we immediately obtain the following proposition.

**Proposition 7.2.15.** (M. Auslander, I. Reiten [21, Proposition 2.2]). *Let $A$ be a ring. Then a homomorphism $f : X \to Y$ of right $A$-modules is right minimal, right almost split, or minimal right almost split if and only if the corresponding morphism $f^{op} : Y \to X$ of left $A^{op}$-modules is left minimal, left almost split, or minimal left almost split respectively.*

As a corollary of this statement we obtain the following proposition.

**Proposition 7.2.16.** *A sequence*

$$0 \longrightarrow L \longrightarrow M \longrightarrow N \longrightarrow 0$$

*of right $A$-modules is almost split if and only if the sequence*

$$0 \longrightarrow N \longrightarrow M \longrightarrow L \longrightarrow 0$$

*of left $A^{op}$-modules is almost split.*

---

[4]See e.g. [145, section 2.1].

**Theorem 7.2.17.** (M. Auslander [15].) *If A is an Artinian ring of finite representation type, then the endomorphism ring of any finitely generated right (left) A-module is also Artinian.*

*Proof.* Since $A$ is an Artinian ring, any finitely generated right $A$-module $M$ decomposes into a direct sum of finitely generated indecomposable right $A$-modules. Without loss of generality we can assume that $M$ is a direct sum of all non-isomorphic indecomposable right $A$-modules $M_r$ $(1 \leq r \leq n)$ with the canonical injections $v_r : M_r \rightarrow M$ and projections $\sigma_r : M \rightarrow M_r$. Let $B = \text{End}_A(M)$ be the endomorphism ring of $M$. Write $v_r \sigma_r = e_r$, that is an idempotent of $B$ for $r = 1, 2, \ldots, n$. Then $1_B = \sum_{r=1}^{n} e_r$ is a decomposition of the identity of $B$ into a sum of pairwise orthogonal idempotents. Therefore the two-sided Peirce decomposition of $B$ has the following form

$$B = \begin{pmatrix} B_{11} & B_{12} & \cdots & B_{1n} \\ B_{21} & B_{22} & \cdots & B_{2n} \\ \vdots & \vdots & \ddots & \vdots \\ B_{n2} & B_{n2} & \cdots & B_{nn} \end{pmatrix}$$

where $B_{ij} = \text{Hom}(M_j, M_i)$. Since $M_i$ is a finitely generated indecomposable module over a right Artinian ring, $B_{ii} = \text{End}_A(M_i)$ is a local Artinian ring. Therefore $B$ is a semiperfect ring with Jacobson radical $J = \sum_r e_r J$ whose two-sided Peirce decomposition is

$$J = \begin{pmatrix} J(B_{11}) & B_{12} & \cdots & B_{1n} \\ B_{21} & J(B_{22}) & \cdots & B_{2n} \\ \vdots & \vdots & \ddots & \vdots \\ B_{n2} & B_{n2} & \cdots & J(B_{nn}) \end{pmatrix}$$

where $J(B_{ii})$ is the Jacobson radical of $B_{ii}$. Since $B_{ii}$ is an Artinian ring, $J(B_{ii})$ is nilpotent for any $i$. Therefore $J$ is also nilpotent by [100, theorem 11.4.1]. Since $B/J$ is a semisimple ring, $B$ is a semiprimary ring. By the Hopkins-Levitzki theorem (see [103, theorem 1.1.22]) a ring is right Artinian if and only if it is semiprimary and right Noetherian, so that it suffices to show that $B$ is a right Noetherian ring. To this end we show that each $e_\alpha J$ is finitely generated as a right $B$-module.

Since $A$ is an Artinian ring of finite representation type for any indecomposable module $M_\alpha$ there exists an almost split extension over $M_\alpha$ by lemma 7.2.12. Let $f : N \rightarrow M_\alpha$ be an almost split extension over $M_\alpha$. If $M_\alpha$ is projective, then $f$ is injective, i.e. $N$ is a submodule of $M_\alpha$. Since $M_\alpha$ is a finitely generated module over an Artinian ring $A$, $N$ is also finitely generated. Therefore it is decomposed into a direct sum of indecomposable $A$-modules. Since $M$ is a direct sum of all indecomposable $A$-modules, $N$ can be embedded in a finite direct sum $M^{(m)}$ of $m$ copies of $M$ as a direct summand. If $M_\alpha$ is not projective, then there is an almost split sequence $0 \rightarrow \text{Ker} f \rightarrow N \rightarrow M_\alpha \rightarrow 0$ where $\text{Ker} f$ is an indecomposable module. Therefore there is $\beta$ such that $\text{Ker} f \simeq M_\beta$. Hence $N$ is an Artinian and

Noetherian module. Since $A$ is an Artinian ring, $N$ is a finitely generated module. So that applying the arguments above $N$ can be embedded in a finite direct sum $M^{(m)}$ of $m$ copies of $M$ as a direct summand.

Let $v : N \to M^{(m)}$ and $\sigma : M^{(m)} \to N$ be homomorphisms with $\sigma v = 1_N$. Then, since $\sigma_\alpha g v_\beta : M_\beta \to M_\alpha$ is not an isomorphism for any $g \in J$, there is a homomorphism $h_\beta : M_\beta \to N$ such that $\sigma_\alpha g v_\beta = f h_\beta$ and so $\sigma_\alpha g v_\beta = f \sigma v h_\beta = f \sigma (\sum_{i=1}^{m} \psi_i \phi_i) v h_\beta = \sum_{i=1}^{m} (f\sigma)(\phi_i v h_\beta)$, where $\psi_i$ and $\phi_j$ are the canonical $i$-th injection $M \to M^{(m)}$ and $j$-th projection $M^{(m)} \to M$, respectively. Therefore $e_\alpha g e_\beta = \sum_{i=1}^{m} (v_\alpha f \sigma \psi_i)(\phi_i v h_\beta \sigma_\beta)$. Clearly $v_\alpha f \sigma \psi_i$ and $\phi_i v h_\beta \sigma_\beta$ belong to $e_\alpha J$ and $B$, respectively. Put $v_\alpha f \sigma \psi_i = w_i$ and $\phi_i v h_\beta \sigma_\beta = b_{\beta,i}$. Then there results that

$$e_\alpha g = \sum_\beta e_\alpha g e_\beta = \sum_{\substack{1 \le i \le m \\ 1 \le \beta \le n}} w_i b_{\beta,i}.$$

Since each $w_i$ does not depend on $g \in J$, this equality shows that $e_\alpha J = \sum_{1 \le i \le m} w_i B$ and hence $J$ is finitely generated as a right $B$-module.

Taking into account that any left $B$-module can be considered as a right module over an opposite ring $B^{op}$ and $J(B) = J(B^{op})$ by [100, Proposition 3.4.5], it suffices to show that $J(B^{op})$ is a finitely generated right $B^{op}$-module. But $B^{op} = (\mathrm{End}_A(M_A))^{op} = \mathrm{End}_{A^{op}}(A^{op} M)$. By proposition 7.1.11, $A^{op}$ is also an Artinian ring of finite representation type. Therefore for a module $M$ considered as a left $A^{op}$-module we can consider the dual almost split sequence which is almost split by proposition 7.2.16. Since in the proof above we used only almost split sequences and projective modules dual of which are injective modules, we can also show that $B^{op}$ is right Artinian, i.e. $B$ is left Artinian. $\square$

## 7.3 Artinian Semidistributive Rings

This section is devoted to the connection between finite dimensional algebras of finite type and semidisributive rings.

**Proposition 7.3.1.** *A local algebra of finite representation type $A$ over an algebraically closed field is uniserial.*

*Proof.* Since $A$ is a local algebra over a field $k$, its radical $R = \mathrm{rad}\, A$ is the unique ideal in it, and $A/R \simeq k$. Consider the algebra $A_1 = A/R^2$. Then $A_1$ is also a local algebra of finite representation type with Jacobson radical $R_1 = R/R^2$ and quiver $Q(A_1) = Q(A)$. Since $R_1^2 = 0$ and $k$ is an algebraically closed field, $Q^b(A_1)$ does not contain multiply arrows by the Gabriel theorem [101, theorem 2.6.1]. Therefore $Q(A_1)$ is either a point or one-pointed graph with one loop. Then, by [100, theorem 12.3.11], $A_1$ is serial. So, by [100, theorem 12.3.10], which asserts that a semiperfect

Noetherian ring is serial if and only if $A/R^2$ is a serial Artinian ring, $A$ is also serial. Since $A$ is local, it is uniserial, as required. $\square$

Let $k$ be a field, and let $A \in M_n(k)$. Consider a $k$-algebra $k[A] = \{f(A) : f(x) \in k[x]\}$. Obviously, $k[A] \simeq k[x]/(\mu_A(x))$, where $\mu_A(x)$ is the minimal polynomial of $A$ and $(\mu_A(x))$ is the principal ideal of $k[x]$ generated by $\mu_A(x)$.

**Lemma 7.3.2.** *Let $k$ be a field, then the algebra $k[a, b]$ with relations*

$$a^2 = b^2 = ab = ba = 0$$

*is of infinite representation type.*

*Proof.* Obviously, any finite dimensional module over the algebra $k[a, b]$ is given by a finite dimensional vector $k$-space $M$ and actions of two linear operators $a$ and $b$ in this space satisfying the equalities:

$$ab = ba = a^2 = b^2 = 0.$$

Let $M$ be a vector space of even dimension $n = 2m$ and let the action of the operators $a$ and $b$ in some basis be given by matrices of the following form:

$$A = \begin{pmatrix} 0 & \mathbf{I}_m \\ 0 & 0 \end{pmatrix}; \quad B = \begin{pmatrix} 0 & J_m(0) \\ 0 & 0 \end{pmatrix},$$

where $\mathbf{I}_m$ is the identity matrix of order $m$, and $J_m(0)$ is a lower Jordan block of order $m$ with zero eigenvalue.

It is obvious that any endomorphism $\varphi$ of $M$ is given by an $n \times n$-matrix $X$ such that, $XA = AX$ and $XB = BX$. View $X$ as

$$X = \begin{pmatrix} X_{11} & X_{12} \\ X_{21} & X_{22} \end{pmatrix},$$

where all four blocks are square of order $m$.

The condition $AX = XA$ implies:

$$\begin{pmatrix} X_{21} & X_{22} \\ 0 & 0 \end{pmatrix} = \begin{pmatrix} 0 & X_{11} \\ 0 & X_{21} \end{pmatrix},$$

whence

$$X = \begin{pmatrix} X_{11} & X_{12} \\ 0 & X_{11} \end{pmatrix}.$$

The second equality $BX = XB$ now implies that:

$$\begin{pmatrix} 0 & J_m(0)X_{11} \\ 0 & 0 \end{pmatrix} = \begin{pmatrix} 0 & X_{11}J_m(0) \\ 0 & 0 \end{pmatrix}.$$

It is easy to see that all matrices $X_{11}$ which commute with $J_m(0)$ generate an $m$-dimensional $k$-algebra $A_m = k[J_m(0)]$ consisting of matrices of the following form:

$$\begin{pmatrix} \alpha_1 & 0 & \cdots & \cdots & \cdots & 0 \\ \alpha_2 & \alpha_1 & 0 & \cdots & \cdots & 0 \\ \alpha_3 & \alpha_2 & \alpha_1 & 0 & \cdots & 0 \\ \vdots & \vdots & \vdots & \ddots & \vdots & \vdots \\ \alpha_{m-1} & \alpha_{m-2} & \cdots & \cdots & \alpha_1 & 0 \\ \alpha_m & \alpha_{m-1} & \alpha_{m-2} & \cdots & \alpha_2 & \alpha_1 \end{pmatrix}.$$

The $k$-algebra $A_m$ is commutative and the ideals in $A_m$ form a descending chain:

$$A_m \supset J_m(0)A_m \supset J_m^2(0)A_m \supset \ldots \supset J_m^{m-1}(0)A_m \supset 0$$

with simple factors, i.e. $A_m$ is a uniserial algebra. Therefore the algebra of endomorphisms of $M$ is local and so $M$ is indecomposable. Thus, for any $m$ there is an indecomposable module of dimension $m^2$. So $k[a, b]$ is of infinite representation type. $\square$

**Remark 7.3.3.** The representations for the more general algebra $k[a, b]$ with relations $ab = ba = 0$ were considered by I.M. Gelfand and V.A. Ponomarev in [83]. Analogous problems were considered by Yu.A. Drozd [60].

**Remark 7.3.4.** Let $A$ be a semiprimary ring with Jacobson radical $R(A) = \mathrm{rad}\, A$ and $R^2(A) = 0$. If $e$ is a non-zero idempotent of $A$ and $B = eAe$ with Jacobson radical $R(B)$, then $R^2(B) = 0$.

Indeed, by [103, proposition 1.1.19], $R(B) = eR(A)e \subset R(A)$. So $R^2(B) = 0$.

**Theorem 7.3.5.** *Let $A$ be a finite dimensional algebra over an algebraically closed field $k$ with radical $R$ and $R^2 = 0$. If $A$ is of finite representation type then $A$ is semidistributive.*

*Proof.* According to the reduction theorem [103, theorem 1.10.10], to show that a semiperfect ring $A$ is semidistributive it is equivalent to show that any minor of the second order of $A$ is semidistributive. Without loss of generality we can consider $A$ to be a basic algebra.

Consider a minor $B$ of the second order of $A$. Since $R^2 = 0$, $R^2(B) = 0$, where $R(B)$ is the Jacobson radical of $B$.

Let $1 = e_1 + e_2$ be a decomposition of $1 \in B$ into a sum of local idempotents and let

$$B = \bigoplus_{i,j=1}^{2} e_i B e_j$$

be the corresponding two-sided Peirce decomposition. We write $B_{ij} = e_i B e_j$ $(i, j = 1, 2)$. By proposition 7.1.10, $B$ and $B_{ii}$ $(i = 1, 2)$ are algebras over an algebraically closed field $k$ of finite representation type. Therefore, by proposition 7.3.1, $B_{ii}$ is a uniserial algebra ($i = 1, 2$).

The radical $R(B)$ of $B$ has the following form:

$$R(B) = \begin{pmatrix} R_1 & B_{12} \\ B_{21} & R_2 \end{pmatrix},$$

where $R_i = R(B_{ii})$ ($i = 1, 2$). Obviously,

$$R^2(B) = \begin{pmatrix} R_1^2 + B_{12}B_{21} & R_1B_{12} + B_{12}R_2 \\ B_{21}R_1 + R_2B_{21} & R_2^2 + B_{21}B_{12} \end{pmatrix}.$$

Consequently, $R_1B_{12} = B_{12}R_2 = 0$, $R_2B_{21} = B_{21}R_1 = 0$, and $B_{12}B_{21} = B_{21}B_{12} = 0$. So, $B_{12}$ and $B_{21}$ are finite dimensional vector spaces over $k$. Let $B_{12} \neq 0$, and $\dim_k B_{12} = m \geq 2$. Then there are elements $x_1, x_2, \ldots x_m \in B_{12}$ which are linear independent over $k$. Consider the elements $a_i = \begin{pmatrix} 0 & x_i \\ 0 & 0 \end{pmatrix}$, $i = 1, \ldots, m$. Obviously, $a_i^2 = a_j^2 = a_i a_j = a_j a_i = 0$ for $i, j = 1, \ldots m$. The subspace generated by elements $x_3, \ldots, x_m$ forms a two-sided ideal $I$ in $B$. It is easy to see that the set

$$T = \begin{pmatrix} R_1 & I \\ B_{21} & R_2 \end{pmatrix}$$

is a two-sided ideal in $B$. Then $S = B/T$ is isomorphic to $k[a_1, a_2]$ with relations $a_1^2 = a_2^2 = a_1 a_2 = a_2 a_1 = 0$, which is of infinite representation type, by lemma 7.3.2. So $B$ is also of infinite type. This contradiction shows that $\dim_k B_{12} = 1$. Analogously we can show that $\dim_k B_{21} = 1$. So $B$ is a semidistributive algebra, by [103, theorem 1.10.9]. And, by [103, theorem 1.10.10], $A$ is a semidistributive algebra, as well.

**Corollary 7.3.6.** *Let $A$ be a finite dimensional algebra of a finite representation type over an algebraically closed field $k$ with Jacobson radical satisfying $R^2 = 0$, then the quiver $Q(A)$ of $A$ is simply laced.*

*Proof.* The proof follows from [100, theorem 14.3.1] and theorem 7.3.5.

**Remark 7.3.7.** The assertion of corollary 7.3.6 is not true for arbitrary algebras. Let

$$A = \begin{pmatrix} \mathbf{R} & \mathbf{C} \\ 0 & \mathbf{R} \end{pmatrix},$$

where $\mathbf{R}$ is the field of real numbers, $\mathbf{C}$ is the field of complex numbers. Then

$$R = \text{rad } A = \begin{pmatrix} 0 & \mathbf{R} \\ 0 & 0 \end{pmatrix} \text{ and } R^2 = 0.$$

The algebra $A$ is of finite representation type and $Q(A)$ has the following form:

$$1 \; \bullet \mathrel{\substack{\longrightarrow \\ \longrightarrow}} \bullet \; 2$$

## 7.4  Artinian Hereditary Semidistributive Rings of Finite Representation Type

Note that by the work of P. Gabriel [79], [80], and V. Dlab and C.M. Ringel [53], a hereditary finite dimensional algebra is of finite representation type if and only if the corresponding diagram is a Dynkin diagram of type $A_n$, $B_n$, $C_n$, $D_n$, $E_6$, $E_7$, $E_8$, $F_4$ or $G_2$ (see Diag. 6.2.4).

The generalization of this result for hereditary Artinian rings was obtained by P. Dowbor, C.M. Ringel and D. Simson in the work [59] using valued graphs with valuations as dimension sequences[5] which were considered in section 6.2.

Let $A$ be an Artinian basic ring with Jacobson radical $R = \operatorname{rad}(A)$. Then, by the Wedderburn-Artin theorem, $B = A/R = \prod_{i=1}^{n} F_i$, where the $F_i$ are division rings, and $R/R^2 = \bigoplus_{i,j} {}_iM_j$ as $B$-bimodule. Then $\mathfrak{L} = (F_i, {}_iM_j)$ is called the **species** of $A$. As was shown at the end of section 6.2 with this species we associate its corresponding valued quiver $(\Gamma_{\mathfrak{L}}, d)$ and its corresponding valued undirected graph $(\Gamma_{\mathfrak{L}}, |d|)$, where $d(i, j)$ is the dimension sequence of the bimodule ${}_iM_j$, and $|d(i, j)|$ is the length of this sequence. We will call $(\Gamma_{\mathfrak{L}}, |d|)$ the **valued graph** of the ring $A$ and denote it $\Gamma(A)$.

**Theorem 7.4.1.** ([59, Theorem 2]). *A hereditary Artinian basic ring $A$ is of finite representation type if and only if $\Gamma(A)$ is a disjoint union of Coxeter diagrams $A_n$, $B_n$, $C_n$, $D_n$, $E_6$, $E_7$, $E_8$, $F_4$, $G_2$ and $H_3$, $H_4$, $I_2(p)$ ($p = 5$ or $p \geq 7$) (see [102, Figure 1.14.1]). In this case the dimension types of the indecomposable $A$-modules form the branch system[6] for $(\Gamma, d)$.*

**Remark 7.4.2.** There exists a hereditary ring $A$ of infinite representation type such that its valued graph $(\Gamma, \mathbf{d}, \Omega)$ is the Dynkin diagram:

$$\bullet \xrightarrow{\ (1,2)\ } \bullet$$

**Remark 7.4.3.** The existence of rings of type $H_3$, $H_4$ and $I_2(p)$ with $p = 5$ and $p \geq 7$ remains open: it depends on rather difficult questions concerning division rings.

In the case when we assume that the division rings $F_i$ are all finitely generated over their centers, then we can exclude these cases immediately, and so we obtain the following theorem:

**Theorem 7.4.4.** ([59, Theorem 3]). *A hereditary Artinian basic ring $A$ with Jacobson radical $R$ and $A/R$ finitely generated over its center is of finite representation type if*

---

[5]See definitions 6.2.20.

[6]see definition 6.2.21.

*and only if* $\Gamma(A)$ *is a disjoint union of Coxeter diagrams* $A_n$, $B_n$, $C_n$, $D_n$, $E_6$, $E_7$, $E_8$, $F_4$, $G_2$.

Let $A$ be a hereditary Artinian semidistributive basic ring. Suppose that $A$ is indecomposable. If $1 = e_1 + e_2 + \cdots + e_n$ is a triangular decomposition of the identity of $A$ into a sum of pairwise orthogonal idempotents then the two-sided Peirce decomposition of $A$ has the following form:

$$A = \begin{pmatrix} D & A_{12} & \cdots & A_{1n} \\ 0 & D & \cdots & A_{2n} \\ \vdots & \vdots & \ddots & \vdots \\ 0 & 0 & \cdots & D \end{pmatrix},$$

where $A_{ij} = e_i A e_j$ is either zero or a one-dimensional left and right vector space over a division ring $D$, and if $A_{ij} \neq 0$ and $A_{jk} \neq 0$ then $A_{ik} \neq 0$ as well. Then the Jacobson radical of $A$ is of the form

$$R = \begin{pmatrix} 0 & A_{12} & \cdots & A_{1n} \\ 0 & 0 & \cdots & A_{2n} \\ \vdots & \vdots & \ddots & \vdots \\ 0 & 0 & \cdots & 0 \end{pmatrix}$$

Therefore $B = A/R = \prod_{i=1}^{n} F_i$, where $F_i = D$ for all $i = 1, \ldots, n$, and $R/R^2 = \bigoplus_{i,j} {}_i M_j$ as $B$-bimodule, where ${}_i M_j = A_{ij}/\sum_{i<k<j} A_{ik}A_{kj}$. To this ring $A$ we can assign the species $\mathfrak{L} = (F_i, {}_i M_j)$ of it. The quiver $\Gamma(\mathfrak{L})$ of $\mathfrak{L}$ is the quiver $Q(A)$ of the ring $A$. Since the quiver $Q(A)$ has no circuits then $A$ is a split ring. By [103, theorem 8.8.6], $A$ is isomorphic to the tensor algebra of the species $\mathfrak{L}$. It follows directly from the definitions that the category of representations of the species $\mathfrak{L}$ and the category of modules of the tensor algebra $\mathfrak{T}(\mathfrak{L})$ are equivalent.

Therefore taking into account theorem 7.4.1 we obtain the following corollary.

**Corollary 7.4.5.** *A hereditary Artinian semidistributive basic ring $A$ is of finite representation type if and only if $\Gamma(A)$ is a disjoint union of Dynkin diagrams $A_n$, $D_n$, $E_6$, $E_7$, $E_8$.*

## 7.5 Notes and References

In [183] A.V. Roiter solved the Brauer-Thrall conjecture for finite dimensional algebras over fields, which states that if the lengths of the finitely generated indecomposable modules are bounded then there is only a finite number of finitely generated indecomposable modules. M. Auslander [15] has proved this theorem for Artinian rings. In [226], Kunio Yamagata not only gave a simpler proof of this theorem, he also showed how to construct all indecomposable modules from simple modules over an Artinian ring of finite representation type. In this chapter we dealt with the Auslander theorem 7.2.9 following to this paper. In the proof of this theorem we use lemma 7.2.7 and lemma 7.2.8, which are due to M. Harada and Y. Sai in [99].

# CHAPTER 8

# *O*-Species and SPSD-Rings of Bounded Representation Type

In chapter 7 some classes of Artinian rings of finite representation type were described, i.e. rings which have a finite number of non-isomorphic indecomposable finitely generated modules. This condition of finiteness of indecomposable finitely generated modules is not natural in general; for instance, for Noetherian rings.

An additive category $M$ is said to be of **finite representation type** if up to isomorphism $M$ has only a finite number of indecomposable objects.

Recall that an $A$-module $M$ is called **finitely presented** if there exists an epimorphism $\varphi : A^{(n)} \to M$ such that $\mathrm{Ker}(\varphi)$ is a finitely generated $A$-module. For any ring $A$ denote by mod-fp$_l$ $A$ (resp. mod-fp$_r$ $A$) the category of finitely presented left (resp. right) $A$-modules. This category is additive and so one can apply the definition of finite representation type to it. So there results the following definition.

**Definition 8.1.** A ring $A$ is said to be of **left** (resp. **right**) **finite representation type** if mod-fp$_l$ $A$ (or mod-fp$_r$ $A$) is of finite representation type. A ring $A$ is of **finite representation type** if it is both of right and left finite representation type.

Suppose that $A$ is an Artinian ring. By the Krull-Remak-Schmidt theorem, every finitely generated $A$-module decomposes uniquely up to isomorphism into a direct sum of finitely generated indecomposable $A$-modules. Since for a right Artinian ring any finitely generated right module is finitely presented, there results the equivalent definition in this case.

A ring $A$, not necessarily Artinian, is said to be of **finite representation type**, if it has only a finite number of non-isomorphic indecomposable finitely presented $A$-modules. For Artinian rings this definition coincides with the earlier one, because in this case each finitely generated $A$-module is finitely presented as well.

For Artinian rings along with the notion of finite representation type the notion of bounded representation type was considered. Recall that a right Artinian ring $A$ is said to be of **bounded representation type** if there is a bound on the length of finitely generated indecomposable right $A$-modules. The first Brauer-Thrall conjecture says that these notions are the same. M. Auslander proved that this conjecture is true for right Artinian rings (see theorem 7.2.13).

Following R.B. Warfield, Jr. a ring $A$ is of **right bounded representation type** if there is an upper bound on the number of generators required for indecomposable finitely presented right $A$-modules. In his paper [222] R.B. Warfield, Jr. put the following question:

*Question* 4. For what semiperfect rings is there an upper bound on the number of generators required for the indecomposable finitely presented modules?

In this chapter there will be described some special classes of semiperfect rings of bounded representation type, which give a some sort of answer to this question.

As showed R.B. Warfield, Jr. there is a serious restriction on the structure of rings of bounded representation type connected with modules of finite Goldie dimension. In section 8.1 it is proved that a semiprimary ring of finite right bounded representation type is left Artinian (see [222]).

In section 8.2 we consider $O$-species which are generalizations of species as introduced by P. Gabriel in [80]. In this section we also consider the tensor algebra $T(\Omega)$ constructed by an $O$-species $\Omega$ and show the equivalence of the category of all representations of an $O$-species $\Omega$ and the category of right $T(\Omega)$-modules.

The connection between right hereditary SPSD-rings and special kinds of $(D, O)$-species is considered in section 8.3.

In section 8.4 we discuss the reduction of representations of $(D, O)$-species to mixed matrix problems over discrete valuation rings and their common skew field of fractions. Some important mixed matrix problems are considered in section 17.5.

Sections 8.6 and 8.7 are devoted to the study of $O$-species of bounded representation type.

In section 8.8 we describe right hereditary SPSD-rings of bounded representation type by reduction to representations of $(D, O)$-species.

## 8.1 Semiperfect Rings of Bounded Representation Type

Let $A$ be a semiperfect ring with Jacobson radical $R$. Let $M$ be a finitely generated $A$-module, and let $S$ be a simple $A$-module. Denote by $\mu(M)$ the number of direct summands in the decomposition of $M/MR$ as a direct sum of simple modules, and by $\mu(M, S)$ the number of these summands which are isomorphic to $S$. These numbers are well-defined by the Krull-Remak-Schmidt theorem.

Let $M$ be a finitely presented $A$-module, and let $f : P \to M$ be its projective cover with kernel $K = \mathrm{Ker}\, f$. Write $\nu(M) = \mu(K)$, and $\nu(M, S) = \mu(K, S)$. These numbers are also well-defined, by [100, lemma 13.1.1].

Following R.B. Warfield, Jr. introduce the following definition.

**Definition 8.1.1.** A ring $A$ is called a ring of (right) **bounded representation type** if there is an upper bound on the number of generators required for indecomposable finitely presented right $A$-modules.

**Remark 8.1.2.** In [101] the notion of bounded representation type for finite dimensional algebras and Artinian rings was introduced. Recall that a finite dimensional algebra $A$ is said to be of bounded representation type if there is a bound on the length of indecomposable finitely dimensional modules. A right Artinian ring $A$ is said to be of bounded representation type if there is a bound on the length of indecomposable finitely generated right $A$-modules. Since for a right Artinian ring $A$ any finitely generated module is finitely presented and a bound on the length of indecomposable modules implies a bound of the number of generators required for indecomposable finitely presented right $A$-modules and vice versa, these notions are the same for right Artinian rings.

**Examples 8.1.3.**

1. Let $A$ be a right Artinian ring. Then $A$ is of finite representation type if and only if it is of bounded representation type, as follows from theorem 7.2.13.
2. Any serial ring is of bounded representation type. This follows from [100, theorem 13.2.1].

Since for a semiperfect ring $A$ every finitely presented $A$-module is a direct sum of indecomposable modules by [103, theorem 9.2.3], there immediately results the following proposition.

**Proposition 8.1.4.** *A semiperfect ring $A$ is Morita equivalent to a ring over which every finitely presented right module is a direct sum of cyclic modules if and only if $A$ is of right bounded representation type.*

As proved R.B. Warfield, Jr. there is a serious restriction on the structure of rings of bounded representation type. This restriction is formulated in terms of modules of finite rank.

**Theorem 8.1.5.** (R.B. Warfield, Jr. [222, Theorem 2.7]). *Let $A$ be a semiperfect ring of right bounded representation type. Then if $I$ is any left ideal of $A$, the quotient ring $A/I$ has finite Goldie dimension, and there is an upper bound for the numbers $\text{rank}(A/I)$. In particular, if $A$ is right semiprimary then $A$ is left Artinian.*

*Proof.* Let $P$ be a projective left $A$-module and let $X$ be a submodule of $P$. If $\text{rank}(P/X) \geq n$, then $P/X$ has a finitely generated submodule $B$ such that $\mu(B) = n$. Let $B_1 = \{y \in P : y + X \in B\}$. Let $R$ be the Jacobson radical of $A$, then, clearly, $B_1/RB_1$ has rank at least $n$, and thus $B_1$ contains a finitely generated submodule $B_2$ such that $\mu(B_2) = n$. $P/B_2$ is an indecomposable finitely presented module with $\mu(P/B_2) = 1$ and $\nu(P/B_2) = n$. If $N$ is the Auslander-Bridger dual to $P/B_2$ then $N$ is an indecomposable finitely presented right $A$-module and $\mu(N) = n$. Since, by hypothesis, $A$ is of right bounded representation type, there is an upper bound on the numbers $\text{rank}(P/X)$, independent of the indecomposable projective $P$ or the submodule $X$.

Let $I$ be any left ideal of $A$, and let $_AA = P_1 \oplus \ldots \oplus P_k$ be a decomposition of the left regular module $A$ into a direct sum of indecomposable projective left $A$-modules. Since $A/I$ is the sum of the submodules $P_i/I \cap P_i$ $(1 \leq i \leq k)$, rank$(A/I) \leq$ rank$(P_1/I \cap P_1) + \ldots +$ rank$(P_k/I \cap P_k)$. The existence of a bound on the numbers rank$(P_i/I \cap P_i)$ therefore implies the existence of a bound on the numbers rank$(A/I)$ (since the number $k$ is fixed).

Finally, if $A$ is semiprimary (by which we mean that $A/R$ is Artinian and $R^n = 0$ for some $n$), then $A$ is left Artinian unless for some integer $m$, $R^m/R^{m+1}$ is an infinite direct sum of simple modules. Clearly, this would imply that $A/R^{n+1}$ was not of finite Goldie dimension, which proves the last statement of the theorem. $\square$

Note that in this proof one can replace "semiprimary" by "right perfect".

**Proposition 8.1.6.** *If a semiperfect ring $A$ is of bounded representation type then any of its minors is of bounded representation type as well.*

*Proof.* Let $P$ be a finitely generated projective $A$-module, $B = \text{End}_A(P)$, and let $M$ be a finitely presented $B$-module. Then there are the exact sequences:

$$0 \to X \to B^n \to M \to 0 \qquad\qquad (8.1.7)$$

$$0 \to Y \to B^m \to X \to 0 \qquad\qquad (8.1.8)$$

Denote by $C(P)$ the full subcategory of the category of all $A$-modules consisting of $A$-modules $M$ such that there exists an exact sequence

$$P^{(J)} \to P^{(I)} \to M \to 0,$$

where $P^{(I)}$ denotes a direct sum of isomorphic modules $P_i$ $(i \in I)$. By the Morita theorem [100, theorem 10.7.2], it follows that the categories $\text{mod}_r B$ and $C(P)$ are equivalent, therefore there is an $A$-module $M' \in C(P)$ such that $M = F(M') = \text{Hom}_A(P, M')$. Therefore there is an exact sequence

$$0 \to X \to B^n \to F(M') \to 0 \qquad\qquad (8.1.9)$$

Applying the exact functor $G = * \otimes_B P$ to the exact sequences (8.1.8) and (8.1.9), we get exact sequences

$$0 \to G(X) \to P^n \to M' \to 0$$

$$0 \to G(Y) \to P^m \to G(X) \to 0$$

Hence $\mu_A(M') = \mu_A(P^n) - \mu_A(G(X)) = ns - \mu_A(G(X))$, where $s = \mu_A(P)$, and $\mu_A(U)$ is a minimal number of generators of an $A$-module $U$.

Since $A$ is a ring of bounded representation type, there exists a number $N$ such that $\mu_A(U) \leq N$ for any $A$-module $U$. Therefore $\mu_A(M') \leq N$ and $\mu_A(G(X)) \leq N$, that is, $ns = \mu_A(M') + \mu_A(G(X)) \leq 2N$, i.e. $n \leq 2N/s$. Writing $2N/s = N_1$ there results that

$$\mu_B(M) \leq n \leq N_1$$

and this is true for any finitely presented $B$-module $M$. Therefore $B$ is a ring of bounded representation type. $\square$

## 8.2 $O$-Species and Tensor Algebras

In this section we consider the notion of $O$-species which generalizes the notion of species as introduced by P. Gabriel in [80] and which are considered in section 6.2. Note that these $O$-species are also a particular case of species as introduced by Yu. Drozd in [64] and which are considered in [103, section 8.8].

Let $\{O_i\}$ be a family of discrete valuation rings (not necessarily commutative) $O_i$ with radicals $R_i$ and skew fields of fractions $D_i$, for $i = 1, 2, \ldots, k$, and let $\{D_j\}$, for $j = k + 1, \ldots, n$, be a family of skew fields. Let $(n_1, n_2, \ldots, n_k)$ be a set of natural numbers. Write

$$H_{n_i}(O_i) = \begin{pmatrix} O_i & O_i & \cdots & O_i \\ R_i & O_i & \cdots & O_i \\ \vdots & \vdots & \ddots & \vdots \\ R_i & R_i & \cdots & O_i \end{pmatrix}$$

which is a subring in the matrix ring $M_{n_i}(D_i)$. It is easy to see that each $H_{n_i}(O_i)$ is a Noetherian serial prime hereditary ring. Write $F_i = H_{n_i}(O_i)$ for $i = 1, 2, \ldots, k$, and $F_j = D_j$ for $j = k + 1, \ldots, n$. Then, by the Goldie theorem, there exists a classical ring of fractions $\tilde{F}_i = M_{n_i}(D_i)$ for $i = 1, 2, \ldots, k$ and $\tilde{F}_j = D_j$ for $j = k + 1, \ldots, n$.

**Definition 8.2.1.** Let $I = \{1, 2, \ldots, n\}$. An $O$-**species** is a family $\Omega = (F_i, {}_iM_j)_{i,j \in I}$, where $F_i = H_{n_i}(O_i)$, for $i = 1, 2, \ldots, k$; $F_j = D_j$ for $j = k + 1, k + 2, \ldots, n$, and moreover ${}_iM_j$ is an $(\tilde{F}_i, \tilde{F}_j)$-bimodule that is finite dimensional both as a left $D_i$-vector space and as a right $D_j$-vector space.

An $O$-species $\Omega$ is called a $(D, O)$-**species** if all $O_i$ have a common skew field of fractions $D$, i.e. all $D_i$ are equal to a fixed skew field $D$ $(i = 1, 2, \ldots, n)$.

An $O$-species $\Omega$ is called a $(k, O)$-**species**, if all the $D_i$ $(i = 1, 2, \ldots, n)$ contain a common central subfield $k$ of finite index in such a way that $\lambda m = m\lambda$ for all $\lambda \in k$ and all $m \in {}_iM_j$ (moreover, each bimodule ${}_iM_j$ is a finite dimensional vector space over $k$). It is a $(k, O)$-**quiver** if $D_i = D$ for each $i$.

One can associate with an $O$-species $\Omega$ a quiver $\Gamma(\Omega)$ which is a directed graph whose vertices are indexed by the numbers $i = 1, 2, \ldots, n$, and there is an arrow from the vertex $i$ to the vertex $j$ if and only if ${}_iM_j \neq 0$. An $O$-species $\Omega$ is called **acyclic** if the quiver $\Gamma(\Omega)$ has no oriented cycles. In this case we can choose the indices so that ${}_iM_j = 0$ for $j \leq i$.

A vertex $i$ is said to be **marked** if $F_i = H_{n_i}(O_i)$. Let $J = \{1, 2, \ldots, k\} \subset I$ be the set of all marked points. An $O$-species $\Omega$ is called **min-marked** if all marked vertices of it are minimal and not connected between themselves in $\Gamma(\Omega)$, i.e. ${}_iM_j = 0$ for $i \in I \setminus J, j \in J$, and ${}_iM_j = 0$ for all $i, j \in J$.

Everywhere in this chapter we will consider min-marked $O$-species.

Similar as for species we can define representations of $O$-species by the following way.

**Definition 8.2.2.** A **representation** $V = (U_i, V_r, {}_j\varphi_i, {}_j\psi_r)$ of an $O$-species $\Omega = (F_i, {}_iM_j)_{i,j \in I}$ is a family of right $F_i$-modules $U_i$ for $i = 1, 2, \ldots, k$, a family of right $D_r$-vector spaces $V_r$ for $r = k + 1, k + 1, \ldots, n$, and $D_j$-linear maps:

$$ {}_j\varphi_i : U_i \otimes_{F_i} {}_iM_j \longrightarrow V_j \qquad (8.2.3) $$

for each $i = 1, 2, \ldots, k; j = k + 1, k + 2, \ldots, n$; and

$$ {}_j\psi_r : V_r \otimes_{D_r} {}_rM_j \longrightarrow V_j \qquad (8.2.4) $$

for each $r, j = k + 1, k + 2, \ldots, n$.

**Definition 8.2.5.** Two representations $V = (U_i, V_r, {}_j\varphi_i, {}_j\psi_r)$ and $V' = (U'_i, V'_r, {}_j\varphi'_i, {}_j\psi'_r)$ are called **equivalent** if there is a family of isomorphisms $\{\alpha_i\}$ of $F_i$-modules from $U_i$ to $U'_i$ and a family of isomorphisms $\{\beta_r\}$ of $D_r$-vector spaces from $V_r$ to $V'_r$ such that for each $i = 1, 2, \ldots, k; r, j = k + 1, k + 2, \ldots, n$ the following equalities hold :

$$ {}_j\varphi'_i(\alpha_i \otimes 1) = \beta_j \cdot {}_j\varphi_i; \qquad (8.2.6) $$

$$ {}_j\psi'_r(\beta_r \otimes 1) = \beta_j \cdot {}_j\psi_r. \qquad (8.2.7) $$

In a natural way one can define the notions of a direct sum of representations and of an indecomposable representation.

The set of all representations of an $O$-species $\Omega = (F_i, {}_iM_j)_{i,j \in I}$ can be turned into a category $\mathrm{Rep}(\Omega)$, whose objects are representations $V = (U_i, V_r, {}_j\varphi_i, {}_j\psi_r)$, and a morphism from object $V = (U_i, V_r, {}_i\varphi_j, {}_i\psi_j)$ to object $V' = (U'_i, V'_r, {}_j\varphi'_i, {}_j\psi'_r)$ is a set of homomorphisms $\alpha_i$ of $H_{n_i}(O_i)$-modules $M_i$ to $M'_i$, and a set of homomorphisms $\beta_r$ of $D_r$-vector spaces $V_r$ to $V'_r$ such that for each $i = 1, 2, \ldots, k; r, j = k + 1, k + 2, \ldots, n$ the equalities (8.2.6) and (8.2.7) hold.

For any $O$-species $\Omega$ one can construct the tensor algebra of bimodules $\mathfrak{T}(\Omega)$. Let $B = \overset{n}{\underset{i=1}{\oplus}} F_i$, $M = \oplus_{i,j} {}_iM_j$. Then $M$ is a $(B, B)$-bimodule and we can define the tensor algebra $\mathfrak{T}_B(M)$ of the bimodule $M$ over the ring $B$ in the following way:

$$ \mathfrak{T}_B(M) = B \oplus M \oplus M^2 \oplus \cdots \oplus M^n \oplus \cdots $$

where $M^n = M \otimes_B M^{n-1}$ for $n > 1$, and multiplication in $\mathfrak{T}_B(M)$ is given by the natural $B$-bilinear map:

$$ M^n \times M^m \to M^n \otimes_B M^m = M^{n+m}. $$

Then $\mathfrak{T}(\Omega) = \mathfrak{T}_B(M)$ is the tensor algebra corresponding to an $O$-species $\Omega$.

**Proposition 8.2.8.** *Let $\Omega$ be an O-species. Then the category* Rep($\Omega$) *of all representations of $\Omega$ and the category* Mod$_r \mathfrak{T}(\Omega)$ *of all right $\mathfrak{T}(\Omega)$-modules are naturally equivalent.*

*Proof.* Form two functors $R : \text{Mod}_r \mathfrak{T}(\Omega) \to \text{Rep}(\Omega)$ and $P : \text{Rep}(\Omega) \to \text{Mod}_r \mathfrak{T}(\Omega)$ in the following way. Let $X_{\mathfrak{T}(\Omega)}$ be a right $\mathfrak{T}(\Omega)$-module. Since $B$ is a subring in $\mathfrak{T}(\Omega)$, $X$ can be considered as a right $B$-module. Then

$$X = ( \overset{k}{\underset{i=1}{\oplus}} U_i) \oplus ( \overset{n}{\underset{r=k+1}{\oplus}} V_r),$$

where $U_i$ is a right $H_{n_i}(O_i)$-module, and $V_r$ is a right $D_r$-vector space. Since $M$ is a $(B, B)$-bimodule, one can define a $B$-homomorphism $\varphi : X \otimes_B M \to X$. Taking into account that $U_i \otimes_B {}_s M_j = 0$ for $i \neq s$, the map $\varphi$ is defined in the following way:

$$\varphi : ( \overset{k}{\underset{i=1}{\oplus}} (U_i \otimes_B {}_i M_j)) \oplus ( \overset{n}{\underset{r=k+1}{\oplus}} (V_r \otimes_B {}_r M_j)) \longrightarrow \overset{n}{\underset{r=k+1}{\oplus}} V_r. \tag{8.2.9}$$

Since $U_i \otimes_B {}_i M_j$ is mapping into $V_j$, and $V_r \otimes_B {}_r M_j$ is mapping into $V_j$, the map $\varphi$ defines a family of $D_j$-homomorphisms:

$${}_j \varphi_i : U_i \otimes_B {}_i M_j = U_i \otimes_{H_{n_i}(O_i)} {}_i M_j \longrightarrow V_j,$$

$${}_j \psi_r : V_r \otimes_B {}_r M_j = V_r \otimes_{D_r} {}_r M_j \longrightarrow V_j,$$

for $i = 1, 2, \ldots, k; r, j = k + 1, \ldots, n$.

Now one can define $R(X_{\mathfrak{T}(\Omega)}) = (U_i, V_r, {}_j \varphi_i, {}_j \psi_r)$. Let $X, Y$ be two right $\mathfrak{T}(\Omega)$-modules, let $\gamma : X \to Y$ be a homomorphism, and let $R(X) = (U_i, V_r, {}_j \varphi_i, {}_j \psi_r)$, $R(Y) = (N_i, W_r, {}_j \tilde{\varphi}_i, {}_j \tilde{\psi}_r)$. Let's define a morphism from $R(X)$ to $R(Y)$. Since $\gamma$ is a $B$-homomorphism, $\gamma(U_i) \subseteq N_i$, $\gamma(V_r) \subseteq W_r$, i.e., $\gamma$ defines a family of $H_{n_i}(O_i)$-homomorphisms $\alpha_i : U_i \to N_i$ and a family of $D_r$-homomorphisms $\beta_r : V_r \to W_r$, which are the restrictions of $\gamma$ to $U_i$ and $V_r$. Therefore one can set $R(\gamma) = \{(\alpha_i), (\beta_r)\}$. Since $\gamma$ is a $\mathfrak{T}(\Omega)$-homomorphism, ${}_j \tilde{\varphi}_i(\alpha_i \otimes 1) = \alpha_j \cdot {}_j \varphi_i$ and ${}_j \tilde{\psi}_r(\beta_r \otimes 1) = \beta_j \cdot {}_j \psi_r$ for $i = 1, 2, \ldots, k; r, j = k + 1, \ldots, n$. Therefore $R(\gamma)$ is a morphism in the category Rep($\Omega$).

Conversely, let $\Omega = (F_i, {}_i M_j)_{i,j \in I}$ and let $V = (U_i, V_r, {}_j \varphi_i, {}_j \psi_r)$ be a representation of $\Omega$. Then one can define $P(V)$ in the following way:

$$P(V) = X = ( \overset{k}{\underset{i=1}{\oplus}} U_i) \oplus ( \overset{n}{\underset{r=k+1}{\oplus}} V_r).$$

We define an action of $B = ( \overset{k}{\underset{i=1}{\oplus}} H_{n_i}(O_i)) \oplus ( \overset{n}{\underset{r=k+1}{\oplus}} D_r)$ on $U_i$ by means of the projection $B \to H_{n_i}(O_i)$ and an action of $B$ on $V_r$ by means of the projection $B \to D_r$. We define an action of $M^n$ on $X$ by $\varphi^{(n)} : X \otimes_B M^n \to X$ by induction as follows:

$$\varphi^{(1)} = \underset{i,j}{\oplus} {}_j \varphi_i \oplus \underset{j,r}{\oplus} {}_j \psi_r : X \otimes_B M = \overset{k}{\underset{i=1}{\oplus}} (U_i \otimes_B {}_i M_j) \overset{n}{\underset{r=k+1}{\oplus}} (V_r \otimes_B {}_r M_j) =$$

$$= \bigoplus_{i=1}^{k} (U_i \otimes_{H_{n_i}(O_i)} {}_iM_j) \bigoplus_{r=k+1}^{n} (V_r \otimes_{D_r} {}_rM_j) \longrightarrow \bigoplus_{r=k+1}^{n} V_r \subseteq X.$$

$$\varphi^{(n+1)} = \varphi(\varphi^{(n)} \otimes 1) : X \otimes_B M^{(n+1)} = (X \otimes_B M) \otimes_B M^n \xrightarrow{\varphi^{(n)} \otimes 1} X \otimes_B M \xrightarrow{\varphi} X.$$

If $\gamma = \{\{\alpha_i\}, \{\beta_r\}\}$ is a morphism of a representation $V = (U_i, V_r, {}_j\varphi_i, {}_j\psi_r)$ to a representation $V' = (N_i, W_r, {}_j\tilde{\varphi}_i, {}_j\tilde{\psi}_r)$, $X = P(V)$, $Y = P(V')$, then

$$\varphi = \bigoplus_i \alpha_i \oplus \bigoplus_r \beta_r : X = \bigoplus_i U_i \oplus \bigoplus_r V_r \longrightarrow \bigoplus_i N_i \oplus \bigoplus_r W_r$$

is a $\mathfrak{T}(\Omega)$-homomorphism and therefore $P(\gamma) = \varphi$.

It is not difficult to show that $R$, $P$ are mutually inverse functors and they give an equivalence of categories $\text{Mod}_r\mathfrak{T}(\Omega)$ and $\text{Rep}(\Omega)$. $\square$

**Definition 8.2.10.** A representation $V = (U_i, V_r, {}_j\varphi_i, {}_j\psi_r)$ of an $O$-species $\Omega = (F_i, {}_iM_j)_{i,j\in I}$ is said to be **finite dimensional** if each $U_i$ is a finitely generated $F_i$-module for $(i = 1, 2, \ldots, k)$ and each $V_r$ is a finite dimensional $F_r$-vector space for $(r = k + 1, k + 1, \ldots, n)$.

Denote by $d(U_i)$ the minimal number of generators of an $H_{n_i}(O_i)$-module $M_i$, and denote by $d(V_r) = \dim_{D_r}(V_r)$ the dimension of a vector space $V_r$ over $D_r$. The dimension of a representation $V = (U_i, V_r, {}_j\varphi_i, {}_j\psi_r)$ is the number

$$d = \dim V = \sum_{i=1}^{k} d(U_i) + \sum_{r=k+1}^{n} d(V_r). \tag{8.2.11}$$

**Definition 8.2.12.** An $O$-species $\Omega$ is said to be of **bounded representation type** if the dimensions of its indecomposable finite dimensional representations have an upper bound.

If $\Omega$ is an $O$-species of bounded representation type, then there exists $N > 0$ such that $\dim V < N$ for any indecomposable finite dimensional representation $V$. Then for any finitely generated $\mathfrak{T}(\Omega)$-module $X$ we have $\mu(X) < N_1$, where $N_1$ is some fixed number depending on $N$, i.e., $\mathfrak{T}(\Omega)$ is a ring of bounded representation type. The converse also holds: if $\mathfrak{T}(\Omega)$ is a ring of bounded representation type, then $\Omega$ is a $O$-species of bounded representation type. So taking into account proposition 8.2.8 we have the following corollary.

**Corollary 8.2.13.** *An $(D, O)$-species $\Omega$ is of bounded representation type if and only if the tensor algebra $\mathfrak{T}(\Omega)$ is of bounded representation type.*

**Corollary 8.2.14.** *Let $\Omega_1$ be a D-species, which is a subspecies of a $(D, O)$-species $\Omega$. If $\Omega$ is of bounded representation type, then $\Omega_1$ is of finite type.*

*Proof.* Since $\Omega$ is of bounded representation type, from by corollary 8.2.13 and proposition 8.1.6 it follows that a tensor algebra $\mathfrak{T}(\Omega_1)$ is of bounded representation type, as well. Since $\Omega_1$ is a $D$-species, $\mathfrak{T}(\Omega_1)$ is an Artinian ring. And so it is of finite representation type, by theorem 7.2.13. Therefore, $\Omega_1$ is also of finite representation type. $\square$

## 8.3 $(D, O)$-Species and Right Hereditary SPSD-Rings

In this section we consider the connection of special kinds of $(D, O)$-species with right hereditary SPSD-rings.

Let $\Omega = (F_i, {}_iM_j)_{i,j\in I}$ be a min-marked $(D, O)$-species, where $F_i = H_{n_i}(O_i)$ for $i = 1, \ldots, k$ and $F_i = D$ for $i = k + 1, \ldots, n$. Each such a species we can assign a quiver $\Gamma(\Omega)$ which is a directed graph defined in the following way:

1. The set of vertices is a finite set $I = \{1, 2, \ldots, n\}$.
2. The finite subset $I_0 = \{1, 2, \ldots, k\}$ of $I$ is a set of marked vertices.
3. The vertex $i$ connects with the vertex $j$ by $t_{ij}$ arrows, where

$$t_{ij} = \frac{1}{n_i n_j} \dim_D({}_iM_j) \times \dim({}_iM_j)_D,$$

if $F_i = H_{n_i}(O_i)$. Moreover, we assume that $n_i = 1$ if $F_i = D$.
A marked vertex will be denoted by $\odot$.

A $(D, O)$-species $\Omega = (F_i, {}_iM_j)_{i,j\in I}$ is said to be **weak** if $\Omega$ is min-marked and all $F_i$ are $O_i$ or $D$, i.e. all $n_i = 1$.

The species $\Omega$ is **acyclic** if $\Gamma(\Omega)$ has no oriented cycles. The species $\Omega$ is called **simply connected** if the underlying graph $\overline{\Gamma(\Omega)}$ of $\Gamma(\Omega)$ has no circuits, i.e. the underlying graph $\overline{\Gamma(\Omega)}$ is a tree. Clearly, a simply connected species is acyclic.

In the previous section we showed that to every $O$-species $\Omega = (F_i, {}_iM_j)_{i,j\in I}$ we may associate the tensor algebra $\mathfrak{T}(\Omega) = \bigoplus_{i=0}^{\infty} T_i$, where $T_0 = \prod_{i=1}^{n} F_i = B, T_{i+1} = T_i \otimes_B M$ and $M = \bigoplus_{i,j=1}^{n} {}_iM_j$.

**Lemma 8.3.1.** *Let $\Omega = (F_i, {}_iM_j)_{i,j\in I}$, where all $F_i = D$, be a simply connected $D$-species of finite representation type. Then the tensor algebra $\mathfrak{T}(\Omega)$ is a hereditary Artinian semidistributive ring.*

*Proof.* Since $\Omega$ is an acyclic species, the tensor algebra $A = \mathfrak{T}(\Omega)$ is Morita equivalent to the algebra of the following form

$$A = \begin{pmatrix} D & & U_{ij} \\ & \ddots & \\ 0 & & D \end{pmatrix}$$

where

$$U_{ij} = \bigoplus_{i=i_0<i_1<\cdots<i_k=j} {}_{i_0}M_{i_1} \otimes {}_{i_1}M_{i_2} \otimes \cdots \otimes {}_{i_{k-1}}M_{i_k}. \tag{8.3.2}$$

Since all ${}_iM_j$ are finitely dimensional right and left $D$-spaces, $A$ is an Artinian ring. From [101, Corollary 2.2.13] it follows that $A$ is a hereditary ring.

Note that the ring

$$\begin{pmatrix} D & V_{12} \\ 0 & D \end{pmatrix}, \tag{8.3.3}$$

where $V_{12}$ is $(D,D)$-bimodule, is of finitely representation type if and only if $V_{12}$ is a one dimensional right and left $D$-vector space. Since $\Omega$ is a $D$-species of finite representation type, the tensor algebra $\mathfrak{T}(\tilde{\Omega})$ is of finite representation type as well, and so it does not contain a minor that is isomorphic to the ring (8.3.3). Therefore $A$ is a semidistributive ring. $\square$

**Lemma 8.3.4.** *Let $O$ be a discrete valuation ring with a skew field of fractions $D$. Then $D$ is an injective torsion-free right and left $O$-module.*

*Proof.* Since $O$ is a discrete valuation ring with a skew field of fractions $D$, $Da = D$ and $aD = D$ for any non-zero element $a \in O$. So $D$ is a divisible torsion-free right and left $O$-module. Therefore, by the Baer criterion [103, theorem 1.5.8], it is sufficient to prove that for any non-zero right ideal $I$ in $O$ and any homomorphism $f \in \mathrm{Hom}_O(I, D)$ there exists an element $q \in D$ such that $f(x) = qx$ for each $x \in I$. A right ideal in $O$ has the form $I = aO$ for some non-zero element $a \in O$. Suppose $f(a) = d$. Since $D$ is a divisible $O$-module, there exists an element $q \in D$ such that $d = qa$. An arbitrary element of the ideal $I$ has the form $ab$, where $b \in O$. Therefore $f(ab) = f(a)b = db = qab = q(ab)$, as required. Therefore $D$ is an injective right $O$-module. Analogously $D$ is an injective left $O$-module. $\square$

Along with a weak $(D, O)$-species $\Omega = (F_i, {}_iM_j)_{i,j\in I}$ we can consider a $D$-species $\tilde{\Omega} = (\tilde{F}_i, {}_iM_j)_{i,j\in I}$, where $\tilde{F}_i = D$, since each ${}_iM_j$ is an $(\tilde{F}_i, \tilde{F}_j)$-bimodule. Let $\mathfrak{T}(\tilde{\Omega})$ be a tensor algebra of $D$-species $\tilde{\Omega}$. Since $\mathfrak{T}(\tilde{\Omega})$ is an Artinian ring, it is of bounded representation type if and only if it is of finite representation type.

**Proposition 8.3.5.** *If $\Omega$ is a weak simply connected $(D, O)$-species of bounded representation type, then $\tilde{\Omega}$ is a $D$-species of finite representation type.*

*Proof.* Let $\Omega$ be a weak simply connected $(D, O)$-species with set of marked vertices $J = \{1, 2, \ldots, k\}$. Then the tensor algebra $A = \mathfrak{T}(\Omega)$ is a basic primely triangular ring whose two-sided Peirce decomposition has the following form

$$A = \left( \begin{array}{ccc|c} O_1 & \cdots & 0 & U_1 \\ \vdots & \ddots & \vdots & \vdots \\ 0 & \cdots & O_k & U_k \\ \hline 0 & \cdots & 0 & T \end{array} \right) \tag{8.3.6}$$

where $U_i$ is an $(D, T)$-bimodule $(i = 1, 2, \ldots, k)$. Moreover the ring $T$ is the tensor algebra of the species $\Omega_1 = (F_i, {}_iM_j)_{i,j\in I\setminus J}$ where all $F_i = D$ for $i \in I \setminus J$.

Since $\Omega$ is a $(D, O)$-species of bounded representation type, then the tensor algebra $\mathfrak{T}(\Omega)$ is also of bounded representation type. Then by proposition 8.1.6, $T$ is also of bounded representation type. Since $\Omega_1$ is a $D$-species, $T$ is an Artinian ring and so it is of finite representation type, as well. Since $\Omega$ is acyclic, $\Omega_1$ is also acyclic. Therefore $T$ is an Artinian hereditary semidistributive ring, by lemma 8.3.1.

Let $\tilde{A}$ be a right classical ring of fractions of $A$. We will use the following notations: if $M$ is a right $A$-module, then $M' = M \otimes_A \tilde{A}$; on the other hand if $M$ is a right $\tilde{A}$-module, then $M'$ is the module $M$ considered as an $A$-module. The length of a composition series of a right $\tilde{A}$-module $X$ will be denoted by $l(X)$.

We now prove that for any right $\tilde{A}$-module $M$ there is a right $\tilde{A}$-module $X$ such that $M'' = M \oplus X$.

We have

$$M'' = M' \otimes_A \tilde{A} = (M \otimes_A \tilde{A}) \otimes_A \tilde{A}.$$

Taking into account (8.3.6) we have that $M = \sum_{i=1}^{k} \oplus M_i \oplus M_0$, where $M_i$ is an $O_i$-module and $M_0$ is a $T$-module. Then

$$M \otimes_A \tilde{A} = (\bigoplus_{i=1}^{k} M_i \oplus M_0) \otimes_A \tilde{A} = \bigoplus_{i=1}^{k} (M_i \otimes_{O_i} D) \oplus M_0$$

$$M'' = (M \otimes_A \tilde{A}) \otimes_A \tilde{A} = \bigoplus_{i=1}^{k} M_i \otimes_{O_i} (D \otimes_{O_i} D) \oplus M_0.$$

By lemma 8.3.4 $D$ is an injective torsion-free $O_i$-module for each $i = 1, \ldots, k$. Therefore the mapping $D \to D \otimes_{O_i} D$ with $d \mapsto 1 \otimes d$ for each $d \in D$, is a monomorphism, i.e. there exist exact sequences of $O_i$-modules:

$$0 \to D \to D \otimes_{O_i} D \to \mathrm{Coker}(\varphi_i) \to 0$$

Since $D$ is injective, these sequences split, i.e. $D \otimes_{O_i} D = D \oplus Y_i$ for $i = 1, \ldots, k$. Therefore

$$M'' = \bigoplus_{i=1}^{k} (M_i \otimes_{O_i} (D \oplus Y_i)) \oplus M_0 = \bigoplus_{i=1}^{k} ((M_i \otimes_{O_i} D) \oplus (M_i \otimes_{O_i} Y_i)) \oplus M_0 =$$

$$= M \oplus X.$$

Now suppose that the ring $A$ is of bounded representation type and the ring $\tilde{A}$ is of infinite representation type. Then for any $N > 0$ there is an indecomposable finitely generated $\tilde{A}$-module $M$ such that $l(M) > N$.

Consider the $A$-module $M'$. It is finitely generated and, by [103, theorem 9.2.3], it decomposes into a direct sum of finitely generated indecomposable $A$-modules:

$$M' = N_1 \oplus \cdots \oplus N_t.$$

Then

$$M'' = N_1' \oplus \cdots \oplus N_t'.$$

Since $M'' = M \oplus X$, and $M''$ is a finitely generated module over an Artinian ring $\tilde{A}$, from the uniqueness of the decomposition it follows that there is a number $i$ such that $M$ is a direct summand of $N_i'$, i.e. there is an $\tilde{A}$-module $P$ such that $N_i' = M \oplus P$. Then we have a chain of inequalities:

$$\mu_A(N_i) = \mu_{\tilde{A}}(N_i') \geq l(N_i') = l(M) + l(P) \geq l(M) > N,$$

which contradicts the assumption that $A$ is of bounded representation type. $\square$

**Proposition 8.3.7.** *If $\Omega$ is a simply connected weak $(D, O)$-species of bounded representation type then the corresponding tensor algebra $\mathfrak{T}(\Omega)$ is a right hereditary SPSD-ring of bounded representation type.*

*Proof.* Since $\Omega$ is simply connected, $_iM_j = 0$ for $j \leq i$ and $i, j \in I$. Therefore the tensor algebra $\mathfrak{T}(\Omega)$ is clearly a triangular ring. Since all $F_i$ are prime rings, $\mathfrak{T}(\Omega)$ is a right Noetherian primely triangular ring with prime radical $\mathcal{J} = \bigoplus_{i=i}^{\infty} T_i$. Moreover, $\mathfrak{T}(\Omega)$ is a semiperfect ring, since it contains a finite number of paiwise orthogonal local idempotents which sum is equal to 1.

Suppose that $_iM_j = 0$ for $i \geq j$. Denote by $e_i$ the identity of the ring $F_i$. Then $1 = e_1 + e_2 + \cdots + e_n$ is a triangular prime decomposition of the identity of the ring $A = \mathfrak{T}(\Omega)$ and

$$A_{ij} = e_i A e_j = \bigoplus_{(k_1,\ldots,k_s)} {}_iM_{k_1} \otimes_{F_{k_1}} \cdots \otimes_{F_{k_s}} {}_{k_s}M_j,$$

where $A_{ii} = F_i$. In particular, $\overline{A}_{ij} = {}_iM_j$. Since all $_iM_j$ are left $D$-modules, $A_{ij}$ is also left $D$-modules for all $i, j \in I$. For each $i = 1, \ldots, n$ the ring $A_{ii} = F_i$ is a prime hereditary Noetherian ring. Since the species $\Omega$ is simply connected, $A$ does not contain the minors of the form (1.3.14). Therefore similar to proposition 1.3.13 we can show that $\mathfrak{T}(\Omega)$ is a right hereditary ring.

Since $\Omega$ is a $(D, O)$-species of bounded representation type, the tensor algebra $\mathfrak{T}(\tilde{\Omega})$ is of finite representation type, and so all its minor of the second order is isomorphic to the ring (8.3.3), where $V_{12} = 0$ or a one dimensional right and left $D$-vector space. Therefore $eAf$ is a uniserial left $eAe$-module and right $fAf$-module for local idempotents $e$ and $f$ of the ring $A$. Hence $A = \mathfrak{T}(\Omega)$ is a semidistributive ring by [103, theorem 1.10.9]. $\square$

We show that the inverse statement is also true.

**Proposition 8.3.8.** *Let $A$ be a right hereditary SPSD-ring such that its right classical ring of fractions $\tilde{A}$ is a basic ring. If $A$ is of bounded representation type then it is isomorphic to the tensor algebra of a simply connected weak $(D, O)$-species of bounded representation type.*

*Proof.* Let $A$ be a right hereditary SPSD-ring. Then $A$ is a primely triangular ring with triangular prime decomposition of the identity $1 = e_1 + e_2 + \cdots + e_n$ such that

$e_i A e_i = O_i$ for $i = 1, \ldots, k$ and $e_j A e_j = D$ for $j = k + 1, \ldots, n$. Since $A$ is a right hereditary primely triangular ring, $A_{ij}$ is a left $\tilde{F}_i$-module. Then the ring $A$ can be assigned a $(D, O)$-species $\Omega_A = (F_i, {}_i M_j)_{i,j \in I}$, by setting $I = \{1, 2, \ldots, n\}$, $F_i = e_i A e_i$ and ${}_i M_j = A_{ij} / \sum_{i < k < j} A_{ik} A_{kj}$, where $A_{ij} = e_i A e_j$ is an $(\tilde{F}_i, \tilde{F}_j)$-bimodule. Since ${}_i M_i = 0$, ${}_i M_j = 0$ if $F_j = O_j$, and $\mathrm{Hom}_A(e_i A, e_j A) \neq 0$ if $\mathrm{Hom}_A(e_j A, e_i A) = 0$, the quiver $\Gamma(\Omega_A)$ is without loops and oriented cycles, and minimal marked vertices of $\Gamma(\Omega_A)$ are points $1, 2, \ldots, k$, i.e. $\Omega_A$ is a weak simply connected species.

We will show that the quiver $\Gamma(\Omega_A)$ is in fact the prime quiver of the right hereditary SPSD-ring $A$.

The two-sided Peirce decomposition of the ring $A$ has the following form:

$$A = \begin{pmatrix} B_1 & X \\ 0 & B_2 \end{pmatrix},$$

where $B_1 = \bigoplus_{i=1}^{k} F_i$, $F_i = H_{n_i}(O_i)$, and $B_2$ is an Artinian hereditary ring, and $X$ is a $(B_1, B_2)$-bimodule.

Let $N$ be the prime radical of $A$, then its two-sided Peirce decomposition has the following form:

$$N = \begin{pmatrix} 0 & X \\ 0 & R_2 \end{pmatrix},$$

where $R_2$ is the Jacobson radical of the ring $B_2$. Then

$$A/N = \begin{pmatrix} B_1 & 0 \\ 0 & B_2/R_2 \end{pmatrix} \simeq \bigoplus_{i=1}^{n} F_i, \quad N^2 = \begin{pmatrix} 0 & X R_2 \\ 0 & R_2^2 \end{pmatrix},$$

$$N/N^2 = \begin{pmatrix} 0 & X/X R_2 \\ 0 & R_2/R_2^2 \end{pmatrix} \simeq \bigoplus_{i,j=1}^{n} {}_i M_j,$$

where ${}_i M_j = A_{ij} / \sum_{i < k < j} A_{ik} A_{kj}$. So that the quiver $\Gamma(\Omega_A)$ is the prime quiver of $A$.

Since $A$ is a right hereditary SPSD-ring, its prime quiver contains no circuits by theorem 2.4.20. Therefore $A$ is a split ring by [103, proposition 8.8.2].

Let $\mathfrak{T}(\Omega_A)$ be the tensor algebra corresponding to the $(D, O)$-species $\Omega_A$. Then by [103, theorem 8.8.6], $A$ is isomorphic to $\mathfrak{T}(\Omega_A)$. $\square$

## 8.4 Representations of $(D, O)$-Species and Mixed Matrix Problems

In this section we show in what way one can reduce the description of representations of weak $(D, O)$-species to some flat mixed matrix problems over discrete valuation rings $O_i$ for $i = 1, \ldots, k$ and their common skew field of fractions $D$.

Let $O$ be a discrete valuation ring (DVR) with a classical division ring of fractions $D$.

By left $O$-elementary transformations of rows of a matrix $\mathbf{T}$ with entries in $D$ we mean transformations of two types:

(a) Multiplying a row on the left by an invertible element of $O$;
(b) Replacing a row by itself plus an arbitrary element of $O$ multiply (on the left) of another row.

In a similar way we can define left $D$-elementary transformations of rows and, by symmetry, right $O$-elementary and right $D$-elementary transformations of columns.

Elementary transformations of the type considered above are given by corresponding elementary matrices which are invertible. An automorphism of a finitely generated module $P$ corresponding to an elementary transformation is an elementary automorphism. Multiplications on the left (right) side of a matrix $\mathbf{T}$ by elementary matrices correspond to elementary row (column) transformations.

Recall, that by [100, Proposition 13.1.3] any invertible matrix $\mathbf{B}$ over a local ring $O$ can be reduced by $O$-elementary row (column) transformations to the identity matrix. By [100, Corollary 13.1.4] the matrix $\mathbf{B}$ can be decomposed into a product of elementary matrices. Moreover by [100, Theorem 13.1.6] any automorphism of a finitely generated projective module $P$ over a semiperfect ring $A$ can be decomposed into a product of elementary automorphisms.

Let $\Delta = \{O_i\}_{i=1,\ldots,k}$ be a family of discrete valuation rings $O_i$ with a common skew field of fractions $D$. We define the general flat matrix problem over $\Delta$ and $D$ in the following way.

Let there be given a block rectangular matrix with entries in $D$:

$$
\mathbf{T} =
\begin{array}{|c|c|c|c|c|}
\hline
\mathbf{T}_{11} & \cdots & \mathbf{T}_{1j} & \cdots & \mathbf{T}_{1m} \\
\hline
\vdots & \ddots & \vdots & \ddots & \vdots \\
\hline
\mathbf{T}_{i1} & \cdots & \mathbf{T}_{ij} & \cdots & \mathbf{T}_{im} \\
\hline
\vdots & \ddots & \vdots & \ddots & \vdots \\
\hline
\mathbf{T}_{n1} & \cdots & \mathbf{T}_{nj} & \cdots & \mathbf{T}_{nm} \\
\hline
\end{array}
$$

partitioned into $n$ horizontal strips $\{\mathbf{T}_i\}_{i=1,\ldots,n}$ and $m$ vertical strips $\{\mathbf{T}^j\}_{j=1,\ldots,m}$ so that each block $\mathbf{T}_{ij}$ is the intersection of the $j$-th vertical strip and the $i$-th horizontal strip, moreover some of these matrices may be empty.

Assume that to the $i$-th horizontal strip $\mathbf{T}_i$ of this matrix $\mathbf{T}$ corresponds the ring $F_{i_s} \in \Delta \cup D$ and to the $j$-th vertical strip $\mathbf{T}^j$ correspondents the ring $F_{j_t} \in \Delta \cup D$.

On the matrix $\mathbf{T}$ one has admissible transformations of the following types:

1. Left $F_{i_s}$-elementary transformations of rows within the strip $\mathbf{T}_i$.
2. Right $F_{j_t}$-elementary transformations of rows within the strip $\mathbf{T}^j$.
3. Additions of rows in the strip $\mathbf{T}_j$ multiplied on the left by elements of $F_r \in \Delta \cup D$ to rows in the strip $\mathbf{T}_i$.

4. Additions of columns in the strip $\mathbf{T}^i$ multiplied on the right by elements of $F_p \in \Delta \cup D$ to columns in the strip $\mathbf{T}^j$.

Indecomposable matrices and equivalent matrices are defined in a natural way.

We say that the flat matrix problem is of **finite type** if there exist only a finitely many non-equivalent indecomposable matrices.

**Definition 8.4.1.** The vector

$$d = d(\mathbf{T}) = (d_1, d_2, \ldots, d_n; d^1, d^2, \ldots d^m), \qquad (8.4.2)$$

where
$d_i$ is the number of rows of the $i$-th horizontal strip of $\mathbf{T}$;
$d^j$ is the number of columns of the $j$-th vertical strip of $\mathbf{T}$
for $i = 1, \ldots, n$; $j = 1, \ldots, m$ is called the **dimension vector** of the stripped matrix $\mathbf{T}$. Also set

$$\dim(\mathbf{T}) = \sum_{i=1}^{n} d_i + \sum_{j=1}^{m} d^j$$

**Definition 8.4.3.** We say that a flat matrix problem is of **bounded representation type** if there is a constant $C$ such that $\dim(\mathbf{X}) < C$ for all indecomposable matrices $\mathbf{X}$. Otherwise it is of **unbounded representation type**.

Let $\Omega = (F_i, {}_iM_j)_{i,j\in I}$, where $F_i = O_i$ for $i = 1, 2, \ldots, k$ and $F_j = D$ for $j = k + 1, \ldots, n$, be a weak $(D, O)$-species of bounded representation type.

Suppose that $V = (M_i, V_r, {}_j\varphi_i, {}_j\psi_r)$ is an indecomposable finite dimensional representation of $\Omega$, then $M_i$ is a finitely generated $F_i$-module for $i = 1, \ldots, k$ and $V_r$ is a finite dimensional $D$-vector space for $r = k + 1, \ldots, n$. Since $F_i = O_i$ is a discrete valuation ring, any $O_i$-module $M_i$ is torsion-free and faithful, by [103, proposition 5.4.18].

Therefore any indecomposable representations of $\Omega$ has the following form:

$$V = (M_i, V_r, {}_j\varphi_i, {}_j\psi_r) \qquad (8.4.4)$$

where $M_i$ is a free $F_i$-module.

Consider the category $\mathcal{R}(\Omega)$ whose objects are representations $V = (M_i, V_r, {}_j\varphi_i, {}_j\psi_r)$, and morphisms from an object $V$ to an object $V' = (M'_i, V'_r, {}_j\varphi'_i, {}_j\psi'_r)$ are sets of homomorphisms $\{\alpha_i\}, \{\beta_r\}$, where $\alpha_i : M_i \rightarrow M'_i$ is a homomorphism of $F_i$-modules, $\beta_r : V_r \rightarrow V'_r$ is a homomorphism of $D$-vector spaces ($r = k + 1, \ldots, n$), and the equalities (8.2.6)-(8.2.7) hold.

Let $V$ be an indecomposable finite dimensional representation of the $(D, O)$-species $\Omega$. So each $M_i$ is a finitely generated free $O_i$-module with basis $\omega_1^{(i)}, \ldots, \omega_{m_i}^{(i)}$ ($i = 1, \ldots, k$); and $V_r$ is a finite dimensional $D$-space with basis $\tau_1^{(r)}, \ldots, \tau_{k_r}^{(r)}$ ($r = k + 1, \ldots, n$).

Suppose

$$j\varphi_i(\omega_s^{(i)} \otimes 1) = \sum_{u=1}^{k_j} \tau_u^{(j)} b_{us}^{(ij)}, \tag{8.4.5}$$

$$j\psi_r(\tau_v^{(r)} \otimes 1) = \sum_{u=1}^{k_i} \tau_u^{(j)} a_{uv}^{(ij)}, \tag{8.4.6}$$

where $a_{uv}^{(ij)}, b_{us}^{(ij)} \in D$. Then the matrices $\mathbf{A}_{ij} = (a_{uv}^{(ij)})$, $\mathbf{B}_{ij} = (b_{us}^{ij})$ define the representation $V$ uniquely by equivalence.

Let $\mathbf{U}_i \in M_{m_i}(F_i)$ be the matrix corresponding to the homomorphism $\alpha_i$, and let $\mathbf{W}_i \in M_{k_i}(D)$ be the matrix corresponding to the homomorphism $\beta_i$, $i \in I$. If $\mathbf{A}'_{ij}, \mathbf{B}'_{ij}$ are the matrices corresponding to a representation $V'$ then equalities (8.2.6)-(8.2.7) have the following matrix form:

$$\mathbf{W}_i \mathbf{B}_{ij} = \mathbf{B}'_{ij} \mathbf{U}_j, \quad (i = 1, \ldots, k; \; j = k+1, \ldots, n) \tag{8.4.7}$$

$$\mathbf{W}_j \mathbf{A}_{jr} = \mathbf{A}'_{jr} \mathbf{W}_r, \quad (j, r = k+1, \ldots, n) \tag{8.4.8}$$

If representations $V$ and $V'$ are equivalent then $\alpha_i$, $\beta_r$ are isomorphisms. Therefore the matrices $\mathbf{U}_i$ and $\mathbf{W}_r$ are invertible and equalities (8.2.6)-(8.2.7) are equivalent to the following equalities:

$$\mathbf{W}_i \mathbf{B}_{ij} \mathbf{U}_j^{-1} = \mathbf{B}'_{ij}, \quad (i = 1, \ldots, k; \; j = k+1, \ldots, n) \tag{8.4.9}$$

$$\mathbf{W}_j \mathbf{A}_{jr} \mathbf{W}_r^{-1} = \mathbf{A}'_{jr}, \quad (j, r = k+1, \ldots, n) \tag{8.4.10}$$

So we obtain the following matrix problem of finding indecomposable finite dimensional representations of a $(D, \mathcal{O})$-species $\Omega$.

**Main mixed matrix problem.**

Let $\Delta = \{\mathcal{O}_i\}_{i=1,\ldots,k}$ be a family of discrete valuation rings $\mathcal{O}_i$ with a common skew field of fractions $D$.

Let there be given a block rectangular matrix $\mathbf{T}$ with entries in $D$ partitioned into $n$ horizontal strips $\{\mathbf{T}_i\}_{i=1,\ldots,n}$ and $m$ vertical strips $\{\mathbf{T}^j\}_{j=1,\ldots,m}$ so that a block $\mathbf{T}_{ij}$ is the intersection of the $j$-th vertical strip and the $i$-th horizontal strip, moreover some of these matrices may be empty.

On the matrix $\mathbf{T}$ one has admissible transformations of the following types:

1. Left $F_{i_s}$-elementary transformations of rows within the strip $\mathbf{T}_i$, where $F_{i_s} \in \Delta \cup D$.
2. Right $F_{j_t}$-elementary transformations of rows within the strip $\mathbf{T}^j$, where $F_{j_t} \in \Delta \cup D$.

The admissible transformations on the matrix $\mathbf{T}$ are given by block-diagonal nonsingular matrices $\mathbf{X} = \mathrm{diag}(\mathbf{X}_1, \ldots, \mathbf{X}_n)$ and $\mathbf{Y} = \mathrm{diag}(\mathbf{Y}_1, \ldots, \mathbf{Y}_m)$, where each matrix $\mathbf{X}_i$, $\mathbf{Y}_j$ is a square invertible matrix for $i = 1, \ldots, n$, $j = 1, \ldots, m$, so that $\mathbf{T} \mapsto \mathbf{XTY}$. Moreover, $\mathbf{X}_i \subset M_{m_i}(F_{i_s})$ and $\mathbf{Y}_j \subset M_{k_j}(F_{j_t})$, where $F_{i_s}, F_{j_t} \in \Delta \cup D$.

Clearly, the matrix $\mathbf{T}$ is indecomposable if and only if the corresponding representation of $\Omega$ is indecomposable. The boundedness of a $(D, O)$-species $\Omega$ is strictly connected with the boundedness of corresponding matrix problems.

It is easy to prove the following statement.

**Lemma 8.4.11.** *A $(D, O)$-species $\Omega$ is of bounded representation type if and only if the corresponding main matrix problem is of bounded representation type.*

## 8.5 Some Mixed Matrix Problems

To describe $(D, O)$-species and right hereditary SPSD-rings of bounded representation type it will be important to consider some special mixed matrix problems. These are discussed in this section.

**Matrix problem I.**
*Let $O$ be a DVR with division ring of fractions $D$, and $\varepsilon_1, \ldots, \varepsilon_{n-1} \in O$. Given a matrix $\mathbf{T}$ with entries in $D$ partitioned into $n$ vertical strips:*

$$\mathbf{T} = \boxed{\begin{array}{|c|c|c|c|} \mathbf{A}_1 & \mathbf{A}_2 & \cdots & \mathbf{A}_n \end{array}}$$

*with admissible transformations of the following forms:*

1. *Left $D$-elementary transformations of rows of $\mathbf{T}$.*
2. *Right $O$-elementary transformations of columns within any vertical strip $\mathbf{A}_i$ $(i = 1, \ldots, n)$.*
3. *Additions of any column of the block $\mathbf{A}_i$ multiplied on the right by an arbitrary element of $O$ to any column of the block $\mathbf{A}_j$ if $i < j$ for $i, j = 1, \ldots, n$.*
4. *Additions of any column of the block $\mathbf{A}_{i+1}$ multiplied on the right by an arbitrary element of $\varepsilon_i O$ to any column of the block $\mathbf{A}_i$ for $i = 1, \ldots, n - 1$.*

**Lemma 8.5.1.** *Let $O$ be a discrete valuation ring with skew field of fractions $D$, $M = \mathrm{rad}\, O = \pi O = O\pi$. Suppose $\varepsilon_i = \pi^2$ for all $i = 1, 2, \ldots, n - 1$. Then for any $n$ matrix problem* I *contains an indecomposable matrix $\mathbf{T}$ such that $\dim(\mathbf{T}) > n$.*

*Proof.* Consider the following matrix

$$\mathbf{T} = \boxed{\pi^{n-1} \mid \pi^{n-2} \mid \cdots \mid \pi^2 \mid \pi \mid 1}$$

Suppose there exists an invertible matrix $\mathbf{B} = (b)$, where $b \in D^*$, $b \neq 0$ and an invertible matrix $\mathbf{A} \in M_n^*(O)$:

$$\mathbf{A} = \begin{pmatrix} a_{11} & a_{12} & \cdots & a_{1n} \\ \pi^2 a_{21} & a_{22} & \cdots & a_{2n} \\ \pi^4 a_{31} & \pi^2 a_{32} & \cdots & a_{3n} \\ \vdots & \vdots & \ddots & \vdots \\ \pi^{2^{n-1}} a_{n1} & \pi^{2^{n-2}} a_{n2} & \cdots & a_{nn} \end{pmatrix},$$

where $a_{ij} \in O$ for $i, j = 1, \ldots, n$; $a_{ii}$ are invertible elements in $O$ for $i = 1, \ldots, n$, such that $\mathbf{BTA}$ is decomposable. Then there is a number $i$ such that

$$b(\pi^{n-1} a_{1i} + \ldots + \pi^{n-i} a_{ii} + \pi^{n-i-1} \pi^2 a_{i+1,i} + \pi^{n-i-2} \pi^4 a_{i+2,i} + \ldots +$$

$$+ \pi \pi^{2^{n-1-i}} a_{n-i,i} + \pi^{2^{n-i}} a_{ni}) = 0.$$

Since $b \neq 0$, we obtain that $a_{ii} \in \pi O = M$, i.e., $a_{ii}$ is not invertible element of $O$. A contradiction. □

**Matrix problem II.**
*Consider a rectangular matrix* $\mathbf{T}$ *with entries in a skew field D partitioned into 3 vertical strips* $\mathbf{A}_1, \mathbf{A}_2, \mathbf{A}_3$

$$\mathbf{T} = \boxed{\mathbf{A}_1 \mid \mathbf{A}_2 \mid \mathbf{A}_3}$$

*On the matrix* $\mathbf{T}$ *one has the following admissible transformations*:

1. *Left elementary O-transformations of rows of* $\mathbf{T}$.
2. *Right D-elementary transformations of columns within each vertical strip* $\mathbf{A}_i$ *$(i = 1, 2, 3)$.*
3. *Additions of any column of the block* $\mathbf{A}_1$ *multiplied on the right by an arbitrary element of D to any column of the block* $\mathbf{A}_2$.

**Lemma 8.5.2.** *Let O be a discrete valuation ring with skew field of fractions D, $R = \mathrm{rad}O = \pi O = O\pi$. Then for any n matrix problem* II *contains an indecomposable matrix* $\mathbf{T}$ *such that* $\dim(\mathbf{T}) > n$.

*Proof.* Consider the stripped matrix $\mathbf{T}$ with the following vertical blocks:

$$
\mathbf{A}_1 = \begin{pmatrix} \begin{array}{|c|} \hline 1 \\ \hline 0 \\ 0 \\ \vdots \\ 0 \\ \hline 0 \\ 0 \\ \vdots \\ 0 \end{array} \end{pmatrix}, \quad
\mathbf{A}_2 = \begin{pmatrix} \begin{array}{|cccc|} \hline 0 & 0 & \cdots & 0 \\ \hline 1 & 0 & \cdots & 0 \\ 0 & 1 & \cdots & 0 \\ \vdots & \vdots & \ddots & \vdots \\ 0 & 0 & \cdots & 1 \\ \hline 0 & 0 & \cdots & 0 \\ 0 & 0 & \cdots & 0 \\ \vdots & \vdots & \ddots & \vdots \\ 0 & 0 & \cdots & 0 \end{array} \end{pmatrix},
$$

$$
\mathbf{A}_3 = \begin{pmatrix} \begin{array}{|cccc|} \hline \pi^{-2n-1} & \pi^{-2n-2} & \cdots & \pi^{-3n} \\ \hline \pi^{-2} & 0 & \cdots & 0 \\ 0 & \pi^{-4} & \cdots & 0 \\ \vdots & \vdots & \ddots & \vdots \\ 0 & 0 & \cdots & \pi^{-2n} \\ \hline 1 & 0 & \cdots & 0 \\ 0 & 1 & \cdots & 0 \\ \vdots & \vdots & \ddots & \vdots \\ 0 & 0 & \cdots & 1 \end{array} \end{pmatrix}.
$$

Then the corresponding matrix $\mathbf{T}$ is indecomposable with $\dim(\mathbf{T}) > n$ since the admissible transformations on the stripped matrix

$$
\mathbf{T}_1 = \begin{array}{|c|c|c|c|} \hline \pi^{-2n-1} & \pi^{-2n-2} & \cdots & \pi^{-3n} \\ \hline \end{array}
$$

satisfy the conditions of matrix problem I. $\square$

## Matrix problem III.

*Let $O_1, O_2$ be distinct discrete valuation rings with a common division ring of fractions $D$. One considers a matrix $\mathbf{T}$ with entries in $D$ with the following admissible transformations:*

1. *Left $O_1$-elementary transformations of rows of the matrix $\mathbf{T}$.*
2. *Right $O_2$-elementary transformations of columns of the matrix $\mathbf{T}$.*

**Lemma 8.5.3.**

*A matrix* **T** *of the matrix problem* III *by admissible transformations can be reduced to the following form*:

$$\mathbf{T} = \left( \begin{array}{c|c} \mathbf{I} & \mathbf{O} \\ \hline \mathbf{O} & \mathbf{O} \end{array} \right), \tag{8.5.4}$$

*where* **I** *is the identity matrix. So the matrix problem* III *is of bounded representation type.*

*Proof.*
1-st step. Let $M_i$ be the maximal ideal of a ring $O_i$. By proposition [103, Proposition 3.6.8] there exists an element $\pi_1 \in M_1 \setminus M_1^2$ and $\pi_1 \in O_2^*$. Analogously there exists an element $\pi_2 \in M_2 \setminus M_2^2$ and $\pi_2 \in O_1^*$.

Thus any element $x \in D$ can be written in two different forms $x = \pi_1^n \varepsilon = \eta \pi_1^n$ and $x = \pi_2^m \mu = \xi \pi_2^m$, where $\varepsilon, \eta \in O_1^*$ and $\mu, \xi \in O_2$. Then $v_1(x) = n$ and $v_2(x) = m$.

Pick out an element $x = \pi_2^m \mu \in D$ in the matrix **T** such that $v_2(x)$ is minimal among all elements of **T** and put it in the place $(1, 1)$ of the matrix **T**. Multiplying the first column of this matrix by $\mu^{-1} \in O_2^*$ we obtain that $t_{11} = \pi_2^m$. By means of this element we can make all elements in the first row zero using the transformations on the columns. So we obtain the equivalent form of the matrix **T**:

$$\left( \begin{array}{c|c} \pi_2^n & \mathbf{O} \\ \hline * & \mathbf{T}_1 \end{array} \right).$$

Since $\pi_2^n \in O_1^*$, multiplying the first row by $\pi_2^{-n} \in O_1^*$, we can reduce this matrix to the form:

$$\left( \begin{array}{c|c} 1 & \mathbf{O} \\ \hline * & \mathbf{T}_1 \end{array} \right)$$

Applying this procedure to the matrix $\mathbf{T}_1$ we obtain the following equivalent form of **T**:

$$\left( \begin{array}{c|c} \mathbf{T}_2 & \mathbf{O} \\ \hline * & \mathbf{O} \end{array} \right)$$

where

$$\mathbf{T}_2 = \begin{pmatrix} 1 & 0 & \cdots & 0 \\ * & 1 & \ddots & \vdots \\ \vdots & \ddots & \ddots & 0 \\ * & \cdots & * & 1 \end{pmatrix}$$

Since $O_1 + O_2 = D$, by [103, proposition 3.8.12], for any nonzero element $x = t_{ij} \in \mathbf{T}_2$ with $i > j$ there exist elements $a \in O_1$ and $b \in O_2$ such that $-x = a + b$. So by transformations on the $i$-th row and the $j$-th column of $\mathbf{T}_2$ we can make this element to be 0. So we reduce the matrix **T** to the following equivalent form:

$$\left(\begin{array}{c|c} I & O \\ \hline T_3 & O \end{array}\right),$$

where $I$ is the $k \times k$ identity matrix.

2-nd step. If $T_3 = 0$ then the proof is finished. Otherwise pick out in the matrix $T_3$ an element $y \in D$ with minimal $v_1(y)$ for all elements in $T_3$ and put it in the place $(k+1, 1)$. Suppose $y = \pi_1^n$. If $n \geq 0$ then by the element $t_{11} = 1$ we can make all elements below the element $y$ in the 1-st column be zero. So we obtain an indecomposable direct summand of $T$ of the form (1). So we can assume $v_1(y) = n < 0$. Since $O_1$ is a total valuation ring, $y^{-1} \in O_1$. By the row transformations on $T_3$ we can make all elements below the element $y$ in the 1-st column be zero. Since $y^{-1} \in O_1$ we can add the $k + 1$-th row of $T$ to the first row of it. Suppose that the elements of the obtained matrix are $a_{ij}$. Then $a_{11} = 0$ and $v_1(a_{1j}) > 0$ for all $j > 1$ since $v_1(y^{-1}t_{k+1,j}) \geq 0$ because $v_1(y)$ is minimal. Using the elements $a_{ii} = 1$ for $i = 2, \ldots, k$ we reduce our matrix $T$ to the following form:

$$\left(\begin{array}{cc|c} 0 & O & \\ \hline O & I & 0 \\ \hline y & * & \\ \hline O & * & 0 \end{array}\right)$$

Since the element $0 \neq y \in D$ can be represented in the form $y = ab$, where $a \in O_1^*$, $b \in O_2^*$, then multiplying the 1-st column by $b^{-1}$ and the $k$-th row by $a^{-1}$ we can make the element $t_{k+1,1}$ to be 1, i.e. we obtain the following matrix:

$$\left(\begin{array}{cc|c} 0 & O & \\ \hline O & I & 0 \\ \hline 1 & * & \\ \hline O & M_3 & 0 \end{array}\right)$$

Now all elements except the first one in the $k + 1$-th row can be made zero, since $O_1 + O_2 = D$. So we obtain the following form:

$$\left(\begin{array}{cc|c} 0 & O & \\ \hline O & I & 0 \\ \hline 1 & O & \\ \hline O & M_3 & 0 \end{array}\right)$$

By changing the first strip row with the third one, and then changing the third strip row with the last one we obtain the matrix

$$\left(\begin{array}{c|c|c} 1 & O & O \\ \hline O & I & O \\ \hline O & M_3 & O \\ \hline O & O & O \end{array}\right)$$

Since the dimension of the matrix $M_3$ is smaller than the previous matrix, by mathematical induction we obtain the required result. □

**Matrix problem IV.**

*Let $O_1, O_2$ be distinct discrete valuation rings with a common division ring of fractions D. Consider a matrix **T** with entries in D with the following admissible transformations:*

    1. *Left $O_1 \cap O_2$-elementary transformations of rows of the matrix **T**.*

    2. *Right D-elementary transformations of columns of the matrix **T**.*

**Lemma 8.5.5.**

*The matrix problem IV is of unbounded representation type.*

*Proof.* By [103, proposition 3.3.11], there exists an element $x \in D$ such that $x \notin O_1 \cap O_2$ and $x^{-1} \notin O_1 \cap O_2$. Then the matrix of the following form:

$$
\mathbf{T} = \begin{pmatrix}
1 & 0 & \cdots & 0 \\
0 & 1 & \cdots & 0 \\
\vdots & \vdots & \ddots & \vdots \\
0 & 0 & \cdots & 1 \\
x & x^2 & \cdots & x^n
\end{pmatrix}
$$

is indecomposable under the admissible transformations 1) and 2) of matrix problem IV for any $n \in N$.

**Matrix problem V.**

*Let $O_1, O_2$ be distinct discrete valuation rings with the common division ring of fractions D. Consider a matrix **T** with entries in D partitioned into two vertical strips:*

$$
\mathbf{T} = \boxed{\ \mathbf{A_1}\ \ \vert\ \ \mathbf{A_2}\ }
$$

*with the following admissible transformations:*

    1. *Left $O_1$-elementary transformations of rows of **T**.*

    2. *Right $O_2$-elementary transformations of columns within the matrix $\mathbf{A_1}$.*

    3. *Right D-elementary transformations of columns within the matrix $\mathbf{A_2}$.*

**Lemma 8.5.6.**

*The matrix problem* V *is of unbounded representation type.*

*Proof.* By admissible transformations we can reduce the matrix **T** to the following form:

| I | O | O | B |
|---|---|---|---|
| O | O | I | O |
| O | O | O | O |

and on the matrix **B** one has the following admissible transformations:
  1. Left $O_1 \cap O_1$-elementary transformations of rows within the matrix **B**.
  2. Right $D$-elementary transformations of columns within the matrix **B**.

Therefore for the matrix **B** we get the matrix problem IV which is of unbounded representation type by lemma 8.5.5. □

**Matrix problem VI.**

*Let $O$ be a discrete valuation ring with a division ring of fractions and Jacobson radical $R = \pi O = O\pi$. One considers matrices **T** with entries in D with the following admissible transformations:*
  1. *Left $O$-elementary transformations of rows of **T**.*
  2. *Right $O$-elementary transformations of columns of **T**.*

**Lemma 8.5.7.**

*A matrix **T** of the matrix problem* VI *by admissible transformations can be reduced to the following form:*

$$
\mathbf{T} =
\left(
\begin{array}{cccc|c}
\pi^{n_1} & 0 & \cdots & 0 & \\
0 & \pi^{n_2} & \cdots & 0 & \mathbf{O} \\
\vdots & \vdots & \ddots & \vdots & \\
0 & 0 & \cdots & \pi^{n_k} & \\
\hline
\multicolumn{4}{c|}{\mathbf{O}} & \mathbf{O}
\end{array}
\right)
\tag{8.5.8}
$$

*So the matrix problem* VI *is of bounded representation type.*

*Proof.* Let $O$ be a DVR ring with Jacobson radical $R = \pi O = O\pi$. Any element of $D$ is of a form $\pi^k \varepsilon_1 = \varepsilon_2 \pi^k$.

Let $\mathbf{T} \in M_{n \times m}(D)$. We carry out the proof by induction on $m + n$. The basis of induction, $m + n = 2$, is trivial.

Suppose that the statement has been already proved for all matrices with dimension $< m+n$. In the matrix $\mathbf{T} \in M_{n \times m}(D)$ we can choose an element $x = \pi^k \varepsilon$ with minimal number $v(x) = k$. By elementary transformations of rows and columns first we can put this element in the place $(1,1)$ of the matrix **T**. Then by means of this element we can make all elements in the first row and the first column zero using $O$-elementary transformations on the rows and columns. So we obtain the equivalent form of the matrix $\mathbf{T}$:

$$\left( \begin{array}{c|c} \pi^k & \mathbf{O} \\ \hline \mathbf{O} & \mathbf{T}_1 \end{array} \right).$$

Since the dimension of the matrix $\mathbf{T}_1$ is less than $m + n$, by induction the matrix $\mathbf{T}_1$ can be transformed to the form (8.5.8) and we obtain the required result. So the proposition is proved. $\square$

## 8.6 $(D, O)$-Species of Unbounded Representation Type

Let $\Omega = \{F_i, {}_iM_j\}_{i,j \in I}$ be a weak $(D, O)$-species, where all $F_i$ are equal to $O_i$ or $D$. Each such species we can assign the quiver $\Gamma(\Omega)$ which is a directed graph as defined in section 8.3.

The **diagram of a $(D, O)$-species** $\Omega$ is a graph $Q(\Omega)$ which obtained from $\Gamma(\Omega)$ by deleting the orientation of all arrows which connected unmarked vertices of $\Gamma(\Omega)$.

In this section we will prove the necessity part of the following main theorem:

**Theorem 8.6.1.** *Let the $O_i$ be discrete valuation rings with a common skew field of fractions $D$. Then a weak $(D, O)$-species $\Omega$ is of bounded representation type if and only if the diagram $Q(\Omega)$ is a finite disjoint union of Dynkin diagrams of the forms $A_n$, $D_n$, $E_6$, $E_7$, $E_8$ and the following diagrams:*

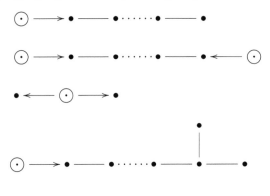

Diag. 8.6.2.

**Lemma 8.6.3.** *Let $O$ be a discrete valuation ring with skew field of fractions $D$ and Jacobson radical $R = \pi O = O\pi$, $I = \{1, 2, 3, 4\}$. Then a weak $(D, O)$-species $\Omega = (F_i, {}_iM_j)_{i,j \in I}$ with diagram*

$$(8.6.4)$$

*is of unbounded representation type.*

*Proof.* It suffices to prove the lemma when $F_1 = O$, $F_2 = F_3 = F_4 = D$, $_1M_2 = {}_1M_3 = {}_1M_4 = {}_O D_D$ and $_iM_j = 0$ for other $i, j \in I$. Write $J = \{1, 2, 3\}$.

Let $X$ be a right $O$-module, $Y_i$ a right $D$-vector space, $\varphi_i : X \otimes_O ({}_1M_i) \to Y_i$ a $D$-linear mapping for $i \in J$.

Consider the category $\text{Rep}(\Omega)$ whose objects are representations $V = (X, Y_i, \varphi_i)_{i \in J}$, and morphisms from an object $V$ to an object $V' = (X', Y'_i, \varphi'_i)_{i \in J}$ are a family of homomorphisms $(\alpha, \beta_i)_{i \in J}$, where $\alpha : X \to X'$ is a homomorphism of $O$-modules, $\beta_i : Y_i \to Y'_i$ is a homomorphism of $D$-vector spaces ($i \in J$), and the following equalities hold:

$$\beta_i \varphi_i = \varphi'_i(\alpha \otimes 1) \quad (i \in J). \tag{8.6.5}$$

We show that the category $\text{Rep}(\Omega)$ is of unbounded representation type. Let $X$ be a finitely generated free $O$-module with basis $\omega_1, \ldots, \omega_n$; and let $Y_i$ be a finite dimensional $D$-space with basis $\tau_1^{(i)}, \ldots, \tau_{k_i}^{(i)}$ ($i \in J$). Suppose

$$\varphi_i(\omega_s \otimes 1) = \sum_{j=1}^{k_i} \tau_j^{(i)} a_{js}^{(i)}, \tag{8.6.6}$$

where $a_{js}^{(i)} \in D$ for $i \in J$, $s = 1, \ldots, n$. Then the matrices $\mathbf{A}_i = (a_{js}^{(i)})$, $i \in J$, define the representation $V$ uniquely up to equivalence.

Let $\mathbf{U} \in M_n(O)$ be the matrix corresponding to the homomorphism $\alpha$, and let $\mathbf{V}_i \in M_{k_i}(D)$ be the matrix corresponding to the homomorphism $\beta_i$, $i \in J$. If $\mathbf{A}'_i$ are matrices corresponding to a representation $V'$ then equalities (8.6.6) have the following matrix form:

$$\mathbf{V}_i \mathbf{A}_i = \mathbf{A}'_i \mathbf{U}, \quad i \in J. \tag{8.6.7}$$

If representations $V$ and $V'$ are equivalent then $\alpha, \beta_i$ are isomorphisms. Therefore the matrices $\mathbf{U}$ and $\mathbf{V}_i$ are invertible and equalities (8.6.7) are equivalent to the following equalities:

$$\mathbf{V}_i \mathbf{A}_i \mathbf{U}^{-1} = \mathbf{A}'_i, \quad i \in J. \tag{8.6.8}$$

So that we obtain the following matrix problem:

Given a block-rectangular matrix

$$\mathbf{T} = \begin{array}{|c|} \hline \mathbf{A}_1 \\ \hline \mathbf{A}_2 \\ \hline \mathbf{A}_3 \\ \hline \end{array}$$

with the following possible admissible transformations:

    1. Right $O$-elementary transformations of columns of **T**.
    2. Left $D$-elementary transformations of rows within each block $\mathbf{A}_i$ ($i \in J$).

Set

$$
\mathbf{A}_1 =
\begin{bmatrix}
1 & 0 & \cdots & 0 & 0 & 0 & \cdots & 0 \\
0 & 1 & \cdots & 0 & 0 & 0 & \cdots & 0 \\
\vdots & \vdots & \ddots & \vdots & \vdots & \vdots & \ddots & \vdots \\
0 & 0 & \cdots & 1 & 0 & 0 & \cdots & 0
\end{bmatrix},
$$

$$
\mathbf{A}_2 =
\begin{bmatrix}
\pi^{-2} & 0 & \cdots & 0 & 1 & 0 & \cdots & 0 \\
0 & \pi^{-4} & \cdots & 0 & 0 & 1 & \cdots & 0 \\
\vdots & \vdots & \ddots & \vdots & \vdots & \vdots & \ddots & \vdots \\
0 & 0 & \cdots & \pi^{-2n} & 0 & 0 & \cdots & 1
\end{bmatrix},
\quad
\mathbf{A}_3 =
\begin{bmatrix}
\pi^{n-1} & 0 \\
\pi^{n-2} & 0 \\
\vdots & \vdots \\
1 & 0
\end{bmatrix},
$$

where $\pi \in R = \mathrm{rad}\, O$, $\pi \neq 0$. By lemma 8.5.1, the matrix **T** is indecomposable and therefore the species (8.6.4) is of unbounded representation type. $\square$

**Lemma 8.6.9.** *Let $O$ be a discrete valuation ring with skew field of fractions $D$ and Jacobson radical $R = \pi O = O\pi$, $I = \{1, 2, 3, 4\}$. Then a weak $(D, O)$-species $\Omega = (F_i, {}_iM_j)_{i,j\in I}$ with diagram*

$$
\bullet \longleftarrow \;\; \odot \;\; \longrightarrow \bullet \longleftarrow \bullet \tag{8.6.10}
$$

*is of unbounded representation type.*

*Proof.* We prove this lemma for the case when $F_1 = O$, $F_2 = F_3 = F_4 = D$, ${}_1M_2 = {}_1M_3 = {}_OD_D$, ${}_4M_3 = {}_DD_D$ and ${}_iM_j = 0$ for other $i, j \in I$. Write $J = \{1, 2, 3\}$.

Let $X$ be a right $O$-module, let $Y_i$ be a right $D$-vector space for $i \in J$, $\varphi_i :$ $X \otimes_O ({}_1M_{i+1}) \to Y_i$ for $i = 1, 2$ and $\varphi_3 : Y_3 \otimes_D ({}_4M_3) \to Y_2$ be $D$-linear mappings.

Consider the category $\mathcal{R}(\Omega)$ whose objects are representations $V = (X, Y_i, \varphi_i)_{i\in J}$, and morphisms from an object $V$ to and object $V' = (X', Y'_i, \varphi'_i)_{i\in J}$ are a family of homomorphisms $(\alpha, \beta_i)_{i\in J}$, where $\alpha : X \to X'$ is a homomorphism of $O$-modules, $\beta_i : Y_i \to Y'_i$ is a homomorphism of $D$-vector spaces ($i \in J$), moreover the following equalities hold:

$$
\beta_i \varphi_i = \varphi'_i(\alpha \otimes 1), \quad (i = 1, 2) \tag{8.6.11}
$$

$$
\beta_2 \varphi_3 = \varphi'_3(\beta_3 \otimes 1). \tag{8.6.12}
$$

We show that the category $\mathrm{Rep}(\Omega)$ is of unbounded representation type. Let $X$ be a finitely generated free $O$-module with basis $\omega_1, \ldots, \omega_n$; let $Y_i$ be a finite dimensional $D$-space with basis $\tau_1^{(i)}, \ldots, \tau_{k_i}^{(i)}$ ($i = 1, 2, 3$). Suppose

$$
\varphi_1(\omega_s \otimes 1) = \sum_{j=1}^{k_1} \tau_j^{(1)} a_{js}^{(1)}, \quad \varphi_2(\omega_s \otimes 1) = \sum_{t=1}^{k_2} \tau_t^{(2)} a_{ts}^{(2)} \tag{8.6.13}
$$

$$\varphi_3(\tau_m^{(3)} \otimes 1) = \sum_{t=1}^{k_2} \tau_t^{(2)} a_{tm}^{(3)}, \tag{8.6.14}$$

where $a_{js}^{(1)}, a_{ts}^{(2)}, a_{tm}^{(3)} \in D$ for $s = 1, \ldots, n$; $j = 1, \ldots, k_1$; $t = 1, \ldots, k_2$; $m = 1, \ldots, k_3$. Then the matrices $\mathbf{A}_1 = (a_{js}^{(1)})$, $\mathbf{A}_2 = (a_{ts}^{(2)})$, $\mathbf{A}_3 = (a_{tm}^{(3)})$ define the representation $V$ uniquely up to equivalence.

Let $\mathbf{U} \in M_n(O)$ be the invertible matrix corresponding to the isomorphism $\alpha$, and let $\mathbf{V}_i \in M_{k_i}(D)$ be the invertible matrix corresponding to the isomorphism $\beta_i, i \in J$. If $\mathbf{A}_i'$ ($i \in J$) are matrices corresponding to a representation $V'$ which equivalent to the representation $V$ then equalities (8.6.13)-(8.6.14) have the following matrix form:

$$\mathbf{V}_i \mathbf{A}_i \mathbf{U}^{-1} = \mathbf{A}_i', \quad (i = 1, 2) \tag{8.6.15}$$

$$\mathbf{V}_2 \mathbf{A}_3 \mathbf{V}_3^{-1} = \mathbf{A}_3' \tag{8.6.16}$$

So that we obtain the following matrix problem:

Given a block-rectangular matrix

$$\mathbf{T} = \begin{array}{|c|c|} \hline \mathbf{A}_1 & \mathbf{0} \\ \hline \mathbf{A}_2 & \mathbf{A}_3 \\ \hline \end{array}$$

On this matrix one admits the following admissible transformations:

1. Right $O$-elementary transformations of columns within the first vertical strip of the matrix $\mathbf{T}$.
2. Right $D$-elementary transformations of columns within the second vertical strip of the matrix $\mathbf{T}$.
3. Left $D$-elementary transformations of rows within each horizontal strip of the matrix $\mathbf{T}$.

Reducing the matrix $\mathbf{A}_3$ to the form

$$\mathbf{A}_3 = \begin{array}{|c|c|} \hline \mathbf{I} & \mathbf{0} \\ \hline \mathbf{0} & \mathbf{0} \\ \hline \end{array}$$

we obtain that the matrix $\mathbf{A}_2$ is reduced to the following form:

$$\mathbf{A}_2 = \begin{array}{|c|} \hline \mathbf{B}_1 \\ \hline \mathbf{B}_2 \\ \hline \end{array}$$

and it is possible to add any row of the matrix $\mathbf{B}_2$ multiplied on the left by elements of $D$ to any row of the matrix $\mathbf{B}_1$. Thus the matrices $\mathbf{B}_2$, $\mathbf{B}_1$ and $\mathbf{A}_1$ form the matrix problem II. Therefore the $(D, O)$-species (8.6.10) is of unbounded representation type. $\square$

Analogously one can prove the following lemma:

**Lemma 8.6.17.** *Let $O$ be a discrete valuation ring with skew field of fractions $D$ and Jacobson radical $R = \pi O = O\pi$, $I = \{1, 2, 3, 4\}$. Then a weak $(D, O)$-species $\Omega = (F_i, {}_iM_j)_{i,j \in I}$ with diagram*

$$\bullet \longleftarrow \odot \longrightarrow \bullet \longrightarrow \bullet \qquad (8.6.18)$$

*is of unbounded representation type.*

**Lemma 8.6.19.** *Let $O$ be a discrete valuation ring with skew field of fractions $D$ and Jacobson radical $R = \pi O = O\pi$, $I = \{1, 2, 3, 4, 5\}$. Then a weak $(D, O)$-species $\Omega = (F_i, {}_iM_j)_{i,j \in I}$ with diagram*

$$
\begin{array}{c}
\odot \\
\downarrow \\
\bullet \longrightarrow \bullet \longleftarrow \bullet \longrightarrow \bullet
\end{array}
\qquad (8.6.20)
$$

*is of unbounded representation type.*

*Proof.* We prove this lemma for the case when $F_1 = O$, $F_2 = F_3 = F_4 = F_5 = D$, ${}_1M_4 = {}_OD_D$, ${}_2M_4 = {}_3M_4 = {}_3M_5 = {}_DD_D$ and ${}_iM_j = 0$ for other $i, j \in I$. Write $J = \{1, 2, 3, 4\}$.

Let $X$ be a right $O$-module, let $Y_i$ be a right $D$-vector space for $i \in J$, and let $\varphi_1 : X \otimes_O ({}_1M_4) \to Y_3$, $\varphi_j : Y_{j-1} \otimes_D ({}_jM_4) \to Y_3$ ($j = 2, 3$), $\varphi_4 : Y_2 \otimes_D ({}_3M_5) \to Y_4$ be $D$-linear mappings.

Consider the category $\mathcal{R}(\Omega)$ whose objects are representations $V = (X, Y_i, \varphi_i)_{i \in J}$, and morphisms from an object $V$ to an object $V' = (X', Y_i', \varphi_i')_{i \in J}$ are a family of homomorphisms $(\alpha, \beta_i)_{i \in J}$, where $\alpha : X \to X'$ is a homomorphism of $O$-modules, $\beta_i : Y_i \to Y_i'$ is a homomorphism of $D$-vector spaces ($i \in J$), and the following equalities hold:

$$\beta_3\varphi_1 = \varphi_1'(\alpha \otimes 1), \quad \beta_3\varphi_2 = \varphi_2'(\beta_1 \otimes 1) \qquad (8.6.21)$$

$$\beta_3\varphi_3 = \varphi_3'(\beta_2 \otimes 1), \quad \beta_4\varphi_4 = \varphi_4'(\beta_2 \otimes 1) \qquad (8.6.22)$$

We show that the category $\mathrm{Rep}(\Omega)$ is of unbounded representation type.

Let $X$ be a finitely generated free $O$-module with basis $\omega_1, \ldots, \omega_n$; and let $Y_i$ be a finite dimensional $D$-space with basis $\tau_1^{(i)}, \ldots, \tau_{k_i}^{(i)}$ ($i \in J$). Suppose

$$\varphi_1(\omega_s \otimes 1) = \sum_{j=1}^{k_3} \tau_j^{(3)} a_{js}, \quad \varphi_2(\tau_m^{(1)} \otimes 1) = \sum_{j=1}^{k_3} \tau_j^{(3)} b_{jm}, \qquad (8.6.23)$$

$$\varphi_3(\tau_t^{(2)} \otimes 1) = \sum_{j=1}^{k_3} \tau_j^{(3)} c_{jt}, \quad \varphi_4(\tau_t^{(2)} \otimes 1) = \sum_{r=1}^{k_4} \tau_r^{(4)} d_{rt}, \tag{8.6.24}$$

where $a_{js}, b_{jm}, c_{jt}, d_{rt} \in D$ for $s = 1, \ldots, n$; $m = 1, \ldots, k_1$; $t = 1, \ldots, k_2$; $j = 1, \ldots, k_3$; $r = 1, \ldots, k_4$. Then the matrices $\mathbf{A}_1 = (a_{js})$, $\mathbf{A}_2 = (b_{jm})$, $\mathbf{A}_3 = (c_{jt})$, $\mathbf{A}_4 = (d_{rt})$ define the representation $M$ uniquely by up to equivalence.

Let $\mathbf{U} \in M_n(O)$ be the matrix corresponding to the homomorphism $\alpha$, and let $\mathbf{V}_i \in M_{k_i}(D)$ be the matrix corresponding to the homomorphism $\beta_i$, $i \in J$. If $\mathbf{A}'_i$ are matrices corresponding to a representation $V'$ then equalities (8.6.23)-(8.6.24) have the following matrix form:

$$\mathbf{V}_3 \mathbf{A}_1 = \mathbf{A}'_1 \mathbf{U}, \quad \mathbf{V}_3 \mathbf{A}_2 = \mathbf{A}'_2 \mathbf{V}_1 \tag{8.6.25}$$

$$\mathbf{V}_3 \mathbf{A}_3 = \mathbf{A}'_3 \mathbf{V}_2, \quad \mathbf{V}_4 \mathbf{A}_4 = \mathbf{A}'_4 \mathbf{V}_2 \tag{8.6.26}$$

If representations $V$ and $V'$ are equivalent then $\alpha$, $\beta_i$ are isomorphisms. Therefore the matrices $\mathbf{U}$ and $\mathbf{V}_i$ are invertible and we obtain the following matrix problem:

Given a block-rectangular matrix $\mathbf{T}$ which partitioned into 2 horizontal strips and 3 vertical strips:

$$\mathbf{T} = \begin{array}{|c|c|c|} \hline \mathbf{A}_1 & \mathbf{A}_2 & \mathbf{A}_3 \\ \hline \mathbf{0} & \mathbf{0} & \mathbf{A}_4 \\ \hline \end{array}$$

On these matrices one allows the following admissible transformations:

1. Right $O$-elementary transformations of columns within the first vertical strip of the matrix $\mathbf{T}$.
2. Right $D$-elementary transformations of columns within the second and the third vertical strips of the matrix $\mathbf{T}$.
3. Left $D$-elementary transformations of rows within each horizontal strip of the matrix $\mathbf{T}$.

Reducing the matrices $\mathbf{A}_2$ and $\mathbf{A}_4$ to the form $\begin{array}{|c|c|} \hline \mathbf{I} & \mathbf{0} \\ \hline \mathbf{0} & \mathbf{0} \\ \hline \end{array}$

we can reduce the matrix $\mathbf{A}_3$ to the following form:

$$\begin{array}{|cccccc|} \hline \mathbf{0} & \mathbf{0} & \mathbf{0} & \mathbf{0} & \mathbf{I} & \mathbf{0} \\ \mathbf{0} & \mathbf{I} & \mathbf{0} & \mathbf{0} & \mathbf{0} & \mathbf{0} \\ \mathbf{0} & \mathbf{0} & \mathbf{0} & \mathbf{0} & \mathbf{0} & \mathbf{0} \\ \mathbf{0} & \mathbf{0} & \mathbf{0} & \mathbf{I} & \mathbf{0} & 0 \\ \mathbf{I} & \mathbf{0} & \mathbf{0} & \mathbf{0} & \mathbf{0} & \mathbf{0} \\ \mathbf{0} & \mathbf{0} & \mathbf{0} & \mathbf{0} & \mathbf{0} & \mathbf{0} \\ \hline \end{array}$$

In according with this reducing the matrix $\mathbf{A}_1$ is divided into 6 horizontal strips and we obtain the following matrix problem.

Given a block-rectangular matrix **B** partitioned into 6 vertical strips

$$\mathbf{B} = \boxed{\mathbf{B}_1 \mid \mathbf{B}_2 \mid \mathbf{B}_3 \mid \mathbf{B}_4 \mid \mathbf{B}_5 \mid \mathbf{B}_6}$$

On these matrices one allows the following admissible transformations:

a. Left $O$-elementary transformations of rows of **B**.
b. Right $D$-elementary transformations of columns within each vertical strip $\mathbf{B}_i$ $(i = 1, 2, .., 6)$.
c. If $\alpha_i \le \alpha_j$ in the poset $S$:

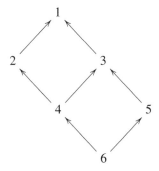

then any column of the vertical strip $\mathbf{B}_i$ multiplied on the right by an arbitrary element of $D$ can be added to any column of the vertical strip $\mathbf{B}_j$.

It is easy to see the blocks $\mathbf{B}_2$, $\mathbf{B}_4$, $\mathbf{B}_5$ form the matrix problem II. Therefore the species (8.6.20) is of unbounded representation type. $\square$

Analogously one can prove the following lemma:

**Lemma 8.6.27.** *Let $O$ be a discrete valuation ring with skew field of fractions $D$ and Jacobson radical $R = \pi O = O\pi$, $I = \{1, 2, 3, 4, 5\}$. Then a weak $(D, O)$-species $\Omega = (F_i, {}_iM_j)_{i,j \in I}$ with diagram*

(8.6.28)

*for an arbitrary orientation of horizontal edges is a species of unbounded representation type.*

**Lemma 8.6.29.** *Let $O_i$ be a discrete valuation ring with skew field of fractions $D$ and Jacobson radical $R_i = \pi_i O_i = O_i \pi_i$ for $i = 1, 2$, and $I = \{1, 2, 3, 4\}$. Then a weak $(D, O)$-species $\Omega = (F_i, {}_i M_j)_{i,j \in I}$ with diagram*

$$(8.6.30)$$

*is of unbounded representation type.*

*Proof.* We will prove the lemma when $F_1 = O_1$, $F_2 = O_2$, $F_3 = F_4 = D$, ${}_1 M_3 = {}_{O_1} D_D$, ${}_2 M_3 = {}_{O_2} D_D$, ${}_4 M_3 = {}_D D_D$ and ${}_i M_j = 0$ for other $i, j \in I$.

Let $X_i$ be a right $O_i$-module for $i = 1, 2$; let $Y_j$ be a right $D$-vector space for $j = 1, 2$; and let $\varphi_i : X \otimes_{O_i} ({}_i M_3) \rightarrow Y_1$ for $i = 1, 2$ and $\varphi_3 : Y_2 \otimes_D ({}_4 M_3) \rightarrow Y_1$ be $D$-linear mappings.

Consider the category $\mathrm{Rep}(\Omega)$ whose objects are representations $V = (X_1, X_2, Y_1, Y_2, \varphi_i)_{i=1,2,3}$, and morphisms from an object $V$ to and object $V' = (X_1', X_2', Y_1', Y_2', \varphi_i')_{i=1,2,3}$ are a family of homomorphisms $(\alpha_1, \alpha_2, \beta_1, \beta_2)$, where $\alpha_i : X_i \rightarrow X_i'$ is a homomorphism of $O_i$-modules, $\beta_i : Y_i \rightarrow Y_i'$ is a homomorphism of $D$-vector spaces for $i = 1, 2$. The following equalities hold:

$$\beta_i \varphi_i = \varphi_i'(\alpha_i \otimes 1), \quad (i = 1, 2) \tag{8.6.31}$$

$$\beta_1 \varphi_3 = \varphi_3'(\beta_2 \otimes 1). \tag{8.6.32}$$

We show that the category $\mathrm{Rep}(\Omega)$ is of unbounded representation type. Let $X_i$ be a finitely generated free $O_i$-module with basis $\omega_1^{(i)}, \ldots, \omega_{m_i}^{(i)}$; let $Y_i$ be a finite dimensional $D$-space with basis $\tau_1^{(i)}, \ldots, \tau_{k_i}^{(i)}$ for $i = 1, 2$.

Suppose

$$\varphi_1(\omega_s^{(1)} \otimes 1) = \sum_{j=1}^{k_1} \tau_j^{(1)} a_{js}^{(1)}, \quad \varphi_2(\omega_t^{(2)} \otimes 1) = \sum_{j=1}^{k_1} \tau_j^{(1)} a_{jt}^{(2)} \tag{8.6.33}$$

$$\varphi_3(\tau_u^{(2)} \otimes 1) = \sum_{j=1}^{k_1} \tau_j^{(1)} a_{ju}^{(3)}, \tag{8.6.34}$$

where $a_{js}^{(1)}, a_{jt}^{(2)}, a_{ju}^{(3)} \in D$ for $s = 1, \ldots, m_1; t = 1, \ldots, m_2; j = 1, \ldots, k_1, u = 1, \ldots, k_2$. Then the matrices $\mathbf{A}_1 = (a_{js}^{(1)})$, $\mathbf{A}_2 = (a_{jt}^{(2)})$; $\mathbf{A}_3 = (a_{ju}^{(3)})$ define the representation $M$ uniquely up to equivalence.

Let $\mathbf{U}_i \in M_{m_i}(O_i)$ be the invertible matrix corresponding to the isomorphism $\alpha_i$, and let $\mathbf{V}_i \in M_{k_i}(D)$ be the invertible matrix corresponding to the isomorphism $\beta_i$ for $i = 1, 2$. If $\mathbf{A}_i'$ are matrices corresponding to a representation $V'$ which is equivalent to the representation $V$ then equalities (8.6.33)-(8.6.34) have the following matrix form

$$V_1 A_i U_i^{-1} = A_i', \quad (i = 1, 2) \tag{8.6.35}$$

$$V_1 A_3 V_2^{-1} = A_3'. \tag{8.6.36}$$

So that we obtain the following matrix problem:
Given a block-rectangular matrix

$$\mathbf{T} = \boxed{\begin{array}{|c|c|c|} A_1 & A_2 & A_3 \end{array}}$$

One allows the following admissible transformations:

1. Left $D$-elementary transformations of rows of $\mathbf{T}$.
2. Right $O_1$-elementary transformations of columns within the matrix $\mathbf{A}_1$.
3. Right $O_2$-elementary transformations of columns within the matrix $\mathbf{A}_2$.
4. Right $D$-elementary transformations of columns within the matrix $\mathbf{A}_3$.

Consider two different cases.

1-st case. Assume that $O_1 = O_2$.

Set

$$\mathbf{A}_1 = \begin{pmatrix} 1 & 0 & \cdots & 0 \\ 0 & 1 & \cdots & 0 \\ \vdots & \vdots & \ddots & \vdots \\ 0 & 0 & \cdots & 1 \end{pmatrix}, \quad \mathbf{A}_2 = \begin{pmatrix} \pi^2 & 0 & \cdots & 0 \\ 0 & \pi^4 & \cdots & 0 \\ \vdots & \vdots & \ddots & \vdots \\ 0 & 0 & \cdots & \pi^{2n} \end{pmatrix}, \quad \mathbf{A}_3 = \begin{pmatrix} 1 \\ \pi \\ \vdots \\ \pi^{n-1} \end{pmatrix}$$

By lemma 8.5.2, the matrix $\mathbf{T}$ is indecomposable. So the corresponding representation $V$ of the species $\Omega$ is indecomposable and the species $\Omega$ with diagram (8.6.30) is of unbounded representation type.

2-nd case. Assume that $O_1 \neq O_2$.

Set $\mathbf{A}_1 = \mathbf{A}_2 = \mathbf{I}$, then

$$\mathbf{T} = \boxed{\begin{array}{|c|c|c|} \mathbf{I} & \mathbf{I} & A_3 \end{array}}$$

and for the matrix $\mathbf{A}_3$ we obtain the matrix problem IV which is of unbounded representation type by lemma 8.5.5. Therefore the species with diagram (8.6.30) is of unbounded representation type. $\square$

**Lemma 8.6.37.** *Let $O_i$ be a discrete valuation ring with skew field of fractions $D$ and Jacobson radical $R_i = \pi_i O_i = O_i \pi_i$ for $i = 1, 2$, and $I = \{1, 2, 3, 4\}$. Then a weak $(D, O)$-species $\Omega = (F_i, {}_i M_j)_{i,j \in I}$ with diagram*

(8.6.38)

*is of unbounded representation type.*

The proof of this lemma is the same as for lemma 8.6.29. □

Note that since we consider only simply connected species, all diagrams which are not presented in diagram 8.6.2 have a subdiagram of one of the types discussed above in this section and hence they are of unbounded representation type.

Therefore the necessity of theorem 8.6.1 follows from lemmas 8.6.3, 8.6.9, 8.6.17, 8.6.19, 8.6.27, 8.6.29, 8.6.37.

## 8.7 $(D, O)$-Species of Bounded Representation Type

In this section we will prove the sufficiency of theorem 8.6.1.

**Lemma 8.7.1.** *Let $O$ be a discrete valuation ring with a skew field of fractions $D$ and Jacobson radical $R = \pi O = O\pi$, and $I = \{1, 2, \ldots, n, n+1\}$. Then a weak $(D, O)$-species $\Omega = (F_i, {}_iM_j)_{i,j \in I}$ with diagram*

$$\odot \longrightarrow \bullet \longrightarrow \bullet \cdots \cdots \bullet \longrightarrow \bullet \qquad (8.7.2)$$

*is of bounded representation type for any arbitrary orientation of edges.*

*Proof.* We will prove the lemma for the case $F_1 = O$, $F_2 = F_3 = \cdots = F_n = F_{n+1} = D$, ${}_1M_2 = {}_OD_D$, ${}_iM_{i+1} = {}_DD_D$ for $i = 2, \ldots, n$ and ${}_iM_j = 0$ for other $i, j \in I$. Write $J = \{1, 2, \ldots, n\}$.

Let $X$ be a right $F_1$-module and let $Y_i$ be a right $D$-vector space for $i \in J$. Let $\varphi_1 : X \otimes_{F_1} ({}_1M_2) \to Y_1$ and $\varphi_i : Y_{i-1} \otimes_D ({}_iM_{i+1}) \to Y_i$ $(i = 2, \ldots, n)$ be $D$-linear mappings.

Consider the category Rep$(\Omega)$ whose objects are representations $V = (X, Y_i, \varphi_i)_{i \in J}$, and morphisms from an object $V$ to and object $V' = (X', Y_i', \varphi_i')_{i \in J}$ are a family of homomorphisms $(\alpha, \beta_i)_{i \in J}$, where $\alpha : X \to X'$ is a homomorphism of $F_1$-modules, $\beta_i : Y_i \to Y_i'$ is a homomorphism of $D$-vector spaces $(i \in J)$, moreover the following equalities hold:

$$\beta_1\varphi_1 = \varphi_1'(\alpha \otimes 1), \quad \beta_i\varphi_i = \varphi_i'(\beta_{i-1} \otimes 1) \quad i = 2, \ldots, n \qquad (8.7.3)$$

We show that the category $\mathcal{R}(\Omega)$ is of bounded representation type. Let $X$ be a finitely generated free $F_1$-module with basis $\omega_1, \ldots, \omega_s$; let $Y_i$ be a finite dimensional

$D$-space with basis $\tau_1^{(i)}, \ldots, \tau_{k_i}^{(i)}$ for $i \in J$. Suppose

$$\varphi_1(\omega_j \otimes 1) = \sum_{u=1}^{k_1} \tau_u^{(1)} b_{uj}, \quad \varphi_i(\tau_{v_i}^{(i-1)} \otimes 1) = \sum_{r=1}^{k_i} \tau_r^{(i)} a_{rv_i}^{(i)}, \qquad (8.7.4)$$

where $a_{rv_i}^{(i)}, b_{uj} \in D$ for $j = 1, \ldots, s; u = 1, \ldots, k_1, v_i = 1, \ldots, k_{i-1}, i = 2, \ldots, n$.

Write $\mathbf{A}_1 = (b_{uj})$, $\mathbf{A}_i = (a_{rv_i}^{(i)})$ for $i = 2, \ldots, n$.

Then the matrices $\mathbf{A}_i$, for $i \in J$ define the representation $V$ uniquely up to equivalence.

Let $\mathbf{U} \in M_s(F_1)$ be the matrix corresponding to the homomorphism $\alpha$, and let $\mathbf{V}_i \in M_{k_i}(D)$ be the matrix corresponding to the homomorphism $\beta_i, i \in J$. If $\mathbf{A}_i'$ are matrices corresponding to a representation $V'$ then equalities (8.7.4) have the following matrix form:

$$\mathbf{V}_1 \mathbf{A}_1 = \mathbf{A}_1' \mathbf{U}, \quad \mathbf{V}_i \mathbf{A}_i = \mathbf{A}_i' \mathbf{V}_{i-1} \quad (i = 2, \ldots, n) \qquad (8.7.5)$$

If representations $V$ and $V'$ are equivalent then $\alpha, \beta_i$ are isomorphisms. Therefore the matrices $\mathbf{U}$ and $\mathbf{V}_i$ are invertible and equalities (8.7.5) are equivalent to the following matrix equalities:

$$\mathbf{V}_1 \mathbf{A}_1 \mathbf{U}^{-1} = \mathbf{A}_1', \quad \mathbf{V}_i \mathbf{A}_i \mathbf{V}_{i-1}^{-1} = \mathbf{A}_i' \quad i = 2, \ldots, n \qquad (8.7.6)$$

So that we obtain the matrix problem of reducing a set of matrices $\mathbf{A}_i$ by means a family of matrices $\mathbf{U}$ and $\mathbf{V}_i$ satisfying equalities (8.7.6) for $i \in J$.

Note that the problem of reducing the family of matrices $\mathbf{A}_i$ by matrices $\mathbf{V}_i$ for $i = 2, \ldots, n$ satisfying equalities (8.7.6) leads to the problem of classifying representations of the quiver $Q$ with diagram $A_{n-1}$:

$$\bullet \!\!-\!\!-\!\!-\!\! \bullet \!\!-\!\!-\!\!-\!\! \bullet \cdots\cdots \bullet \!\!-\!\!-\!\!-\!\! \bullet \qquad (8.7.7)$$

By the Gabriel theorem 6.1.32, this quiver has $\dfrac{n(n-1)}{2}$ indecomposable representations. Then according to these representations the matrix $\mathbf{A}_1$ is partitioned into $2n - 2$ vertical strips, and the partial ordering relation between these strips is linear. Therefore the matrix problem (8.7.6) leads to reducing the following matrix problem.

Given a block-rectangular matrix $\mathbf{T}$ with entries in a skew field $D$ which is partitioned into $2n - 2$ vertical strips:

$$\mathbf{T} = \begin{array}{|c|c|c|c|c|} \hline \mathbf{T}_1 & \mathbf{T}_2 & \cdots & \mathbf{T}_{2n-3} & \mathbf{T}_{2n-2} \\ \hline \end{array}$$

On this matrix one allows the admissible transformations of the following types:

1. Right $O$-elementary transformations of rows of **T**.
2. Left $D$-elementary transformations of columns within each vertical strip $\mathbf{T}_i$.
3. Additions of columns of the $i$-th vertical strip $\mathbf{T}_i$ multiplied on the right by elements of $D$ to columns of the $j$-th vertical strip $\mathbf{T}_j$, if $i \le j$.

By means of these transformations the matrix **T** can be reduced to the form such that any block $\mathbf{T}_i$ has the following form:

$$\mathbf{T}_i = \begin{array}{|c|c|} \hline \mathbf{I} & \mathbf{O} \\ \hline \mathbf{O} & \mathbf{O} \\ \hline \end{array}$$

such that we have the zero matrices up and down of the matrix **I** in the matrix **T**. This means that the matrix $\mathbf{A}_1$ is decomposed into a direct sum of matrices of the form

Thus, for any indecomposable representation $V$ the corresponding matrix has a finite number of elements distinct from zero, and this number depends only on $n$. Therefore the species $\Omega$ with diagram (8.7.2) is of bounded representation type. $\square$

**Lemma 8.7.8.** *Let $O_i$ be discrete valuation rings with a common skew field of fractions $D$ and Jacobson radicals $R_i = \pi_i O_i = O_i \pi_i$, for $i = 1, 2$; and $I = \{1, 2, \ldots, n, n + 1\}$. Then a weak $(D, O)$-species $\Omega = (F_i, {}_iM_j)_{i,j \in I}$ with diagram $S$:*

$$(8.7.9)$$

*is of bounded representation type for an arbitrary orientation of edges.*

*Proof.* Renumber the vertices of the diagram (8.7.9) in such a way that the vertex $1 \in S$ corresponds to $F_1 = O_1$, and the vertex $2 \in S$ corresponds to $F_2 = O_2$, so that we obtain the following diagram:

$$\overset{\textstyle 1}{\odot} \longrightarrow \overset{\textstyle 3}{\bullet} \underline{\quad\quad} \bullet \cdots\cdots \bullet \underline{\quad\quad} \overset{\textstyle n+1}{\bullet} \longleftarrow \overset{\textstyle 2}{\odot}$$

We will prove the lemma for the following case: $F_i = O_i$, for $i = 1, 2$, $F_3 = F_4 = \cdots = F_n = F_{n+1} = D$, ${}_1M_3 = {}_{O_1}D_D$ is a left $O_1$-module and right $D$-module, ${}_2M_{n+1} = {}_{O_2}D_D$ is a left $O_2$-module and right $D$-module, ${}_iM_{i+1} = {}_DD_D$ for $i = 3, \ldots, n$ and ${}_iM_j = 0$ for other $i, j \in I$. Write $J = \{1, 2, \ldots, n - 1\}$.

Let $X_i$ be an $F_i$-module ($i = 1, 2$), and let $Y_j$ be right $D$-vector space for $j \in J$. Let $\varphi_1 : X_1 \otimes_{F_1} (_1M_3) \to Y_1$, $\varphi_2 : X_2 \otimes_{F_2} (_2M_{n+1}) \to Y_{n-1}$, $\varphi_{j+2} : Y_j \otimes_D (_{j+2}M_{j+3}) \to Y_{j+1}$ ($j = 1, \ldots, n - 2$) are $D$-linear mappings.

Consider the category $\text{Rep}(\Omega)$ whose objects are representations $V = (X_1, X_2, Y_i, \varphi_i, \varphi_n)_{i \in J}$, and a morphism from an object $V$ to an object $V' = (X_1', X_2', Y_i', \varphi_i', \varphi_n')_{i \in J}$ is a family of homomorphisms $(\alpha_1, \alpha_2, \beta_j)_{j \in J}$, where $\alpha_i : X_i \to X_i'$ is a homomorphism of $F_i$-modules for $i = 1, 2$; $\beta_j : Y_j \to Y_j'$ is a homomorphism of $D$-vector spaces ($j \in J$). Moreover, the following equalities hold:

$$\beta_1\varphi_1 = \varphi_1'(\alpha_1 \otimes 1), \quad \beta_{n-1}\varphi_2 = \varphi_2'(\alpha_2 \otimes 1) \tag{8.7.10}$$

$$\beta_j\varphi_{j+1} = \varphi_{j+1}'(\beta_{j-1} \otimes 1) \quad (j = 2, \ldots, n - 1) \tag{8.7.11}$$

We show that the category $\text{Rep}(\Omega)$ is of bounded representation type. Let $X_i$ be a finitely generated free $F_i$-module with basis $\omega_1^{(i)}, \ldots, \omega_{s_i}^{(i)}$ ($i = 1, 2$); let $Y_j$ be a finite dimensional $D$-space with basis $\tau_1^{(j)}, \ldots, \tau_{k_j}^{(j)}$ for $j \in J$. Suppose homomorphisms $\varphi_i$ are given by matrices $\mathbf{A}_i$ with entries in $D$ for $i = 1, \ldots, n$. Then these matrices define the representation $V$ uniquely up to equivalence.

Suppose to the isomorphism $\alpha_i$ is given by the invertible matrix $\mathbf{U}_i$ with entries in $F_i$ for $i = 1, 2$, and the isomorphism $\beta_j$ is given by the invertible matrix $\mathbf{V}_j$ with entries in $D$ for $j \in J$.

If $\mathbf{A}_i'$ ($i = 1, 2, \ldots, n$) are matrices corresponding to a representation $V'$ which is equivalent to the representation $V$ then equalities (8.7.10)-(8.7.11) have the following matrix form:

$$\mathbf{V}_1\mathbf{A}_1\mathbf{U}_1^{-1} = \mathbf{A}_1', \quad \mathbf{V}_{n-1}\mathbf{A}_2\mathbf{U}_2^{-1} = \mathbf{A}_2' \tag{8.7.12}$$

$$\mathbf{V}_j\mathbf{A}_{j+1}\mathbf{V}_{j-1}^{-1} = \mathbf{A}_{j+1}', \quad (j = 2, \ldots, n - 1) \tag{8.7.13}$$

So that we obtain the matrix problem of reducing matrices $\mathbf{A}_i$ by matrices $\mathbf{U}_1, \mathbf{U}_2$ and $\mathbf{V}_i$, for $i = 1, 2, \ldots, n - 1$, satisfying equalities (8.7.12)-(8.7.13).

Note that the reducing the family of matrices $\mathbf{A}_1, \mathbf{A}_3, \ldots, \mathbf{A}_n$ by admissible transformations (8.7.14)-(8.7.15) leads to the matrix problem of classifying representations of a $(D, O)$-species with diagram (8.7.2) described in lemma 8.7.1. Then with accordance to this reduction of matrices $\mathbf{A}_1, \mathbf{A}_3, \ldots, \mathbf{A}_n$ the matrix $\mathbf{A}_2$ is partitioned into $n$ vertical strips. Thus we obtain the following matrix problem.

Given a block-rectangular matrix $\mathbf{T}$ with entries in a skew field $D$ that is partitioned into $n$ vertical strips:

$$\mathbf{T} = \boxed{\begin{array}{|c|c|c|c|} \hline \mathbf{T}_1 & \mathbf{T}_2 & \cdots & \mathbf{T}_n \\ \hline \end{array}}$$

On this matrix one allows admissible transformations of the following types:

1. Left $O_2$-elementary transformations of rows of $\mathbf{T}$.
2. Right $O_1$-elementary transformations of columns within vertical strip $\mathbf{T}_k$ for some $1 \leq k \leq n$.

3. Right $D$-elementary transformations of columns within each vertical strip $\mathbf{T}_i$, if $i \neq k$.
4. Additions of columns in the vertical strip $\mathbf{T}_i$ multiplied on the right by elements of $D$ to columns in the vertical strip $\mathbf{T}_j$, if $i \leq j$.

Using these transformations and taking into account lemma 8.5.3 and lemma 8.5.7 the matrix $\mathbf{T}$ can be reduced to the form in which every block $\mathbf{T}_i$ has one of the following forms:

$$
\begin{array}{|c|c|}
\hline
\mathbf{I} & \mathbf{O} \\
\hline
\mathbf{O} & \mathbf{O} \\
\hline
\end{array}
\quad \text{or} \quad
\begin{array}{|c|c|}
\hline
\pi^m \mathbf{I} & \mathbf{O} \\
\hline
\mathbf{O} & \mathbf{O} \\
\hline
\end{array}
$$

and up and down, on the left and on the right of the matrix $\mathbf{I}$ (or $\pi^m \mathbf{I}$) in the matrix $\mathbf{T}$ we have the zero matrices. This means that the matrix $\mathbf{A}_2$ is decomposed into a direct sum of matrices of these forms.

Thus, the species with diagram 8.7.9 is of bounded representation type. $\square$

**Lemma 8.7.14.** *Let $O$ be a discrete valuation ring with a skew field of fractions $D$ and Jacobson radical $R = \pi O = O\pi$, and $I = \{1, 2, 3\}$. Then a weak $(D, O)$-species $\Omega = (F_i, {}_iM_j)_{i,j \in I}$ with diagram*

$$\bullet \longleftarrow \odot \longrightarrow \bullet \tag{8.7.15}$$

*is of bounded representation type.*

*Proof.* We will prove the lemma for the case $F_1 = O$, $F_2 = F_3 = D$, ${}_1M_2 = {}_1M_3 = {}_OD_D$ is a left $F_1$-module and right $D$-module, and ${}_iM_j = 0$ for other $i, j \in I$. Write $J = \{1, 2\}$.

Let $X$ be an $F_1$-module and let $Y_i$ be right $D$-vector space for $i = 1, 2$. Let $\varphi_i : X \otimes_{F_1} ({}_1M_{i+1}) \to Y_i$ ($i = 1, 2$) be $D$-linear mappings.

Consider the category $\mathrm{Rep}(\Omega)$ whose objects are representations $V = (X, Y_i, \varphi_i)_{i \in J}$, and a morphism from an object $V$ to an object $V' = (X', Y_i', \varphi_i')_{i \in J}$ is a family of homomorphisms $(\alpha, \beta_i)_{i \in J}$, where $\alpha : X \to X'$ is a homomorphism of $F_1$-modules, $\beta_i : Y_i \to Y_i'$ is a homomorphism of $D$-vector spaces ($i = 1, 2$). Moreover the following equalities hold:

$$\beta_i \varphi_i = \varphi_i'(\alpha \otimes 1), \quad (i = 1, 2). \tag{8.7.16}$$

We show that the category $\mathrm{Rep}(\Omega)$ is of bounded representation type. Let $X$ be a finitely generated free $O$-module with basis $\omega_1, \ldots, \omega_s$; let $Y_i$ be a finite dimensional $D$-space with basis $\tau_1^{(i)}, \ldots, \tau_{k_i}^{(i)}$ for $i = 1, 2$. Suppose the homomorphisms $\varphi_1, \varphi_2$ are given by the matrices $\mathbf{A}_1, \mathbf{A}_2$ with entries in $D$. Then these matrices define the representation $V$ uniquely up to equivalence.

Let $\mathbf{U} \in M_s(F_1)$ be the invertible matrix corresponding to the isomorphism $\alpha$, and let $\mathbf{V}_i \in M_{k_i}(D)$ be the invertible matrix corresponding to the isomorphism $\beta_i$

($i = 1, 2$). If $\mathbf{A}'_1$, $\mathbf{A}'_2$ are matrices corresponding to a representation $V'$ which is equivalent to $V$ then equalities (8.7.16) have the following matrix forms:

$$\mathbf{V}_i\mathbf{A}_i\mathbf{U}^{-1} = \mathbf{A}'_i, \quad (i = 1, 2). \tag{8.7.17}$$

So that the reduction of the matrices $\mathbf{A}_1$, $\mathbf{A}_2$ by the matrices $\mathbf{U}$, $\mathbf{V}_1$, $\mathbf{V}_2$ leads to the following matrix problem:

Given a matrix $\mathbf{T}$ with entries in $D$ partitioned into 2 vertical strips:

$$\mathbf{T} = \boxed{\mathbf{T}_1 \mid \mathbf{T}_2}$$

On these matrices one allows admissible transformations of the following types:

1. Right $D$-elementary transformations of columns within each vertical strips.
2. Left $O$-elementary transformations of rows of $\mathbf{T}$.

Using these transformations and taking into account lemma 8.5.7 the matrix $\mathbf{T}$ can be reduced to the direct sum of the following matrices:

$$1.\boxed{1 \mid 0} \qquad 2.\boxed{0 \mid 1}$$

$$3.\boxed{1 \mid 1} \qquad 4.\begin{array}{|c|c|} \hline 1 & \pi^k \\ \hline 0 & 1 \\ \hline \end{array}$$

Thus, a species with diagram 8.7.15 is of bounded representation type. $\square$

**Lemma 8.7.18.** *Let $O$ be a discrete valuation ring with a skew field of fractions $D$ and Jacobson radical $R = \pi O = O\pi$, and $I = \{1, 2, \ldots, n, n + 1\}$. Then a weak $(D, O)$-species $\Omega = (F_i, {}_iM_j)_{i,j\in I}$ with diagram*

$$\tag{8.7.19}$$

*is of bounded representation type for an arbitrary orientation of edges in the diagram.*

*Proof.* We will prove the lemma for the case $F_1 = O$, $F_2 = F_3 = \cdots = F_n = F_{n+1} = D$, ${}_1M_2 = {}_OD_D$ is a left $F_1$-module and right $D$-module, ${}_iM_{i+1} = {}_{n-1}M_{n+1} = {}_DD_D$ for $i = 2, \ldots, n - 1$ and ${}_iM_j = 0$ for other $i, j \in I$. Write $J = \{1, 2, \ldots, n\}$.

Let $X$ be a right $F_1$-module and let $Y_i$ be a right $D$-vector space for $i \in J$, and let $\varphi_1 : X \otimes_{F_1} (_1M_2) \to Y_1$, $\varphi_j : Y_{j-1} \otimes_D (_jM_{j+1}) \to Y_j$ ($j = 2, \ldots, n-1$), $\varphi_n : Y_{n-2} \otimes_D (_{n-1}M_{n+1}) \to Y_n$ be $D$-linear mappings.

Consider the category $\mathrm{Rep}(\Omega)$ whose objects are representations $V = (X, Y_i, \varphi_i)_{i \in J}$, and morphisms from an object $V$ to and object $V' = (X', Y_i', \varphi_i')_{i \in J}$ is a family of homomorphisms $(\alpha, \beta_i)_{i \in J}$, where $\alpha : X \to X'$ is a homomorphism of $F_1$-modules, $\beta_i : Y_i \to Y_i'$ is a homomorphism of $D$-vector spaces ($i \in J$). Moreover the following equalities hold:

$$\beta_1\varphi_1 = \varphi_1'(\alpha \otimes 1), \quad \beta_i\varphi_i = \varphi_i'(\beta_{i-1} \otimes 1) \quad (i = 2, \ldots, n-1) \tag{8.7.20}$$

$$\beta_n\varphi_n = \varphi_n'(\beta_{n-2} \otimes 1) \tag{8.7.21}$$

We show that the category $\mathrm{Rep}(\Omega)$ is of bounded representation type. Let $X$ be a finitely generated free $O$-module, and let $Y_i$ be a finite dimensional $D$-space ($i \in J$). Suppose the homomorphism $\varphi_i$ is given by a matrix $\mathbf{A}_i$ with entries in $D$. Then the matrices $\mathbf{A}_i$, for $i \in J$ define the representation $V$ uniquely up to equivalence.

Let $\mathbf{U} \in M_s(F_1)$ be the invertible matrix corresponding to the isomorphism $\alpha$, and let $\mathbf{V}_i \in M_{k_i}(D)$ be the invertible matrix corresponding to the isomorphism $\beta_i$, $i \in J$. If $\mathbf{A}_i'$ are matrices corresponding to a representation $V'$ which equivalent to the representation $V$ then equalities (8.7.20)-(8.7.21) have the following matrix form:

$$\mathbf{V}_1\mathbf{A}_1\mathbf{U}^{-1} = \mathbf{A}_1', \quad \mathbf{V}_i\mathbf{A}_i\mathbf{V}_{i-1}^{-1} = \mathbf{A}_i' \quad (i = 2, \ldots, n-1) \tag{8.7.22}$$

$$\mathbf{V}_n\mathbf{A}_n\mathbf{V}_{n-2}^{-1} = \mathbf{A}_n' \tag{8.7.23}$$

So that we obtain the matrix problem of reducing a family of matrices $\mathbf{A}_i$ by means of a family of matrices $\mathbf{U}$ and $\mathbf{V}_i$, for $i \in J$, satisfying equalities (8.7.22)-(8.7.23).

Note that the reduction of matrices $\mathbf{A}_i$ by matrices $\mathbf{V}_i$, for $i = 2, \ldots, n$, satisfying equalities (8.7.22)-(8.7.23) leads to the problem of classifying representations of the following quiver with diagram $D_n$:

By the Gabriel theorem, this quiver has $(n-1)n$ indecomposable representations. In accordance with these representations the matrix $\mathbf{A}_1$ is partitioned into $2n - 1$ vertical strips:

$$\mathbf{A}_1 = \boxed{\mathbf{B}_1 \mid \mathbf{B}_2 \mid \cdots \mid \mathbf{B}_{2n-1}}$$

The admissible transformations on the columns of the matrices $\mathbf{B}_i$ are given by the poset $S = \{\delta_1, \delta_2, \ldots, \delta_{2n-1}\}$ whose diagram has the following form:

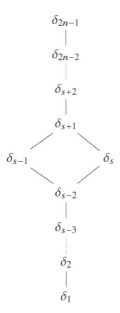

That is, if $\delta_i \leq \delta_j$ in $S$, then any column in the block $\mathbf{B}_i$ multiplied by an arbitrary element of $D$ can be added to any column of the block $\mathbf{B}_j$.

So that on the matrix $\mathbf{A}_1$ one allows admissible transformations of the following types:

1. Left $O$-elementary transformations of rows of $\mathbf{A}_1$.
2. Right $D$-elementary transformations of columns within each vertical strip $\mathbf{B}_i$.
3. Additions of columns of the vertical strip $\mathbf{B}_i$ multiplied on the right by elements of $D$ to columns of the vertical strip $\mathbf{B}_j$, if $\delta_i \leq \delta_j$ in $S$.

Using these transformations the matrix $\mathbf{A}_1$ can be reduced to the form in which the corresponding matrix $\mathbf{B}_i$ is decomposed into a direct sum of the matrices of the following form:

1. 
| 1 | 0 |
|---|---|

2. 
| 1 | 1 |
|---|---|

3. 
| 1 | $\pi^k$ |
|---|---------|
| 0 | 1 |

Thus, a species with diagram (8.7.19) is of bounded representation type. □

Now the sufficiency of theorem 8.6.1 follows from lemmas 8.7.1, 8.7.8, 8.7.14, 8.7.18.

## 8.8 Right Hereditary SPSD-Rings of Bounded Representation Type

In section 8.3 we showed that the description of right hereditary SPSD-rings of bounded representation type can be reduced to the description of $(D, O)$-species of bounded representation type.

Let $A$ be a right hereditary SPSD-ring. Then as was shown in section 2.4 and 8.3 the prime quiver $PQ(A)$ of $A$ is an acyclic simply connected quiver which coincides with the diagram of some finite poset $S$ with weights $H_{n_i}(O_i)$ or $D$, where the $O_i$ are discrete valuation rings with a common classical division ring of fractions $D$, and all points with weights $H_{n_i}(O_i)$ correspond to the minimal elements of $S$. Let $\overline{PQ(A)}$ be the graph obtained from the diagram $PQ(A)$ by deleting the orientation of all arrows except those connected the points with weights $H_{n_i}(O_i)$. If the right classical ring of fractions $\tilde{A}$ of $A$ is a basic ring then $PQ(A)$ is a simply connected weak quiver.

Taking into account proposition 8.3.8 and theorem 8.6.1 we immediately obtain the following theorem which describes right hereditary rings SPSD-ring of bounded representation type.

**Theorem 8.8.1.** *Let $A$ be a right hereditary SPSD-ring which classical ring of fractions $\tilde{A}$ is a basic ring. Then $A$ is of bounded representation type if and only if the diagram $\overline{PQ(A)}$ of the prime quiver $PQ(A)$ is a finite disjoint union of Dynkin diagrams of the forms $A_n$, $D_n$, $E_6$, $E_7$, $E_8$ and the following diagrams:*

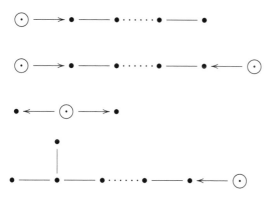

Diag. 8.8.2.

**Remark 8.8.3.** A diagram $\Gamma^*(\Omega)$ is said to be an **extension** of a undirected diagram $\overline{\Gamma(\Omega)}$ if it is obtained from $\overline{\Gamma(\Omega)}$ by changing each marked vertex in $\overline{\Gamma(\Omega)}$ by an infinite chain beginning in this vertex (and which does not have any added connections with other vertices of the diagram). In this case all diagrams 8.8.2 will be infinite of the following form:

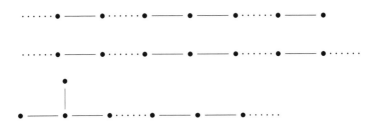

Using this definition theorem 8.8.1 can be rewritten in the following simple form:

**Theorem 8.8.4.** *Let $A$ be a right hereditary SPSD-ring which classical ring of fractions $\tilde{A}$ is a basic ring. Then $A$ is of bounded representation type if and only if the diagram $\overline{\Gamma(\Omega)} = \overline{PQ(A)}$ of the prime quiver $PQ(A)$ is a finite disjoint union of Dynkin diagrams $A_n$, $D_n$, $E_6$, $E_7$, $E_8$ or any connected finite part of the extended diagram $\Gamma^*(\Omega)$ of the diagram $\overline{\Gamma(\Omega)}$ is one of the Dynkin diagrams $A_n$, $D_n$.*

## 8.9 Notes and References

The matrix problems, i.e. the problems of reducing a family of matrices by some family of admissible transformations, arise in many problems of mathematics. Many problems in the theory of representations were solved by reducing to some matrix problems (see eg. [137], [161], [162], [163], [164]), [231]).

The general definition of a matrix problem over a field was first given by A.V. Roiter in [184]. More general definition of a matrix problem over a ring was given by Yu.A. Drozd in [61].

The definition of a flat matrix problem of mixed type over a discrete valuation ring and its skew field of fractions was given by A.G. Zavadskii and U.S. Revitskaya in [233]. Earlier such matrix problems were considered by N.M. Gubareni in [87], [88], and A.G. Zavadskii and V.V. Kirichenko in [230], [231]. Some flat mixed matrix problems were also considered in [94].

$(D, O)$-species were first introduced and considered by N.M. Gubareni in [88]. Right hereditary SPSD-rings and $D, O$-species of bounded representation type were studied by N.M. Gubareni in [87], [88], [91], [92], [93].

The notion of $(D, O)$-species considered in this section was generalized to the case where $O$ is a Noetherian hereditary ring without zero divisors and $D$ is its skew field of fractions in [65].

# REFERENCES

[1] Abrams G. 1994. Group gradings and recovery results for generalized incidence rings. Journal of Algebra 164: 859-876.

[2] Achkar H. 1974. Sur les anneaux arithmétiques à gauche. C.R. Acad. Sci. Paris 278(5): A307-A309.

[3] Angeleri Hügel L. 2006. An Introduction to Auslander-Reiten theory. *In*: Advanced School on Representation Theory and Related Topics, ICTP. Trieste.

[4] Alfaro R. and A.I. Kelarev. 2006. On cyclic codes in incidence rings. Studia Scientiarum Math. Hungarica 43(1): 69-77.

[5] Andrunakievich V.A. 1984. On completely prime ideals of a ring. Math. USSR-Sbornik 49(2): 285-290.

[6] Andrunakievich V.A. and Yu.M. Ryabukhin. 1968. Rings without nilpotent elements, and completely simple ideals. Dokl. Akad. Nauk SSSR 180: 9-11 (in Russian); English translation: 1968. Sov. Math. Dokl. 9: 565-568.

[7] Anh P.N. and M. Siddoway. 2010. Divisibility theory of semihereditary rings. Proc. Amer. Math. Soc. 138(12): 4231-4242.

[8] Arnold D.A. 2000. Abelian groups and representations of finite partially ordered sets. Springer-Verlag, New York.

[9] Assem I., M.I. Platzeck, M.J. Redondo and S. Trepode. 2003. Simply connected incidence algebras. Discrete Math. 269(1): 333-356.

[10] Assem I., D. Simson and A. Skowroński. 2006. Elements of the Representation Theory of Associative Algebras. vol. I. Techniques of Representation Theory. Cambridge Univ. Press, London Math. Soc. Student Texts 65, Cambridge.

[11] Atiyah M. 1956. On the Krull-Schmidt theorem with application to sheaves. Bull. Soc. Math. France 84: 307-317.

[12] Auslander M. 1962. Modules over unramified regular local rings. pp. 230-233. *In*: Proc. of Int. Congress of Math. Stockholm.

[13] Auslander M. and M. Bridger. 1969. Stable module theory. Mem. Amer. Math. Soc. 94.

[14] Auslander M. 1971. Representation dimension of Artin algebras. Queen Mary College Notes, London.

[15] Auslander M. 1974. Representation theory of Artin algebras II. Comm. Algebra 1: 269-310.

[16] Auslander M. 1977. Existence theorems for almost split sequences, Ring Theory II. *In*: Proc. of the Second Oklahoma Conf., Marcel Dekker, New York and Basel.

[17] Auslander M. 1978. Functors and morphisms determined by objects. Lecture Notes in Pure and Appl. Math. 37: 1-244.

[18] Auslander M. 1982. A functorial approach to representation theory. Lecture Notes in Math. 944: 105-179.

[19] Auslander M. 1986. A survey of existence theorems for almost split sequences. Representations of algebras. London Math. Soc. Lecture Notes Ser. 116 Cambridge Univ. Press, Cambridge 81-89.

[20] Auslander M. and I. Reiten. 1975. Representation theory of Artin algebras III. Almost split sequences. Comm. Algebra 3(3): 239-294.

[21] Auslander M. and I. Reiten. 1977. Representation theory of Artin algebras IV. Invariants given by almost split sequences. Comm. Algebra 5(5): 443-518.

[22] Auslander M., I. Reiten and S.O. Smalø. 1995. Representation Theory of Artin Algebras. Cambridge University Press, Cambridge.

[23] Baer D., W. Geigle and H. Lenzig. 1987. The preprojective algebra of tame hereditary Artin algebra. Comm. Algebra 15: 425-457.

[24] Belding W. 1973. Incidence rings of pre-ordered sets. Notre Dame J. Formal Logic 14: 481-509.

[25] Bell A.D. and K.R. Goodearl. 1995. Algebras of bounded finite dimensional representation type. Glazgow Math. Journal 37: 289-302.

[26] Behrens E.A. 1960. Distributive Darstellbar Ringe I. Math. Z. 73(5): 409-432.

[27] Behrens E.A. 1961. Distributive Darstellbar Ringe II. Math. Z. 76(4): 367-384.

[28] Bernstein J., I. Gel'fand and V. Ponomarev. 1973. Coxeter functors and Gabriel's theorem. Russian Math. Surveys 28: 17-32.

[29] Blair R.I. 1953. Ideal lattice and the structure of rings. Trans. Amer. Math. Soc. 75(1): 136-153.

[30] Bondarenko V.M., N.M. Gubareni, M.A. Dokuchaev, V.V. Kirichenko and M.A. Khibina. 2008. Representations of primitive posets (in Russian). Fudam. Prikl. Mat. 14(6): 41-74; English translation: 2010. Journal of Math. Sci. 164(1): 26-48.

[31] Bourbaki N. 1968. Croups et algèbres de Lie. Chaptires 4,5,6. Paris, Hermann.

[32] Brenner S. 1974. On four subspaces of a vector space. J. Algebra 29: 587-599.

[33] Brungs H.H. 1976. Rings with distributive lattice of right ideals. J. Algebra 40(2): 392-400.

[34] Brungs H.H. 1986. Bèzout domains and rings with a distributive lattice of right ideals. Canad. J. Math. 38(2): 286-303.

[35] Brungs H.H. 1987. Ideal theory in rings with a distributive lattice of right ideals. Wiss. Beitr. M.-Luther-Univ. Halle, Wittenberg 48: 73-80.

[36] Brungs H.H. and J.Gräter. 1991. Value groups and distributivity. Canad. J. Math. 43(6): 1150-1160.

[37] Bunina E.I. and A.S. Dobrokhotova-Maykova. 2010. Elementary equivalence of generalized incidence rings. Journal of Mathematical Sciences 164(2): 178-181.

[38] Caballero C.E. 1986. Self-duality and *l*-hereditary semidistributive rings. Comm. Algebra 14(10): 1821-1843.

[39] Camillo V. 1975. Distributive modules. J. Algebra 36(1): 16-25.

[40] Cartan H. and S. Eilenberg. 1956. Homological algebra. Princeton University Press, Princeton, New York.

[41] Chapman S.T. and S. Glaz (eds.). 2000. Non-Noetherian Commutative Ring Theory. Math. and its Appl., Kluver Acad. Publ.

[42] Chatters A.W. 1972. A decomposition theorem for Noetherian hereditary rings. Bull. London Math. Soc. 4: 125-126.

[43] Chen X. 2009. Quivers and Representations. pp. 507-561. *In*: M.Hazewinkel (ed.), Handbook of Algebra, Vol.6.

[44] Cicala D. 2014. Matrix problems and their relation to the representation theory of quivers and posets. Thesis M.Sc., University of Ottawa, Canada, Ottawa.

[45] Cohn P.M. 1959. On the free product of associative rings. Math. Zeitschr. 71: 380-398.

[46] Couchot F. 2001. Commutative local rings of bounded module type. Comm. Algebra 29(3): 1347-1355.

[47] Couchot F. 2005. Local rings of bounded module type are almost maximal valuation rings. Comm. Algebra 33(8): 2851-2855.

[48] Danlyev Kh.M., V.V. Kirichenko, Z.P. Haletskaja and Yu.V. Yaremenko. 1995. Weakly prime semiperfect 2-rings and modules over them pp. 5-32. *In*: Studies in Algebra, Math. Inst. Ukr. Akad. Nauk, Kiev.

[49] De la Pena J.A. and D. Simson. 1992. Preinjective modules, reflections functors, quadratic forms, and Auslander-Reiten sequences. Trans. Amer. Math. Soc. 329: 733-753.

[50] Diestel R. 2005. Graph theory, Springer.

[51] Dilworth R.P. 1950. A decomposition theorem for partially ordered sets. Annals of Mathematics 51(1): 161-166.

[52] Dlab V. and C.M. Ringel. 1974. Representations of graphs and algebras. Carleton Math. Lecture Notes 8. Carleton Univ., Ottawa.

[53] Dlab V. and C.M. Ringel. 1975. On algebras of finite representation type. J. Algebra 33(2): 306-394.

[54] Dlab V., C.M. Ringel. 1976. Indecomposable representations of graphs and algebras. Mem. Amer. Math. Soc. 173.

[55] Dokuchaev M.A., N. Gubareni. 2010. Rings related to finite posets. Scientific Research of the Institute of Mathematics and Computer Science 9(2): 25-36.

[56] Dokuchaev M.A., V.V. Kirichenko, B.V. Novikov and A.P. Petravchuk. 2007. On incidence modulo ideal rings. Journal of Algebra and Its Applications 6(4): 553-586.

[57] Donovan P. and M.R. Freislich. 1973. The representation theory of finite graphs and associated algebras. Carleton Math. Lecture Notes 5. Carleton Univ., Ottawa.

[58] Dowbor P. and D. Simson. 1980. Quasi-Artin species and rings of finite representation type. J. Algebra 63: 435-443.

[59] Dowbor P., C.M. Ringel and D. Simson. 1980. Hereditary Artinian rings of finite representation type. Representation Theory II. Lecture Notes in Math. 832: 232-241.

[60] Drozd Yu.A. 1972. Representations of commutative algebras (in Russian). Funkcional. Anal. i Prilozen. 6(4): 41-43.

[61] Drozd Yu.A. 1972. Matrix problems and categories of matrices (in Russian). Zapiski Nauch. Sem. Leningrad. Otdel. Mat. Inst. Steklov. (LOMI) 28: 144-153; English translation: 1975. J. Soviet Math. 3(5): 692-699.

[62] Drozd Yu.A. 1974. Coxeter transformations and representations of partially ordered sets (in Russian). Funk. Anal. Prilozh. 8(3): 34-42.

[63] Drozd Yu.A. 1980. Tame and wild problems. pp. 242-258. *In*: V.Dlab and P.Gabriel (eds.). Representation Theory II. Lecture Notes in Math. 832, Springer.

[64] Drozd Yu.A. 1980. The structure of hereditary rings (in Russian). Math. Sbornik 113(155), N.1(9): 161:172; English translation: 1982. Math. USSR Sbornik 41(1): 139-148.

[65] Drozd Yu.A. and L.V. Izjumchenko. 1992. Representations of $\mathcal{D}$-species. Ukr. Math. Journal 44: 572-574.

[66] Drozd Yu.A. and E. Kubichka. 2004. Dimensions of finite type for representations of partially ordered sets. Algebra and Discrete Mathematics 3: 21-37.

[67] Dynkin E.B. 1946. Classification of the simple Lie groups (in Russian). Mat. Sbornik 18(60): 347-352.

[68] Dynkin E.B., 1947. The structure of semi-simple Lie algebras (in Russian). Uspekhi Mat. Nauk 2: 59-127; English translation: 1962. American Math. Soc. Transl. 9(1): 328-469.

[69] Dynkin E.B. 1952. Semi-simple subalgebras of semi-simple Lie algebras (in Russian). Mat. Sbornik 30: 349-462; English translation: 1957. Amer. Math. Soc. Trans. Ser. 6(2): 111-244.

[70] Eckmann B. and P. Hilton 1960. Homotopy Groups of Maps and Exact Sequences. Comment. Math. Helv. 34: 271-304.

[71] Eisenbud D. and P. Griffith. 1971. The structure of serial rings. Pacific J. Math. 36: 173-182.

[72] Enochs E.E. and O.M.G. Jenda. 2000. Relative Homological Algebra, Walter de Gruyter, Berlin-New York.

[73] Farkas D.R. 1974. Radicals and primes in incidence algebras, Discrete Math. 10: 257-268.

[74] Feinberg R.B. 1977. Characterization of incidence algebras. Discrete Math. 17: 47-70.

[75] Finch P.D. 1970. On the Möbius function of a nonsingular binary relation. Bull. Austr. Math. Soc. 3: 155-162.

[76] Fitting H. 1935. Primärkomponentenzerlegung in nichtkommutative Ringe. Mathematische Annalen 111: 19-41.

[77] Fuchs L. 1949. Über die Ideale arithmetischer ringe. Comment. Math. Helv. 23: 334-341.

[78] Gabriel P. 1972-1973. Representations indecomposables des ensembles ordonnes. Sem. P. Dubreil 13: 1301-1304.

[79] Gabriel P. 1972. Unzerlegbare Darstellungen I. Manuscripta Math. 6: 71-103.

[80] Gabriel P. 1973. Indecomposable represenations. II. Ist. Naz. Alta Mat. Symposia Mathematica 11. Academic Press, London, New York, 81-104.

[81] Gabriel P. 1973. Représentations indécomposables des ensembles ordonnés. (in French) D'après L.A. Nazarova et A. V. Roiter. 1972. Zap. Naucn. Sem. Leningrad. Otdel. Mat. Inst. Steklov. 28: 5-31. Séminaire P. Dubreil (26e année: 1972/73), Algèbre, Exp. 13 Secrétariat Mathématique, Paris.

[82] Gabriel P. and A.V. Roiter 1997. Representations of finite-dimensional algebras. With a chapter by B. Keller. Springer-Verlag, Berlin.

[83] Gel'fand I.M. and V.A. Ponomarev. 1968. Indecomposable representations of the Lorentz group (in Russian). Uspekhi Mat. Nauk 23(2): 3-60; English translation: 1968. Russian Math. Surveys 23(2): 1-58.

[84] Gel'fand I.M. and V.A. Ponomarev. 1970. Problems of linear algebra and classification of quadruples of subspaces in a finite-dimensional vector space. pp. 167-237. Colloq. Math. Soc. Janos Bolyai 5. Hilbert space operators and operator algebras. Tihany, Hungary.

[85] Gel'fand I.M. and V.A. Ponomarev. 1979. Model algebras and representations of graphs (in Russian). Funk. Anal. i Prilozen. 13: 1-12; English translation: 1980. Funct. Anal. Appl. 13: 157:166.

[86] Goodearl K.R. and R.B. Warfield, Jr. 1989. An introduction to noncommutative Noetherian rings. London Mathematical Society Student Texts 16, Cambridge Univ. Press, Cambridge.

[87] Gubareni N.M. 1977. Right hereditary rings of module bounded type (in Russian). Preprint-148. Inst. Electrodynamics Akad. Nauk Ukrain. SSR, Kiev.

[88] Gubareni N.M. 1978. Semiperfect right hereditary rings of module bounded type (in Russian). Preprint IM-78.1. Inst. Mat. Akad. Nauk Ukrain. SSR, Kiev.

[89] Gubareni N.M., V.V. Kirichenko and U.S. Revitskaja. 1999. Semiperfect semidistributive semihereditary rings of modular restricted type. Proc. Gomel State Univ. Problems in Algebra 1(15): 29-47.

[90] Gubareni N. 2010. Finitely presented modules over right hereditary SPSD-rings. Scientific Research of the Institute of Mathematics and Computer Science 9(2): 49-58.

[91] Gubareni N. 2012. Structure of finitely generated modules over right hereditary SPSD-rings. Scientific Research of the Institute of Mathematics and Computer Science 11(3): 45-55.

[92] Gubareni N. 2012. On right hereditary SPSD-rings of bounded representation type. I. Scientific Research of the Institute of Mathematics and Computer Science 11(3): 57-70.

[93] Gubareni N. 2012. On right hereditary SPSD-rings of bounded representation type. II. Scientific Research of the Institute of Mathematics and Computer Science 11(4):: 53-63.

[94] Gubareni N. 2013. Some mixed matrix problems over several discrete valuation rings. Journal of Applied Mathematics and Computational Mechanics 12(4): 47-58.

[95] Haack J.K. 1980. Incidence rings with self-duality. Proc. Amer. Math. Soc. 78: 165-169.

[96] Haack J.K. 1984. Isomorphisms of incidence rings. Illinois Journal of Mathematics 28(4): 676-683.

[97] Hanlon P. 1981. Algebras of acyclic type. Can. J. Math. 33: 129-141.

[98] Happel D. 1988. Triangulated categories in the Representation Theory of Finite Dimensional Algebras. London Mat. Soc. Lecture Notes 119. Cambridge Univ. Press.

[99] Harada M. and Y. Sai. 1970. On categories of indecomposable modules I. Osaka J. Math. 8: 323-344.

[100] Hazewinkel M., N. Gubareni and V.V. Kirichenko. 2004. Algebras, rings and modules. Volume 1. Kluwer Acad. Publ.

[101] Hazewinkel M., N. Gubareni and V.V. Kirichenko. 2007. Algebras, rings and modules. Volume 2. Springer.

[102] Hazewinkel M., N. Gubareni and V.V. Kirichenko. 2010. Algebras, rings and modules: Lie algebras and Hopf algebras. AMS, Providence.

[103] Hazewinkel M. and N. Gubareni. 2016. Algebra, Rings and Modules: Non-commutative Algebras and Rings. CRC Press, Taylor & Francis Group, Boca Raton, London, New York.

[104] Iyama O. 2003. Symmetry and duality on $n$-Gorenstein rings. J. Algebra 269: 528-535.

[105] Jacobson N. 1943. The Theory of Rings. American Mathematical Society Mathematical Surveys. vol. I. American Mathematical Society, New York.

[106] Jacobson N. 1943. The theory of rings. Math. Surveys II. Amer. Math. Soc., Providence, R.I., New York.

[107] Jaffard P., 1960. Les systèmes d'idéaux. Paris.

[108] Jans J.P. 1957. On the indecomposable representations of algebras. Ann. Math. 66(3): 418-429.

[109] Jans J.P. 1963. On finitely generated modules over Noetherian rings. Trans. Amer. Math. Soc. 106: 330-340.

[110] Jans J.P. 1964. Rings and Homology. Holt, Rinehart and Winston. New York,

[111] Jensen Chr. U. 1963. On characterizations of Prüfer rings. Math. Scand. 13: 90-98.

[112] Jensen Chr. U. 1966. Arithmetical rings. Acta Math. Acad. Sci. Hung. 17: 115-123.

[113] Jensen Chr. U. 1964. A remark on arithmetical rings. Proc. Amer. Math. Society 15: 951-954.

[114] Jondrup S. and D. Simson. 1981. Indecomposable modules over semiperfect rings. J. Algebra 73(1): 23-29.

[115] Kaplansky I. 1970. Commutative algebra. Univ. Chicago Press, Chicago.

[116] Kasjan S. and D. Simson. 1996. Tame prinjective type and Tits form of two-peak posets I. J. Pure Appl. Math. 106: 307-330.

[117] Kasjan S. and D.Simson. 1997. Tame prinjective type and Tits form of two-peak posets II. J. Algebra 187: 71-96.

[118] Kerner O. 1981. Partially ordered sets of finite representation type. Comm. Algebra 9(8): 783-809.

[119] Kelarev A.V. 2002. Rings constructions and applications. Series in Algebra 9, World Scientific, River Edge, New Jork.

[120] Kelarev A.V. 2003. On incidence rings of group automata. Bull. Austr. Math. Soc. 67: 407-411.

[121] Kelarev A.V. 2004. On the structure of incidence rings of group automata. International Journal of Algebra and Computation 14(4): 505-511.

[122] Kelarev A.V. and D.S. Passman. 2008. A description of incidence rings of group automata. Contemporary Mathematics 456: 27-34.

[123] Kelly G.M. 1964. On the radical of a category. J. Australian Math. Soc. 4: 299-307.

[124] Kirichenko V.V. 1978. Classification of the pairs of mutually annihilating operators in a graded space and representations of a dyad of generalized uniserial algebras (in Russian). Rings and linear groups. Zap. Naucn. Sem. Leningrad. Otdel. Mat. Inst. Steklov. (LOMI) 75: 91-109.

[125] Kirichenko V.V. and M.A. Khibina. 1993. Semiperfect semidistributive rings pp. 457-480. *In*: Infinite Groups and Related Algebraic Topics, Institute of Mathematics NAS Ukraine.

[126] Kirichenko V.V., V.V. Mogileva, E.M. Pirus and M.A. Khibbina. 1995. Semiperfect rings and piecewise domains pp. 33-65. *In*: Algebraic Researches, Institute of Mathematics NAS Ukraine.

[127] Kirichenko V.V. 2000. Semiperfect semidistributive rings. Algebras and Representation theory 3: 81-98.

[128] Kirichenko V.V. and Yu.V. Yaremenko. 2001. Semiperfect semidistributive rings. Math. Notes 69(1): 134-137.

[129] Kleiner M.M. 1972. Partially ordered sets of finite type (in Russian). Zapiski Nauch. Semin. LOMI 28: p. 32-41; English translation: 1975. J. Soviet. Math.3: 607-615.

[130] Kleiner M.M. 1972. On the exact representations of partially ordered sets of finite type (in Russian). Zapiski Nauch. Semin. LOMI 28: 42-59; English translation: 1975. J. Soviet. Math. 3: 616-628.

[131] Kleiner M.M. 1988. Pairs of partially ordered sets of tame representation type. Linear Algebra Appl. 104: 103-115.

[132] Köthe G. 1935. Verallgemeinerte Abelsche Gruppen mit hypercomplexem Operatorenring. Math. Z. 39: 31-44.

[133] Krause H. 1992. Stable equivalences preserves representation type. Comment. Math. Helv. 72: 266-284.

[134] Krause H. 2010. Representations of quivers via reflection functors. arXiv preprint, arXiv: 0804.1428v2, 31 Aug.

[135] Krause H. 2014. Krull-Schmidt categories and projective covers. arXiv preprint, arXiv: 1410.2228v1, 10 Oct.

[136] Kriegl A. 1981. A characterization of reduced incidence algebras. Discrete Math. 34: 141-144.

[137] Krugljak S.A. 1972. Representations of algebras with zero square radical (in Russian). Zap. Nauchn. Sem. LOMI 28: 60-68; English translation: 1975. J. Soviet. Math. 3(5).

[138] Krylov P.A. and A.A. Tuganbaev. 2008. Modules over Discrete Valuation Rings. Walter de Gruyter GmbH & Co, Berlin, Germany.

[139] Lam T.Y. 1991. A First Course in Noncommutative rings. Graduate Texts in Mathematics 131. Springer-Verlag, Berlin-Heidelberg-New York.

[140] Lam T.Y. 2001. Lectures on Module and Rings. Springer.

[141] Lang S. 1965. Algebra. Addison-Wesley Publ. Company.

[142] Lemay J. 2012. Valued graphs and the representation theory of Lie algebras. Axioms 1: 111-148.

[143] Leroux P. and J.Sarraillé. 1981. Structure of incidence algebras of graphs. Comm. Alg. 9(15): 1479-1517.

[144] MacLane S. 1963. Homology. Springer. 1963.

[145] MacLane S. 1998. Categories for the working mathematician. Springer.

[146] Martsinkovsky A. and J.R. Strooker. 2004. Linkage of modules. J. Algebra 271: 587-626.

[147] Marubayashi H., M. Haruo and U. Akira. 1997. Non-commutative Valuation Rings and Semi-hereditary Orders. Springer-Verlag.

[148] May J.P. 2011. Notes on Tor and Ext, http://www.math.uchicago.edu/ may/MISC/ TorExt.pdf

[149] Mazurek R. 1991. Distributive rings with Goldie dimension one. Comm. Algebra 19(3): 931-944.

[150] Mazurek R. and E.R. Puczyłowski. 1977. On semidistributive rings. Comm. Algebra 25(11): 3463-3471.

[151] Mazurek R. and M. Ziembowski. 2006. On Bèzout and distributive generalized power series rings. J. Algebra 306: 397-411.

[152] Mazurek R. and M. Ziembowski. 2009. Duo, Bèzout and Distributive Rings of Skew Power Series. Publicacions matematiques 53(2): 257-271.

[153] McCoy N.H. 1973. Completely prime and completely semi-prime ideals. Colloq. Math. Soc. János Bolyai 6: 147-152.

[154] McGovern, W.W. 2008. Bèzout rings with almost stable range 1. J. Pure Appl. Algebra 212(2): 340-348.

[155] Menzel W. 1960. Über den Untergruppenverband einer Abelschen Operatorgruppe. Teil II. Distributive und M-Verbande von Untergruppen einer Abelschen Operatorgruppe. Math. Z. 74(1): 52-65.

[156] Menzel W. 1961. Ein Kriterium für Distributivität des Untergruppenverband einer Abelschen Operatorgruppe. Math. Z. 75(3): 271-276.

[157] Mirsky L. 1971. A dual of Dilworth's decomposition theorem. Amer. Math. Monthly 78(2): 876-877.

[158] Nachev N.A. 1977. On incidence rings. Vest. Moscow Univ. Mat. (Moscow Univ. Math. Bull.) 32: 29-34.

[159] Nachev N.A. 1981. The global dimension of incidence rings I (in Bulgarian). Plovdiv Univ. Nauchn. Trud. 18: 19-41.

[160] Nachev N.A. 1981. The global dimension of incidence rings II (in Bulgarian). Plovdiv Univ. Nauchn. Trud. 18: 43-63.

[161] Nazarova L.A. 1961. Integral representations of Klein's four group (in Russian). Dokl. Akad. Nauk SSSR 140: 1111-1014; English translation: 1961. Soviet Mathematics 2(5): 1304-1308.

[162] Nazarova L.A. 1967. Representations of a tetrad (in Russian). Izv. Akad. Nauk SSSR Ser. Mat. 31: 1361-1378; English translation: 1967. Math. USSR - Izvestija 1(6): 1305-1321.

[163] Nazarova L.A. and A.V. Roiter. 1972. Representations of partially ordered sets (in Russian). Zap. Nauchn. Sem. LOMI 28: 5-31; English translation: 1975. J. Soviet Math. 3: 585-606.

[164] Nazarova L.A. 1973. Representations of quivers of infinite type (in Russian). Izv. Akad. Nauk SSSR, Ser. Mat. 37: 752-791; English translation: 1973. Math. USSR Izv. 7: 749-792.

[165] Nazarova L. and A.V. Roiter. 1975. Kategorielle Matrizen-Probleme und die Brauer-Thrall-Vermutung. Mitt. Math. Sem. Geissen 115: 1-153.

[166] Nazarova L.A. 1974. Partially ordered sets of infinite type pp. 244-252. In: 1974. Lecture Notes in Math., Springer-Verlag 488.

[167] Nazarova L.A. 1975. Representations of partially ordered sets of infinite type. Izv. Akad. Nauk SSR, Ser. Mat. 39: 963-991.

[168] Nazarova L.A. and A.G. Zavadskij. 1977. Partially ordered sets of tame type. pp. 122-143 (in Russian). In: Matrix problems, Kiev.

[169] Nishida K. 2009. Linkage and duality of modules. Math. J. Okayama Univ. 51: 71-81.

[170] O'Donnell C.J. 1996. Maximal and minimal prime ideals of incident algebras. Comm. Alg. 24(5): 1823-1840.

[171] Oppermann S. 2005. Auslander-Reiten Theory of Representation-Directed Artinian Rings. Univ. Stuttgart, Fachbereich Math.

[172] Ovsienko S.A. and A.V. Roiter. 1977. Bilinear forms and categories of representations. pp. 71-80 (in Russian). *In*: 1977. Matrix problems, Kiev.

[173] Peskine C. and L. Szpiro. 1974. Liaison des variétés algébriques. I. Invent. Math. 26: 271-302.

[174] Reiten I. 1997. Dynkin diagrams and the representation theory of algebras. Notices of the AMS 44(5): 546-556.

[175] Rickard J. 1989. Derived categories and stable equivalence. J. Pure and Appl. Alg. 61: 303-317.

[176] Ringel C.M. 1976. Representations of $K$-species and bimodules. J. Algebra 41: 269-302.

[177] Ringel C.M. 1978. Finite dimensional hereditary algebras of wild representation type. Math. Z. 161: 235-255.

[178] Ringel C.M. 1996. The preprojective algebra of quiver. pp. 467-487. *In*: 1998. Algebras and Modules II (Geiranger, 1996). CMS Conf. Proc. 24. Amer. Math. Soc., Providence, RI.

[179] Ringel C.M. 2008. The First Brauer-Thrall Conjecture. pp. 371-376. *In*: 2008. R.Göbel, B. Goldsmith (eds.). Models, Modules and Abelian Groups. Walter de Cruyter, Berlin.

[180] Ringel C.M. 2010. Gabriel-Roiter inclusions and Auslander-Reiten theory. J. Algebra 324: 3579-3590.

[181] Robson J.C. 1967. Pri-rings and ipri-rings. Quart. J. Math. Oxford Ser. 18(2): 125-145.

[182] Robson J.C. 1967. Rings in which finitely generated right ideals are principal. Proc. London Math. Soc. 17(3): 617-628.

[183] Roiter A.V. 1968. Unbounded dimensionality of indecomposable representations of an algebra with an infinite number of indecomposable representations. Math. USSR Izv. 2(6): 1223-1230.

[184] Roiter A.V. 1972. Matrix problems and representations of bisystems (in Russian). Zapiski Nauch. Sem. Leningrad. Otdel. Mat. Inst. Steklov. (LOMI), 28: 130-143.

[185] Rota G.-C. 1964. On the foundations of combinatorial theory I. Theory of Möbius functions. Z. Wahr Scheinlichkeits Therie und Verw. Gebiete 2: 340-368.

[186] Rotman J.J. 1979. An introduction to homological algebra. Academic Press, New York.

[187] Rowen L.H. 1986. Finitely presented modules over semiperfect rings. Proc. Amer. Math. Soc. 97: 1-7.

[188] Sands A.D. 1990. Radicals of structural matrix rings. Quaestiones Math. 13: 77-81.

[189] Sato M. 1979. On equivalences between module categories. J. Algebra 59: 412-420.

[190] Shmatkov V.D. 1994. Isomorphisms and Automorphisms of Incidence Rings and Algebras (in Russian), Ph.D. Thesis, Moscow.

[191] Simson D. 1985. Special Schurian vector space categories and l-hereditary right QF-2 artinian rings, Commentationes Math. 25: 135-147.

[192] Simson D. 1985. Vector space categories, right peak rings and their socle projective modules. J. Algebra 92: 532-571.

[193] Simson D. 1992. Linear Representations of Partially Ordered Sets and Vector Categories. Algebra, Logic and Appl. 4. Gordon and Breach.

[194] Simson D. 1993. Posets of finite prinjective type and class of orders. J. Pure Appl. Algebra 90: 77-103.

[195] Simson D. 1993. On representation types of module subcategories and orders. Bull. Polish Acad. Sci. 41: 77-93.

[196] Simson D., I. Assem and A. Skowroński. 2006. Elements of representation theory of associative algebras. vol. I, Cambridge Univ. Press, London Math. Soc. Student Texts 65.

[197] Simson D. and A. Skowroński. 2007. Elements of the Representation Theory of Associative Algebras. vol. II. Tubes and Concealed Algebras of Euclidian type. Cambridge Univ. Press, London Math. Soc. Student Texts 71.

[198] Simson D. and A. Skowroński. 2007. Elements of the Representation Theory of Associative Algebras. vol. III. Representation-Infinite Tilted Algebras. Cambridge Univ. Press, London Math. Soc. Student Texts 72.

[199] Smith K.C. and L.van Wyk. 1994. An internal characterization of structural matrix rings. Comm. in Algebra 22(14): 5599-5622.

[200] Spiegel E. and Ch.J. O'Donnell. 1997. Incidence algebras. Monographs and Textbook in Pure and Applied Mathematics, 206. Marcel Dekker, Inc..

[201] Spiegel E. 1994. Radicals of incidence algebras. Comm. Alg. 22(2): 139-149.

[202] Stanley R.P. 1970. Structure of incidence algebras and their automorphism groups. Amer. Math. Soc. Bull. 76: 1236-1239.

[203] Stephenson W. 1974. Modules whose lattice of submodules is distributive. Proc. London Math. Soc. 28(2): 291-310.

[204] Tainiter M. 1971. Incidence algebras on generalized semigroups. J. Combin. Theory 11: 170-177.

[205] Thierrin G. 1960. On duo rings. Canad. Math. Bull. 3: 167-172.

[206] Tuganbaev A.A. 1984. Distributive rings. Mat. Zametki 35(3): 329-332.

[207] Tuganbaev A.A. 1985. Right distributive rings. Izv. Vuzov, Mat. 1: 46-51.

[208] Tuganbaev A.A. 1988. Distributive rings with finiteness conditions. Izv. Vuzov, Mat. 10: 50-55.

[209] Tuganbaev A.A. 1989. Modules with distributive lattice of submodules. Uspekhi Mat. Nauk 44(1): 215-216.

[210] Tuganbaev A.A. 1995. Distribute semiprime rings. Mat. Zametki 58(5): 736-761.

[211] Tuganbaev A.A. 1996. Direct sums of distributive modules Mat. Sb. 187 (12): 137-156.

[212] Tuganbaev A.A. 1997. Semidistributive hereditary rings. Uspekhi Mat. Nauk 52(4): 215-216.

[213] Tuganbaev A.A. 1998. Semidistributive modules and rings. Kluver Acad. Publ..

[214] Tuganbaev A.A. 1999. Distributive modules and related topics. CRC Press.

[215] Tuganbaev A.A. 2000. Modules with Distributive Submodules Lattice. pp. 399-416. In: 2000. M. Hazewinkel (ed.). Handbook of Algebra 2. Elsevier Science, Amsterdam.

[216] Tuganbaev A.A. 2000. Serial and Semidistributive Modules and Rings. pp. 417-437. In: 2000. M. Hazewinkel (ed.). Handbook of Algebra 2. Elsevier Science, Amsterdam.

[217] Varadajan K. 1989. The incidence ring of a poset. Indian J. of Math. 31: 59-64.

[218] Voss E.R. 1980. On the isomorphism problem for incidence rings. Illinois J. Math. 24: 624-638.

[219] Warfield R.B., Jr. 1969. Purity and algebraic compactness for modules. Pacific Journal of Mathematics 28(3): 699:719.

[220] Warfield R.B., Jr. 1969. Krull-Schmidt theorem for infinite sums of modules. Proc. Amer. Math. Soc. 22(22): 460-465.

[221] Warfield R.B., Jr. 1970. Decomposibility of finitely presented modules. Proc. Amer. Math. Soc. 25: 167-172.

[222] Warfield R.B., Jr. 1975. Serial rings and finitely presented modules. J. Algebra 37(2): 187-222.

[223] Weibel Ch.A. 1994. An introduction to homological algebra. Cambridge Univ. Press, Cambridge.

[224] van Wyk L. 1988. Maximal left ideals in structural matrix rings. Comm. Algebra 16: 399-419.

[225] van Wyk L. 1988. Special radicals in structural matrix rings. Comm. Algebra 16: 421-435.

[226] Yamagata K. 1978. On Artinian Rings of Finite Representation Type. J. Algebra 50: 276-283.

[227] Zanardo P. 1989. Valuation domains of bounded module type. Arch. Math. 53: 11-19.

[228] Zanardo P. 2002. Modules over Archimedean valuation domains and the problem of bounded module type. Commun. Algebra 30(4): 1979-1993.

[229] Zarisky O. and P. Samuel. 1958. Commutative algebra. vol. I.

[230] Zavadskij A.G. and V.V. Kirichenko. 1976. Torsion-free modules over primary rings (in Russian). Zapiski Nauchn. Sem. LOMI AN SSSR 57: 100-116; English translation: 1979. J. Soviet Math. 11(4): 598-612.

[231] Zavadskij A.G. and V.V. Kirichenko. 1977. Semimaximal rings of finite type (in Russian). Mat. Sb. 103(145), N3: 323-345; English translation: 1977. Math. USSR-Sb. 32: 273-291.

[232] Zavadskij A.G. 1977. Differentiation with respect to a pair of places (in Russian). pp.115-121. *In*: 1977. Matrix problems, Akad. Nauk Ukrain. SSR Inst. Mat., Kiev.

[233] Zavadskij A.G. and U.S. Revitskaya. 1999. Sbornik: Mathematics 190(6): 835-858.

[234] Zavadskij A.G. 2005. On two-point differentiation and its generalization. pp. 413-436. *In*: 2005. Algebraic structures and their representations, Contemp. Math. 376. Amer. Math. Soc., Providence, RI.

[235] Zhaoyong H. 1999. On a generalization of the Auslander-Bridger transpose. Communications in Algebra 27(12): 5791-5812.

[236] Zimmermann W. 1983. Existenz von Auslander-Reiten Folgen. Arch. Math. 40: 40-49.

[237] Zimmermann W. 1988. Auslander-Reiten sequences over Artinian rings. J. Algebra 119: 366-392.

[238] Zhaoyong H. 1999. On a generalization of the Auslander-Bridger transpose. Communications in Algebra 27(12): 5791-5812.

# Glossary of Ring Theory

**(0,1)-order**

A tiled order $A = \{O, \mathcal{E}(A)\}$ is called a $(0, 1)$-*order* if $\mathcal{E}(A)$ is a $(0, 1)$-matrix.

**$\alpha$-critical module**

A non-zero module is a $\alpha$-*critical module* if K.dim $M = \alpha$ and K.dim$(M/N) < \alpha$ for some ordinal $\alpha$ and each non-zero submodule $N$.

**algebra** (over a field $K$)

An *algebra* over a field $K$ (or $K$-algebra) is a set $A$ which is both a ring and a vector space over $K$ in such a manner that the additive group structures are the same and the axiom $(\lambda a)b = a(\lambda b) = \lambda(ab)$ satisfies for all $\lambda \in K$ and $a, b \in A$

**algebra of bounded representation type**

A finite dimensional algebra $A$ is said to be of *bounded representation type* if there is a bound on the length of the indecomposable finite dimensional $A$-modules.

**algebra of finite type**

An algebra of finite representation type is also called an *algebra of finite type.*

**algebra of finite representation type**

An algebra $A$ is said to be of *finite representation type* if $A$ has only a finite number of non-isomorphic finite dimensional indecomposable representations up to isomorphism.

**algebra of infinite representation type**

An algebra which is not of finite representation type is called an *algebra of infinite representation type.*

**algebra of tame representation type**

An algebra $A$ is said to be of *tame representation type* if it is of infinite type but all families of indecomposable representations are 1-parametric, i.e. for any $r$ there are $(A, K[x])$-bimodules $M_1, \ldots, M_n$ (where the natural number $n$ may depend on $r$), which are finitely generated and free over $K[x]$ such that any indecomposable $A$-module of dimension $r$ is isomorphic to some $A$-module of the form $M_i \otimes K[x]/(x - \lambda)$ for some $\lambda \in K$.

**algebra of wild representation type**

An algebra $A$ is said to be of *wild representation type* if there is an $(A, k\langle x, y\rangle)$-bimodule $M$ which is finitely generated and free over $K\langle x, y\rangle$ and such that the functor $M \otimes_{K\langle x,y\rangle} *$ sends non-isomorphic finite dimensional $K\langle x, y\rangle$-modules to non-isomorphic $A$-modules.

**almost split homomorphism**

A homomorphism $g : M \to N$ is called *almost split* if it is either right almost split or left almost split.

**arithmetical ring**

A commutative ring $A$ is called *arithmetical* if the ideals of the local ring $A_M$ are totally ordered by inclusion for all maximal ideals $M$ of $A$.

**arrow ideal** (of a path algebra)

Let $Q$ be a quiver and let $K$ be a field. The two-sided ideal of a path algebra $KQ$ generated by the set of all arrows of $Q$ is called the *arrow ideal* of $KQ$.

**Artin algebra**

Let $C$ be a commutative Artinian ring. A ring $A$ is called an *Artin algebra* if $A$ is finitely generated as $C$-module.

**Artinian module**

A module $M$ is *Artinian* if it satisfies descending chain condition on submodules of $M$.

**Artinian ring**

A ring is called *Artinian* if it is both a right and left Artinian ring.

**ascending chain condition** (on submodules)

A module $M$ satisfies the *ascending chain condition* (or a.c.c.) if every ascending chain of submodules of $M$

$$M_1 \subseteq M_2 \subseteq M_3 \subseteq \cdots$$

contains only a finite number of elements, i.e. there exists an integer $n$ such that $M_n = M_{n+1} = M_{n+2} = \cdots.$

**associative ring**

A ring $A$ is called *associative* if the multiplication satisfies the associative law, that is, $(a_1 a_2)a_3 = a_1(a_2 a_3)$ for all $a_1, a_2, a_3 \in A$.

**Auslander-Bridger transpose**

Let $A$ be a semiperfect ring, and let $M$ be a right $A$-module with minimal projective resolution

$$P_1 \xrightarrow{\varphi} P_0 \longrightarrow M \longrightarrow 0.$$

If $^* = \mathrm{Hom}_A(-, A)$, define $\mathrm{Tr}(M)$ as $\mathrm{Coker}(\varphi^*)$ where

$$0 \longrightarrow M^* \longrightarrow P_0^* \xrightarrow{\varphi^*} P_1^* \xrightarrow{\omega} \mathrm{Tr}(M) \longrightarrow 0$$

The correspondence $M \mapsto \mathrm{Tr}(M)$ defines a duality functor, which is called *Auslander-Bridger transpose* between the projective stable categories of right and left modules:

$$\mathrm{Tr} : \underline{\mathrm{mod}}_r A \longleftrightarrow \underline{\mathrm{mod}}_l A^{op}$$

**Baer ring**
A ring is called *Baer* if it is both right and left Baer.

**basic ring**
A semiperfect ring $A$ is called *basic* if the quotient ring $A/R$, where $R$ is the Jacobson radical of $A$, is a direct sum of division rings.

**Bézout domain**
A commutative integral domain is called a *Bézout domain* if every one of its finitely generated ideals is principal.

**Bézout ring**
A ring is called a *Bézout ring* if it is both a right and left Bézout ring.

**bounded representation type** (for an algebra)
A finite dimensional algebra $A$ is said to be of *bounded representation type* if there is a bound on the length of the indecomposable finite dimensional $A$-modules.

**bounded representation type** (for an Artinian ring)
An Artinian ring $A$ is said to be of *bounded representation type* if there is a bound on the length of the finitely generated indecomposable right (left) $A$-modules.

**cancellable module**
A module $M$ is called *cancellable* if $M \oplus X \cong M \oplus Y$ implies $X \cong Y$ for any pair of modules $X$ and $Y$.

**cardinal sum** (of poset)
Let $(X, \leq_1)$ and $(Y, \leq_2)$ be two (disjoint) posets. The *cardinal sum* $X \sqcup Y$ (or *disjoint union* $X \cup Y$) of $X$ and $Y$ is the set of all $x \in X$ and $y \in Y$. The relations $x \leq_1 x_1$ and $y \leq_2 y_1$ ($x, x_1 \in X$; $y, y_1 \in Y$) have the same meanings and there are no another relations in $X \sqcup Y$.

**Cartan matrix**
An (abstract) **Cartan matrix** is a square matrix $\mathbf{C} = (c_{ij})$ with integer entries satisfying the conditions:
  1. $c_{ii} = 2$ for all $i$.
  2. $c_{ij} < 0$ for $i \neq j$.

3. $c_{ij} = 0$ if and only if $c_{ji} = 0$.
4. **C** is a positive definite matrix.

### central idempotent
An idempotent $e \in A$ is called *central* if $ea = ae$ for all $a \in A$.

### centrally primitive idempotent
An idempotent is called *centrally primitive* if it cannot be written as a sum of two non-zero orthogonal central idempotents.

### classical Krull dimension
*Classical Krull dimension* of a ring $A$, denoted by cl.K.dim($A$), is defined by transitive recursion as follows:
1. cl.K.dim($A$) = $-1$ if and only if $A = 0$.
2. Define $\mathcal{R}_{-1}(A) = \emptyset$. For each ordinal $\alpha \geq 0$, if $\mathcal{R}_\beta(A)$ has been defined for each $\beta < \alpha$, define $\mathcal{R}_\alpha(A)$ to be the set of prime ideals $P$ such that all prime ideals $Q \supset P$ are contained in $\bigcup_\beta \mathcal{R}_\beta(A)$ (e.g. $\mathcal{R}_0(A)$ is the set of maximal ideals of $A$). If some $\mathcal{R}_\gamma(A)$ contains all prime ideals of $A$, then cl.K.dim($A$) is the smallest such $\gamma$.

### classical left ring of fractions
*Classical left ring of fractions* $Q$ of a ring $A$ is a ring $Q \supseteq A$ such that the following conditions hold:
1. All regular elements of the ring $A$ are invertible in the ring $Q$.
2. Each element of the ring $Q$ has the form $b^{-1}a$, where $a, b \in A$ and $b \in C_A(0)$.
Here $C_A(0)$ is a set of regular elements of $A$.

### classical right ring of fractions
*Classical right ring of fractions* $Q$ of a ring $A$ is a ring $Q \supseteq A$ such that the following conditions hold:
1. All regular elements of the ring $A$ are invertible in the ring $Q$.
2. Each element of the ring $Q$ has the form $ab^{-1}$, where $a, b \in A$ and $b \in C_A(0)$.

### classical ring of fractions
*Classical ring of fractions* $Q$ of a ring $A$ is a ring $Q \supseteq A$ which is both classical right ring of fractions and classical left ring of fractions of $A$.

### clean ring
A ring is called *clean* if every its element is a sum of a unit and an idempotent.

### closed submodule
An essentially closed submodule is also called a *closed submodule*.

### commutative ring
A ring $A$ is called *commutative* if the multiplication satisfies the commutative law, that is, $a_1 a_2 = a_2 a_1$ for all $a_1, a_2 \in A$.

**compatible ideals**

Let $B_1$, $B_2$ be valuation rings of a division ring $D$. Then there exists a minimal valuation ring $B_{1,2}$ of $D$ containing both $B_1$ and $B_2$. A right ideal $I_1$ of $B_1$ and a right ideal $I_2$ of $B_2$ are called *compatible* if $I_1 B_{1,2} = I_2 B_{1,2}$.

**complement**

A *complement* to a submodule $N$ in a module $M$ is a submodule $C$ which is maximal with respect to the property that $N \cap C = 0$.

**complementary submodule**

A submodule $C$ in $M$ is called *complementary* if there exists a submodule $N$ of $M$ such that $C$ is a complement to $N$ in $M$.

**complete set of idempotents**

A finite set of orthogonal idempotents $e_1, e_2, \ldots, e_m \in A$ is called *complete* if

$$e_1 + e_2 + \ldots + e_m = 1 \in A.$$

**completely prime ideal**

An ideal $I$ of a ring $A$ is called *completely prime* if $xy \in I$ implies that either $x \in I$ or $y \in I$ for any $x, y \in A$.

**completely reducible module**

A module is called *completely reducible* if it can be decomposed into a direct sum of simple modules.

**composition series**

A finite chain of submodules of a module $M$:

$$0 = M_0 \subset M_1 \subset \cdots \subset M_n = M$$

is called a *composition series* for a module $M$ if all quotient modules $M_{i+1}/M_i$ are simple for $i = 0, 1, \ldots, n-1$.

**Coxeter diagram**

A *Coxeter diagram* (or *Coxeter graph*) is a finite undirected graph with a set of vertices $S$ whose edges are labelled with integers $\geq 3$ or with the symbol $\infty$.

**critical ideal**

A non-zero ideal of a ring $A$ is called *critical* if it is critical as a right $A$-module.

**critical module**

A non-zero module is called *critical* if it is $\alpha$-critical for some $\alpha$.

**critical subposet**

The posets of infinite representation type:

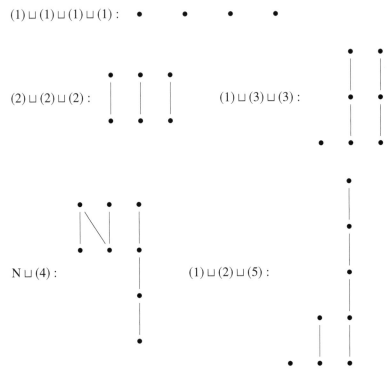

$(1) \sqcup (1) \sqcup (1) \sqcup (1) :$

$(2) \sqcup (2) \sqcup (2) :$

$(1) \sqcup (3) \sqcup (3) :$

$N \sqcup (4) :$

$(1) \sqcup (2) \sqcup (5) :$

in the list of the Kleiner theorem are called the *critical subposets*.

**cyclic module**

An $A$-module is called *cyclic* if it is generated by one element, i.e. there is an element $m_0 \in M$ such that every element $m \in M$ has the form $m = m_0 a$ for some $a \in A$.

**decomposable module**

A module which is isomorphic to a direct sum $M_1 \oplus M_2$ of non-zero modules $M_1$ and $M_2$ is called *decomposable*.

**decomposable ring**

A ring which can be decomposed into a direct product of two non-zero rings is called *decomposable*.

**Dedekind domain**

A commutative hereditary integral domain is called a *Dedekind domain*.

## Dedekind-finite module

An $A$-module $M$ such that $M = M_1 \oplus N$, where $N$ is an $A$-module and $M \cong M_1$, implies that $N = 0$, is called *Dedekind-finite*. In other words, a Dedekind-finite module is not isomorphic to any proper direct summand of itself.

## Dedekind-finite ring

A ring $A$ such that $ba = 1$ implies $ab = 1$ for $a, b \in A$ is called *Dedekind-finite*.

## Dedekind-infinite ring

A ring which is not Dedekind-finite ring is called *Dedekind-infinite*.

## descending chain condition (on submodules)

A module $M$ satisfies the *descending chain condition* (or d.c.c.) if every ascending chain of submodules of $M$

$$M_1 \supseteq M_2 \supseteq M_3 \subseteq \cdots$$

contains only a finite number of elements, i.e. there exists an integer $n$ such that $M_n = M_{n+1} = M_{n+2} = \cdots$.

## diagonal of a ring

The quotient ring $A/Pr(A)$, where $Pr(A)$ is a the prime radical of a ring $A$, is called the *diagonal of a ring $A$*.

## directly finite module

A Dedekind-finite module is also called *directly finite*.

## directly finite ring

A Dedekind-finite ring $A$ is also called *directly finite*.

## discrete valuation

A valuation is called *discrete* if its totally ordered value group is discrete.

## discrete valuation domain

A subring $A$ of a field $K$ is called a *discrete valuation domain* if there is a discrete valuation $v : K \to \mathbf{Z} \cup \{\infty\}$ such that

$$A = \{x \in K \ : \ v(x) \geq 0\}.$$

## discrete valuation ring

A ring $A$ which can be embedded into a division ring $D$ with discrete valuation $v : D \to \mathbf{Z} \bigcup\{\infty\}$ of $D$ such that

$$A = \{x \in D : v(x) \geq 0\}$$

is called a *discrete valuation ring*.

### distributive module
A module $M$ such that

$$K \cap (L + N) = K \cap L + K \cap N$$

for all submodules $K, L, N$ of $M$ is called *distributive*.

### division ring
A non-zero ring $A$ with identity such that every non-zero element of $A$ is invertible, i.e. $U(A) = A\backslash\{0\}$, is called a *division ring*.

### domain
A non-zero ring $A$ which has neither right nor left zero divisors, i.e. $ab \neq 0$ for any non-zero elements $a, b \in A$, is called a *domain*.

### Dubrovin valuation ring
A subring $A$ of a simple Artinian ring $S$ is called a *Dubrovin valuation ring* if there is an ideal $M$ of $A$ such that
  1. $A/M$ is a simple Artinian ring.
  2. For each $s \in S \setminus A$ there are $a_1, a_2 \in A$ such that $sa_1 \in A \setminus M$ and $a_2 s \in A \setminus M$.

### duo ring
A ring is called a *duo ring* if every one-sided ideal is two-sided.

### Dynkin diagram
The following graphs are called the **Dynkin diagrams**:

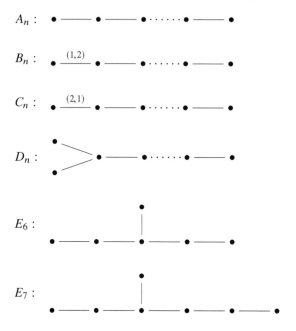

$E_8$ :

$F_4$ :  $(1,2)$

$G_2$ :  $(1,3)$

### essential epimorphism

An epimorphism $f : M \to N$ is called *essential* if for each sequence of $A$-modules

$$X \xrightarrow{\varphi} M \xrightarrow{f} N$$

such that $f\varphi$ is an epimorphism, $\varphi$ is also an epimorphism.

### essential extension

A module $M$ is called an *essential extension* of a submodule $N$ if $N$ has a non-zero intersection with every non-zero submodule of $M$, i.e. $N$ is an essential submodule in $M$.

### essential monomorphism

A monomorphism $f : N \to M$ is called *essential* if for each sequence of $A$-modules

$$N \xrightarrow{f} M \xrightarrow{\varphi} X$$

such that $\varphi f$ is a monomorphism, $\varphi$ is also a monomorphism.

### essential submodule

A submodule $N$ of $M$ which has non-zero intersection with every non-zero submodule of $M$ is called *essential*. Hence a right ideal is essential if its intersection with any other right ideal is non-zero. Similarly for left ideals.

### essentially closed submodule

A submodule $C$ in $M$ which has no proper essential extensions inside $M$ is called *essentially closed*.

### essentially finitely generated module

A module $M$ over a ring $A$ is called *essentially finitely generated* if there exists a finitely generated submodule $N \subseteq M$ which is essential in $M$.

### exchange property

Given a cardinal $\mathfrak{N}$, an $A$-module $M$ is said to have the $\mathfrak{N}$-*exchange property* if for any $A$-module $X$ and any two decompositions of $X$:

$$X = M_1 \oplus N = \bigoplus_{i \in I} X_i$$

with $M_1 \simeq M$ and $|I| \leq \mathfrak{N}$ it follows that there are submodules $Y_i \subseteq X_i$ such that

$$X = M_1 \oplus \left( \bigoplus_{i \in I} Y_i \right).$$

A module $M$ has the *exchange property* if $M$ has the $\mathfrak{N}$-exchange property for any cardinal $\mathfrak{N}$. If $M$ has the $\mathfrak{N}$-exchange property for any finite cardinal $\mathfrak{N}$, $M$ is said to have the *finite exchange property*. If $M$ has the $\mathfrak{N}$-exchange property for $|\mathfrak{N}| = n$, $M$ is said to have the *n-exchange property*.

**exchange ring**

A ring is called *exchange* if every right (left) regular module has the finite exchange property, i.e. has *n*-exchange property for any natural number $n$.

$\mathrm{Ext}^1$**-quiver** (of a finite dimensional algebra)

Let $A$ be a finite dimensional basic algebra over a field $K$ with Jacobson radical $R$. Let $P_1, P_2, \ldots, P_n$ be the pairwise non-isomorphic principal (i.e. projective indecomposable) right $A$-modules, and let $P_i/P_i R = S_i$ be right simple $A$-modules for $i = 1, \ldots, n$. Then the $\mathrm{Ext}^1$-*quiver* $Q(A)$ of $A$ is a quiver whose vertices are $\{1, 2, \ldots, n\}$ and the vertex $i$ joins with the vertex $j$ by $t_{ij}$ arrows, where

$$t_{ij} = \dim_K \mathrm{Ext}^1(S_i, S_j)$$

The $\mathrm{Ext}^1$-quiver is the same as the Gabriel quiver.

**faithful module**

A right $A$-module $M$ is called *faithful* if

$$\mathrm{ann}_M(A) = \{m \in M \mid ma = 0 \text{ for all } a \in A\} = 0.$$

**FBN-ring**

A ring is called an *FBN-ring* if it is both a left and right FBN-ring.

**FD-ring**

A finitely decomposable ring is called an *FD-ring*.

**FDD-ring**

A finitely decomposable diagonal ring is called an *FDD-ring*.

**FDI-ring**

A ring $A$ is called an *FDI-ring* if there exists a decomposition of the identity $1 \in A$ into a finite sum $1 = e_1 + e_2 + \cdots + e_n$ of pairwise orthogonal primitive idempotents $e_i$.

**field**

A commutative division ring is called a *field*. Every finite division ring is a field.

**finitely decomposable ring**

A ring which can be expressed as a direct product of a finite number of indecomposable rings is called a *finitely decomposable ring*.

## finite exchange property

An $A$-module $M$ has *finite exchange property* if $M$ has $n$-exchange property for any natural number $n$.

## finite representation type

An algebra $A$ is said to be of *finite representation type* if $A$ has only a finite number of non-isomorphic finite dimensional indecomposable representations up to isomorphism.

A ring $A$ is said to be of *finite representation type* if $A$ has up to isomorphism only a finite number of indecomposable finitely presented $A$-modules.

## finitely generated module

An $A$- module $M$ is called *finitely generated* if it has a finite set of generators, i.e. there exists a set of elements $X = \{m_1, m_2, \ldots, m_n\} \subset M$ such that every element $m \in M$ can be written as $m = \sum_{i=1}^{n} m_i a_i$ for some $a_i \in A$.

## finitely presented module

An $A$-module $M$ is called *finitely presented* if the following equivalent conditions hold:

1. There is an exact sequence

$$A^m \longrightarrow A^n \longrightarrow M \to 0,$$

for some $m, n \in N$.

2. $M$ is finitely generated and there is an epimorphism $\varphi : P \longrightarrow M$ with $P$ projective, such that $\mathrm{Ker}(\varphi)$ is a finitely generated module.

## finitely related module

An $A$-module $M$ is called *finitely related* if there exists an exact sequence $0 \longrightarrow K \longrightarrow F \longrightarrow M \longrightarrow 0$, where $F$ is a free $A$-module of arbitrary rank and $K$ is a finitely generated $A$-module.

## flat dimension

*Flat dimension* of a module $X$ is the length of an exact sequence

$$0 \longrightarrow F_n \xrightarrow{d_n} F_{n-1} \xrightarrow{d_{n-1}} \ldots \xrightarrow{d_2} F_1 \xrightarrow{d_1} F_0 \xrightarrow{\varepsilon} X \longrightarrow 0,$$

where all $F_i$ are flat and there is no shorter such sequence.

## flat module

An $A$-module $X$ is called *flat* if $X \otimes_A *$ is an exact functor.

## formal triangular matrix ring

A ring of the following form

$$A = \begin{pmatrix} A_1 & X_{12} & \cdots & X_{1n} \\ X_{21} & A_2 & \cdots & X_{2n} \\ \vdots & \vdots & \ddots & \vdots \\ X_{n1} & X_{n2} & \cdots & A_n \end{pmatrix}$$

if $X_{ij} = 0$ for all $1 \le j < i \le n$ (resp. $1 \le i < j \le n$), is called *formal triangular matrix ring*.

## fractional ideal

A *fractional ideal* of a ring $A$ in the field $K$ is an $A$-module $M \subset K$ such that $xM \subset A$ for some element $0 \ne x \in A$.

## free module

An $A$-module $M$ is called *free* if it is isomorphic to a direct sum of (right) regular modules, i.e. $M \simeq \bigoplus_{i \in I} M_i$ where $M_i \simeq A_A$ for all $i \in I$.

## Gabriel quiver (of a finite dimensional algebra)

Let $A$ be a finite dimensional basic algebra over a field $K$ with Jacobson radical $R$. Let $P_1, P_2, \ldots, P_n$ be the pairwise non-isomorphic principal (i.e. projective indecomposable) right $A$-modules, and let $P(P_iR) \cong \bigoplus_{k=1}^{s} P_k^{t_{ik}}$ where $P(P_iR)$ is a projective cover of $R_i = P(P_iR)$. Then the *Gabriel quiver* $Q(A)$ of $A$ is a quiver whose vertices are $\{1, 2, \ldots, n\}$ and the vertex $i$ joins with the vertex $j$ by $t_{ij}$ arrows, where $t_{ij}$ is equal to the number of principal modules $P_j$ in the projective cover $P(P_iR)$.

The Gabriel quiver is the same as the $\mathrm{Ext}^1$-quiver.

## generalized matrix ring

A formal triangular matrix ring is also called a *generalized matrix ring*.

## global dimension

*Global dimension* of a Noetherian ring $A$ is the common value of projective global and weak dimensions of $A$.

## Goldie dimension

The uniform dimension of a ring (module) is also called *Goldie dimension*.

## Goldie ring

A ring which is both a right and left Goldie ring is called a *Goldie ring*.

## hereditary ring

A ring $A$ which is both right and left hereditary is called *hereditary*.

**IBN-ring**

An *IBN-ring* is a ring with invariant basis number property.

**idempotent**

An element $e$ for which $e^2 = e$ is called an *idempotent*.

**idempotent-lifting**

It is said that idempotents may be *lifted modulo an ideal $\mathcal{I}$* of a ring $A$ if from the fact that $g^2 - g \in \mathcal{I}$, where $g \in A$, it follows that there exists an idempotent $e^2 = e \in A$ such that $e - g \in \mathcal{I}$.

**incidence ring**

The *incidence ring* $I(S, A)$ of a locally finite poset $S$ with a partial order relation $\leq$ over an associative (not necessary commutative) ring $A$ with identity is the set of functions $f : S \times S \to A$ such that $f(x, y) = 0$ if $x \nleq y$ with operations:

$$(f + g)(x, y) = f(x, y) + g(x, y),$$

$$(fg)(x, y) = \sum_{x \leq z \leq y} f(x, z)g(z, y)$$

$$(af)(x, y) = a(f(x, y))$$

for all $x, y, z \in S$ and $a \in A$.

**incomparable rings**

Two valuation rings $B_1$ and $B_2$ of a field $K$ are called *incomparable* if $B_1 \not\subseteq B_2$ and $B_2 \not\subseteq B_1$.

**indecomposable module**

A module which is not isomorphic to a direct sum $M_1 \oplus M_2$ of non-zero modules $M_1$ and $M_2$ is called *indecomposable*.

**indecomposable ring**

A ring $A \neq 0$ which cannot be decomposed into a direct product of two non-zero rings is called *indecomposable*.

**independent rings**

Two valuation rings $B_1$ and $B_2$ of a field $K$ is called *independent* if their least overing in $K$ is $K$ itself.

**injective dimension**

*Injective dimension* of a module $X$ is the length of a injective resolution

$$0 \longrightarrow X \overset{i}{\longrightarrow} Q_0 \overset{d_0}{\longrightarrow} \ldots \longrightarrow Q_{n-1} \overset{d_{n-1}}{\longrightarrow} Q_n \longrightarrow 0$$

if there is no shorter one.

### injective envelope (hull)

A module $M$ which is both an essential extension of $M$ and an injective module is called the *injective envelope* (or *injective hull*).

### injective module

A module $Q$ is called *injective* if for any monomorphism $\varphi : M \longrightarrow N$ and any homomorphism $\psi : M \longrightarrow Q$ there exists a homomorphism $h : N \longrightarrow Q$ such that $\psi = h\varphi$.

### injective resolution

*Injective resolution* of an $A$-module $M$ is an exact sequence of $A$-modules

$$0 \longrightarrow M \xrightarrow{i} Q_0 \xrightarrow{d_0} Q_1 \xrightarrow{d_1} Q_2 \longrightarrow \cdots$$

for which the $Q_n$ are injective for all $n \geq 0$.

### integral domain

A commutative domain is called an *integral domain*.

### invariant basis number property

A ring $A$ is said to have *invariant basis number property* if for all positive integers $m, n$, an isomorphism $A^n \simeq A^m$ as right $A$-modules implies $n = m$.

### invariant valuation ring

A subring $A$ of a division ring $D$ is called an *invariant valuation ring* if it satisfies the following two conditions:

  1. For every $x \in D^*$, $x \in A$ or $x^{-1} \in A$.
  2. For every $d \in D^*$, $dAd^{-1} = A$.

### invertible ideal

A fractional ideal $M$ of an integral domain $A$ in the field $K$ is called *invertible* if there exists a fractional ideal $M^{-1}$ such that $MM^{-1} = A$.

### Jacobson radical of a ring

The *Jacobson radical* of a ring $A$ is the intersection of all maximal right ideals in a ring $A$. It is also the intersection of all maximal left ideals.

### Kronecker algebra

A four-dimensional algebra over $K$

$$A = \begin{pmatrix} K & K \oplus K \\ 0 & K \end{pmatrix}$$

is called the *Kronecker algebra*. It is isomorphic to the path algebra $KQ$ where $Q$ is the quiver with two vertices and two arrows:

$$1 \underset{\beta}{\overset{\alpha}{\rightrightarrows}} 2$$

i.e. $Q_0 = \{1, 2\}$ and $Q_1 = \{\alpha, \beta\}$. The algebra $KQ$ has a basis $\{\varepsilon_1, \varepsilon_2, \alpha, \beta\}$.

## Krull dimension
*Krull dimension* of a right $A$-module $M$ (denoted by K.dim $M$ if it exists) is defined by transfinite recursion as follows:
1. K.dim $M = -1$ if and only if $M = 0$.
2. If $\alpha$ is an ordinal and K.dim $M \neq \delta$ for any ordinal $\delta < \alpha$, then K.dim $M = \alpha$ if for every countable descending chain $M_0 \supseteq M_1 \supseteq \ldots$ of submodules $M_i$ of $M$ one has that K.dim$(M_{i-1}/M_i) < \alpha$ for all but finitely many indices $i$.

## Krull-Schmidt category
An additive category $\mathfrak{C}$ is said to be a *Krull-Schmidt category* (also called a Krull-Remak-Schmidt category) provided the endomorphism ring End$(X)$ of any indecomposable object $X$ of $\mathfrak{C}$ is a local ring. In a Krull-Schmidt category any object is isomorphic to a finite direct sum of indecomposable objects, and such a decomposition is unique up to isomorphism and permutation of summands.

## l-hereditary ring
An Artinian ring is called *l-hereditary* if it satisfies the following condition: given any pair of indecomposable projective left $A$-modules $P$ and $Q$ and given any $A$-homomorphism $\varphi : P \to Q$, either $\varphi = 0$ or $\varphi$ is a monomorphism.

## left almost split homomorphism
Let $M$ and $N$ be $A$-modules. A homomorphism $g : M \longrightarrow N$ is said to be *left almost split* if
1. It is not a split monomorphism.
2. For any homomorphism $h : M \longrightarrow Y$ which is not a split monomorphism, there is a homomorphism $h_1 : N \longrightarrow Y$ such that $h = gh_1$.

## left Artinian ring
A ring $A$ is *left Artinian* if the left regular module $_AA$ is Artinian.

## left Baer ring
A ring is *left Baer* if any right annihilator in $A$ is of the form $eA$ for some idempotent $e \in A$.

## left Bézout ring
A ring is called *left Bézout* if every one of its finitely generated left ideals is principal.

## left bounded ring
A left Noetherian ring is called *left bounded* if every one of its essential left ideals contains a non-zero two-sided ideal.

## left coherent ring
A ring is called *left coherent* if it satisfies the following equivalent conditions:
1. The direct product of any family of flat right $A$-modules is flat.
2. The direct product of any family of copies of $A$ is flat as a right $A$-module.
3. Any finitely generated left ideal in $A$ is finitely presented.

**left FBN-ring**

A left Noetherian ring $A$ is called an *FBN-ring* if $A/P$ is left bounded for every prime ideal $P$.

**left finite-dimensional ring**

A ring $A$ is called *left finite-dimensional* if it contains no infinite sum of non-zero left ideals of $A$, i.e. the left regular module $_AA$ is finite-dimensional.

**left finite representation type**

A left Artinian ring $A$ is of *left finite representation type* if $A$ has up to isomorphism only a finite number of indecomposable finitely generated left $A$-modules.

**left global dimension**

*Left global dimension* of a ring $A$ is the common value of left projective and left injective global dimension of $A$.

**left Goldie ring**

A ring $A$ is called a *left Goldie ring* if it satisfies the following conditions:
    1. $A$ satisfies the ascending chain condition on left annihilators.
    2. $A$ contains no infinite direct sum of non-zero left ideals.

**left injective global dimension**

*Left injective global dimension* of a ring $A$ is supremum of injective dimensions of left $A$-modules.

**left hereditary ring**

A ring $A$ is called *left hereditary* if each of its left ideals is a projective $A$-module.

**right Kasch ring**

A *left Kasch ring* is a ring $A$ for which every simple left $A$-module is isomorphic to a left ideal of $A$.

**left minimal morphism**

An $A$-module homomorphism $f : M \rightarrow N$ is called *left minimal* if each $\varphi \in \mathrm{End}_A N$ with $\varphi f = f$ is an automorphism.

**left $n$-chain ring**

A ring $A$ is called *left $n$-chain* if for any $n + 1$ elements $x_0, x_1, \ldots, x_n \in A$ there is $i$ such that

$$x_i \in \sum_{k \neq i} A x_k .$$

**left Noetherian ring**

A ring $A$ is called *left Noetherian* if the left regular module $_AA$ is Noetherian.

**left nonsingular ring**

A ring $A$ is called *left nonsingular* if $\mathcal{Z}(_AA) = 0$.

## left order
A ring $A \subseteq Q$ is a *left order* if $Q$ is a classical left ring of fractions of a ring $A$.

## left Ore condition
Let $A$ be a ring with the nonempty set $C_A(0)$ of all regular elements in $A$. For any element $a \in A$ and any element $r \in C_A(0)$ there exists a regular element $y \in C_A(0)$ and an element $b \in A$ such that $ya = br$.

## left Ore domain
A ring which is both a domain and a left Ore ring is called a *left Ore domain*.

## left Ore ring
A ring $A$ satisfying the left Ore conditions, i.e. for any element $a \in A$ and any regular element $r \in A$ there exists a regular element $y \in A$ and an element $b \in A$ such that $ya = br$, is called a *left Ore ring*.

## left perfect ring
A ring $A$ with Jacobson radical $R$ is called *left prefect* if $A/R$ is semisimple and $R$ is left $T$-nilpotent.

## left primitive ring
A ring $A$ that has a faithful simple left $A$-module is called a *left primitive ring*.

## left projective global dimension
*Left projective global dimension* of a ring $A$ is supremum of projective dimensions of left $A$-modules.

## left quiver of a ring
Let $A$ be a semiperfect left Noetherian ring, $P_1, ..., P_s$ be all pairwise non-isomorphic principal left $A$-modules. Let $P(RP_i) = \overset{s}{\underset{j=1}{\oplus}} P_j^{t_{ij}}$ be projective cover of $R_i = RP_i$ $(i = 1, ..., s)$. The *left quiver* $Q(A)$ is a graph which vertices $1, .., s$ associate with principal modules $P_1, ..., P_s$ and the vertex $i$ joins with the vertex $j$ by $t_{ij}$ arrows.

## left regular element
An element $x \in A$ is called *left regular* if $ax \neq 0$ for any non-zero element $a \in A$.

## left regular module
*Left regular module* $_AA$ is a ring $M = A$ together with a map $\varphi : A \times M \to M$ which is the usual multiplication $\varphi(a, m) = am \in M$.

## left Rickart ring
A ring $A$ is called *left Rickart* if every one of its principal left ideals is projective.

## left semidistributive ring
A ring $A$ is *left semidistributive* if the left regular module $_AA$ is semidistributive.

**left semihereditary ring**

A ring $A$ is called *left semihereditary* if each of its left finitely generated ideals is a projective $A$-module.

**left self-injective ring**

A ring $A$ is called *left self-injective* if the left regular module $_AA$ is injective.

**left serial ring**

A ring is called *left serial* if it is a left serial module over itself.

**left singular ideal**

The ideal

$$Z(_AA) = \{a \in A \ : \ I a = 0 \ \text{ for some left ideal } \ I \subseteq_e A\}$$

is called the *left singular ideal* of a ring $A$.

**left socular ring**

A ring $A$ is called *left socular* if every non-zero left $A$-module has a non-zero socle

**left stable range one ring**

A ring $A$ has *left stable range one* whenever if $a, b \in A$ and $Aa + Ab = A$, there exists an element $y \in A$ such that $a + yb$ is invertible in $A$.

**left suitable ring**

A ring $A$ is *left suitable* if for any element $x \in A$ there exists an idempotent $e^2 = e \in Ax$ such that $1 - e \in A(1 - x)$.

**left T-nilpotent ideal**

A left ideal $\mathcal{J}$ is called *left T-nilpotent* if for any sequence $a_1, a_2, \ldots, a_n \ldots$ of elements $a_i \in \mathcal{J}$ there exists a positive integer $k$ such that $a_1 a_2 \ldots a_k = 0$.

More generally left T-nilpotence as defined here makes sense for any multiplicatively closed set, including right ideals.

**left uniserial ring**

A ring is called *left uniserial* if it is a left uniserial module over itself, i.e. the lattice of left ideals is linearly ordered.

**left uniform ring**

A ring is called *left uniform* if it is uniform as a left module over itself.

**left weak dimension**

*Left weak dimension* of a ring $A$ is supremum of flat dimensions of left $A$-modules.

**length of a module**

The length of the composition series of a module $M$ is called the *length* of $M$.

**local dual module**

A right $A$-module

$$M^+ = \text{Hom}_S(_SM, _SQ)$$

is called the *local dual* of $M$, where $M$ is a finitely presented right $A$-module with endomorphism ring $S = \text{End}_A M$, which is a local ring with unique maximal ideal $J = \text{rad}(S)$, and $Q$ is an injective hull of the left $S$-module $S/J$.

**local idempotent**

An idempotent $e \in A$ is called *local* if $eAe$ is a local ring.

**local ring**

A ring $A$ which has a unique maximal right ideal is called *local*.

**locally invariant ring**

A total valuation ring $A$ of a division ring $D$ with maximal ideal $M$ is called *locally invariant* if

$$xP(x) = P(x)x$$

for all $x \in M$, where $P(x)$ is the minimal completely prime ideal of $A$.

**locally finite poset**

A poset $X$ is *locally finite* if each its interval, i.e. each subset $[x, y] = \{z \in S : x \le z \le y\}$, is finite.

**Manis valuation pair**

A pair $(A, P)$ where $A$ is a *Manis valuation ring* and $P = \{x \in K : v(x) > 0\}$ is called a *Manis valuation pair*.

**Manis valuation ring**

A subring $A = \{x \in K : v(x) \ge 0\}$ of a commutative ring $K$ with a valuation $v : K \rightarrow G \cup \{\infty\}$ is called a *Manis valuation ring* if $v$ satisfies the following conditions:

  i. $v(ab) = v(a) + v(b)$ for all $a, b \in K$.
  ii. $v(a + b) \ge \min(v(a), v(b))$ for all $a, b \in K$.
  iii. $v(1) = 0, v(0) = \infty$,
  where $G$ is a totally ordered additive group.

**Marot ring**

A commutative ring $A$ is called a *Marot ring* if every one of its regular ideals is generated by a set of regular elements of $A$.

**minimal injective presentation**

A *minimal injective presentation* of a module $M$ is the same as a minimal injective resolution of $M$.

**minimal injective resolution**
An exact sequence of the form

$$0 \longrightarrow M \xrightarrow{h_0} E_0 \xrightarrow{h_1} E_1,$$

where $E_0$ is an injective hull of $M$, and $E_1$ is an injective hull of $\operatorname{Im} h_0$ is called a *minimal injective resolution* of $M$.

**minimal left almost split homomorphism**
A homomorphism $g : M \to N$ is called *minimal left almost split* if it is both left minimal and left almost split.

**minimal monomorphism**
A homomorphism $A \to B$ is *minimal* if $\operatorname{Ker}(f)$ is small in $A$.

**minimal prime ideal**
A prime ideal of a ring $A$ which does not properly contain any other prime ideal is called *minimal*.

**minimal projective presentation**
An exact sequence

$$P_1 \xrightarrow{\pi_1} P_0 \xrightarrow{\pi_0} M \longrightarrow 0$$

is called a *minimal projective presentation* of a module $M$ if $P_0 \xrightarrow{\pi_0} M$ and $P_1 \xrightarrow{\pi_1} \operatorname{Ker}(\pi_0)$ are projective covers.

**minimal projective resolution**
A *minimal projective resolution* is the same as a minimal projective presentation.

**minimal right almost split morphism**
A homomorphism $f : N \longrightarrow M$ is called *minimal right almost split* if it is both right minimal and right almost split.

**minor of a ring**
The endomorphism ring $B = \operatorname{End}_A(P)$ is called a *minor* of order $n$ of a ring $A$, where $P$ is a finitely generated projective $A$-module which can be decomposed into a direct sum of $n$ indecomposable modules.

**Morita equivalent rings**
Rings whose categories of modules are equivalent are called *Morita equivalent*.

**$n$-chain ring**
A ring which is both a right and left $n$-chain ring is called *$n$-chain*.

**$n$-exchange property**
An $A$-module $M$ has *$n$-exchange property* if for any $A$-module $X$ and any two decompositions of $X$:

$$X = M_1 \oplus N = \bigoplus_{i \in I} X_i$$

with $M_1 \simeq M$ and $|I| \le n$ it follows that there are submodules $Y_i \subseteq X_i$ such that

$$X = M_1 \oplus \left( \bigoplus_{i \in I} Y_i \right).$$

### *n*-finite ring

A ring $A$ is called *n-finite* if it satisfies the equivalent conditions:
1. $M_n(A)$ is a Dedekind-finite ring.
2. If $A^n \simeq A^n \oplus M$ for some $A$-module $M$ then $M = 0$.
3. Any module epimorphism $\varphi : A^n \longrightarrow A^n$ is an isomorphism.

### Nakayama permutation

A permutation $\nu$ is called a *Nakayama permutation* if it satisfies the following conditions:
1. $\mathrm{soc} P_k = P_{\nu(k)}/P_{\nu(k)}R$.
2. $\mathrm{soc} P_{\nu(k)} = Q_k/RQ_k$,

for each $k \in \{1, 2, \dots, s\}$, where $A_A = P_1^{n_1} \oplus \cdots \oplus P_s^{n_s}$ ($_A A = Q_1^{n_1} \oplus \cdots \oplus Q_s^{n_s}$) is the canonical decomposition of a semiperfect ring $A$ into a direct sum of right (left) principal modules.

### nil-ideal

An ideal of a ring $A$ all of whose elements are nilpotent is called a *nil-ideal*.

### nilpotent element

An element $x$ of a ring $A$ is called *nilpotent* if $x^n = 0$ for some natural number $n \in \mathbf{N}$.

### nilpotent ideal

An ideal $I$ is called *nilpotent* if $I^n = 0$ for some natural number $n \in \mathbf{N}$. In this case $x_1 x_2 \cdots x_n = 0$ for any elements $x_1, , x_2, \dots, x_n \in I$. Every nilpotent ideal is a nil-ideal. There exist nil-ideals which are not nilpotent.

### Noetherian module

A module $M$ which satisfies the ascending chain condition on submodules of $M$ is called *Noetherian*.

### Noetherian ring

A ring $A$ which is both a right and left Noetherian ring is called *Noetherian*.

### non-Dedekind-finite ring

A ring which is not Dedekind-finite is called *non-Dedekind-finite*.

### nonsingular module

An $A$-module $M$ such that $\mathcal{Z}(M) = 0$ is called *nonsingular*.

### nonsingular ring

A ring which is both a right and left nonsingular ring is called *nonsingular*.

## normal ring
A ring whose all idempotents are central is called a *normal ring*.

## order
A ring $A$ in $Q$ which is both a right and left order in $Q$ is called an *order*.

## orderable group
A group $G$ is called *orderable* if there is a total order on it making this group a totally ordered group.

## ordinal sum (of posets)
An *ordinal sum* of posets $X$ and $Z$, denoted by $X < Z$, is a disjoint union of posets $X \cup Z$ with additional relation $x < z$ for all $x \in X$ and for all $z \in Z$.

## Ore domain
A ring which is both a domain and an Ore ring is called an *Ore domain*.

## Ore extension
A skew polynomial ring is also called an *Ore extension*.

## Ore ring
A ring which is both a right and left Ore ring is called an *Ore ring*.

## orthogonal idempotents
Idempotents $e, f$ such that $ef = fe = 0$ are called *orthogonal*.

## orthogonally finite ring
A ring which does not contain an infinite set of pairwise orthogonal non-zero idempotents is called *orthogonally finite*.

## perfect ring
A ring which is both a right and left perfect is called *perfect*.

## piecewise domain
A ring $A$ with a complete set of pairwise orthogonal idempotents $\{e_1, e_2, \ldots, e_n\}$ such that $xy = 0$ implies $x = 0$ or $y = 0$ whenever $x \in e_i A e_j$ and $y \in e_j A e_k$ for $1 \leq i, j, k \leq n$ is called a *piecewise domain*.

## Peirce decomposition
A decomposition of a ring $A$ into the following form:

$$A = \bigoplus_{i,j=1}^{n} e_i A e_j,$$

where $1 = e_1 + e_2 + \ldots + e_n$ with pairwise orthogonal idempotents $e_1, e_2,\ldots,e_n$ of a ring $A$ is called a *Peirce decomposition*.

## primary ring

A ring $A$ with Jacobson radical $R$ is called *primary* if the quotient ring $A/R$ is a simple Artinian ring.

## prime block

A semiperfect ring is called a *prime block* if its prime quiver $PQ(A)$ is connected.

## prime radical

*The prime radical* of a ring $A$ is the intersection of all prime ideals in $A$.

## prime ring

A ring $A$ is called *prime* if the product of any two non-zero two-sided ideals of $A$ is not equal to zero.

## prime triangular decomposition of identity

*Prime triangular decomposition of identity* is a triangular decomposition of identity $1 = e_1 + e_2 + \cdots + e_n$ of a ring $A$ if $e_i A e_i$ is a prime ring for all $i = 1, \ldots, n$.

## primely triangular ring

A ring $A$ is called a *primely triangular ring* if there is a prime triangular decomposition of identity, i.e. the two-sided Peirce decomposition of $A$ has the following form:

$$
A = \begin{pmatrix}
A_1 & A_{12} & \cdots & A_{1,n-1} & A_{1n} \\
0 & A_2 & \cdots & A_{2,n-1} & A_{2n} \\
\vdots & \vdots & \ddots & \vdots & \vdots \\
0 & 0 & \cdots & A_{n-1} & A_{n-1,n} \\
0 & 0 & \cdots & 0 & A_n
\end{pmatrix},
$$

where $e_i A e_i$ is a prime ring for all $i = 1, 2, \ldots, n$.

## primitive idempotent

An idempotent $e \in A$ which has no decomposition $e = e_1 + e_2$ into a sum of non-zero orthogonal idempotents $e_1, e_2 \in A$ is called *primitive*.

## primitive poset

A finite poset $\mathcal{P}$ is called *primitive* if it is a cardinal sum of linearly ordered sets $L_1, \ldots, L_m$.

## primitive ring

A ring which is both right primitive and left primitive is a *primitive ring*. Every primitive ring is prime.

**principal ideal domain**
A principal ideal ring which is a domain is called a *principal ideal domain*.

**principal ideal ring**
A ring is called a *principal ideal ring* if all its right and left ideals are principal.

**principal left module**
An indecomposable projective left module over a semiperfect ring is called a *principal left module*.

**principal right module**
An indecomposable projective right module over a semiperfect ring is called a *principal right module*.

**projective cover**
A projective module $P$ of a module $M$ is called a *projective cover* and denoted by $P(M)$ if there is an epimorphism $\varphi : P \longrightarrow M$ such that $\operatorname{Ker}\varphi$ is a small submodule in $P$.

**projective dimension**
*Projective dimension* of a module $X$ is the length of a projective resolution

$$0 \longrightarrow P_n \xrightarrow{d_n} \dots \xrightarrow{d_2} P_1 \xrightarrow{d_1} P_0 \xrightarrow{\varepsilon} X \longrightarrow 0$$

if there is no shorter one. It is equal to $\infty$ if there is no finite length projective resolution.

**projective module**
A module $P$ is called *projective* if for any epimorphism $\varphi : M \longrightarrow N$ and for any homomorphism $\psi : P \longrightarrow N$ there is a homomorphism $h : P \rightarrow M$ such that $\psi = \varphi h$.

**projective resolution**
*Projective resolution* of an $A$-module $M$ is an exact sequence of $A$-modules

$$\cdots \longrightarrow P_2 \xrightarrow{d_2} P_1 \xrightarrow{d_1} P_0 \xrightarrow{\pi} M \longrightarrow 0$$

for which the $P_n$ are projective for all $n \geq 0$.

**projective stable category** (of $A$-modules)
The quotient category

$$\underline{\operatorname{mod}}_r A = \operatorname{mod}_r A/\mathcal{P}$$

is called the *projective stable category* of $\operatorname{mod}_r A$ modulo projectives. The objects of this category are the same as those in $\operatorname{mod}_r A$, and the group of morphisms $\underline{\operatorname{Hom}}_A(M, N)$ of morphisms from $M$ to $N$ in $\underline{\operatorname{mod}}_r A$ is defined as the quotient group

$$\underline{\operatorname{Hom}}_A(M, N) = \operatorname{Hom}_A(M, N)/\mathcal{P}(M, N)$$

of $\operatorname{Hom}_A(M, N)$ with the composition of morphisms induced by a composition in $\operatorname{mod}_r A$. Here $\mathcal{P}_A(M, N)$ is the subset of $\operatorname{Hom}_A(M, N)$ consisting of all homomorphisms which factor through a finitely generated projective $A$-modules.

**projectively equivalent modules**
Two modules $X$ and $Y$ over a ring $A$ are said to be *projectively equivalent* if there exist projective $A$-modules $P$ and $Q$ such that $X \oplus P \simeq Y \oplus Q$.

**Prüfer domain**
A commutative semihereditary integral domain is called a *Prüfer domain*.

**Prüfer ring**
A commutative ring is called a *Prüfer ring* if every one of its regular finitely generated ideals is invertible.

**pure monomorphism**
A *pure monomorphism* is a monomorphism $L \longrightarrow M$ whose image is a pure submodule of $M$.

**pure submodule**
Let $M$ be a right $A$-module. A submodule $L$ of $M$ is said to be *pure* if for the natural embedding $i : L \hookrightarrow M$ the induced sequence

$$0 \to L \otimes_A X \xrightarrow{i \otimes 1_X} M \otimes_A X$$

is exact for any left $A$-module $X$.

**PWD**
A piecewise domain is also called a PWD.

**QF-ring**
A quasi-Frobenius ring is also called a *QF-ring*.

**quasi-Baer ring**
A ring is called *quasi-Baer* if the right annihilator of every right ideal of it is generated by an idempotent as a right ideal.

**quasi-Frobenius ring**
A ring is called *quasi-Frobenius* if it is a self-injective Artinian ring.

**quiver of a ring**
The right (or left) quiver of a semiperfect Noetherian ring $A$ is called the *quiver of a ring* $A$.

**$R$-Prüfer ring**
A subring $B$ of a ring $A$ with regard to a set of fixed elements $R$ of $A$ is called an *R-Prüfer ring* if $B_{[P]}$ is a valuation ring for every maximal ideal $P$ of $B$. Here $B_{[P]} = \{b \in R \mid db \in B \text{ for some } d \in B \backslash P\}$.

**radical of a module**
The intersection of all maximal submodules of a module $M$ is called the *radical* of the module $M$.

**radical of a quadratic form**

The set $\text{rad}(q) = \{\alpha \in \mathbf{Z}^n : (\alpha, \beta) = 0 \text{ for all } \beta \in \mathbf{Z}^n\}$ is called the *radical* of a quadratic form $Q$.

**reduced rank**

*Reduced rank* of a right $A$-module $M$ over of a semiprime right Goldie ring $A$ with classical right ring of fractions $Q$, which is semisimple, is

$$\rho(M) = \text{length}_Q(M \otimes_A Q) = \text{u.dim}(M \otimes_A Q).$$

**reduced ring**

A ring is called *reduced* if it has no non-zero nilpotent elements.

**refinement**

*Refinement* of a finite chain of submodules of $M$

$$M = M_0 \supseteq M_1 \supseteq \cdots \supseteq M_k = 0 \qquad (*)$$

is a submodule series

$$M = N_0 \supseteq N_1 \supseteq \cdots \supseteq N_t = 0 \qquad (**)$$

if all the terms of $(*)$ occur in the series $(**)$.

**reflexive module**

An $A$-module $M$ is called *reflexive* if

$$\delta_M : M \longrightarrow M^{**}$$

defined by

$$\delta_M(m)(f) = f(m)$$

for $m \in M$ and $f \in M^*$ is an isomorphism, where $M^* = \text{Hom}_A(M, A)$.

**regular element**

An element $y$ of a ring $A$ is called *regular* if $ay \neq 0$ and $ya \neq 0$ for any non-zero element $a \in A$.

**regular ideal**

An ideal of a commutative ring which contains a regular element is called a *regular ideal*.

**Rickart ring**

A ring which is right and left Rickart is called a *Rickart ring*.

**right almost split homomorphism**

Let $M$ and $N$ be $A$-modules. A homomorphism $f : N \longrightarrow M$ is said to be *right almost split* if

    1. It is not a split epimorphism.

2. For any homomorphism $g : X \longrightarrow M$ which is not a split epimorphism, there is a homomorphism $h : X \longrightarrow N$ such that $g = fh$.

### right Artinian ring
A ring $A$ is called *right Artinian* if the right regular module $A_A$ is Artinian.

### right Baer ring
A ring is called *right Baer* if any right annihilator in $A$ is of the form $eA$ for some idempotent $e \in A$.

### right Bézout ring
A ring is called *right Bézout* if every its finitely generated right ideal is principal.

### right bounded representation type
A ring $A$ is said to be of *right bounded representation type* if there is an upper bound on the number of generators required for indecomposable finitely presented right $A$-modules.

### right bounded ring
A right Noetherian ring is called *right bounded* if every one of its essential right ideals contains a non-zero two-sided ideal.

### right coherent ring
A ring is called *right coherent* if it satisfies the following equivalent conditions:

1. The direct product of any family of flat left $A$-modules is flat.
2. The direct product of any family of copies of $A$ is flat as a left $A$-module.
3. Any finitely generated right ideal in $A$ is finitely presented.

### right FBK-ring
A right fully bounded ring having right Krull dimension is called a *right FBK-ring*.

### right FBN-ring
A ring which is both right fully bounded and right Noetherian is called a *right FBN-ring*.

### right finite-dimensional ring
A ring is called a *right finite-dimensional ring* if it contains no infinite sum of non-zero right ideals of $A$, i.e. the right regular module $A_A$ is finite-dimensional.

### right finite representation type
A right Artinian ring $A$ is of **right finite representation type** if $A$ has up to isomorphism only a finite number of indecomposable finitely generated right $A$-modules.

### right fully bounded ring
A ring $A$ is a *right fully bounded ring* if $A/P$ is right bounded for each prime ideal $P$ in $A$.

**right global dimension**

*Right global dimension* of a ring $A$ is the common value of right projective and right injective global dimension of $A$.

**right Goldie ring**

A ring $A$ is called a *right Goldie ring* if it satisfies the following conditions:
1. $A$ satisfies the ascending chain condition on right annihilators.
2. $A$ contains no infinite direct sum of non-zero right ideals.

**right injective global dimension**

*Right injective global dimension* of a ring $A$ is supremum of injective dimensions of right $A$-modules.

**right hereditary ring**

A ring $A$ is *right hereditary* if each of its right ideals is a projective $A$-module.

**right FBN-ring**

A right Noetherian ring $A$ is a *right FBN-ring* if $A/P$ is right bounded for every prime ideal $P$.

**right Kasch ring**

A *right Kasch* ring is a ring $A$ for which every simple right $A$-module is isomorphic to a right ideal of $A$.

**right Krull dimension**

*Right Krull dimension* of a module $M$ is defined by transfinite recursion as follows:
1. K.dim $M = -1$ if and only if $M = 0$.
2. If $\alpha$ is an ordinal and K.dim $M \neq \delta$ for any ordinal $\delta < \alpha$, then K.dim $M = \alpha$ if for every countable descending chain $M_0 \supseteq M_1 \supseteq \ldots$ of submodules $M_i$ of $M$ one has that K.dim$(M_{i-1}/M_i) < \alpha$ for all but finitely many indices $i$.

**right Krull dimension**

*Right Krull dimension* of a ring is the Krull dimension of the right regular module $A_A$ (if it exists).

**right Ore condition**

Let $A$ be a ring with a nonempty set $C_A(0)$ of all regular elements in $A$. For any element $a \in A$ and any element $r \in C_A(0)$ there exists a regular element $y \in C_A(0)$ and an element $b \in A$ such that $ay = rb$.

**right minimal morphism**

An $A$-module homomorphism $f : M \to N$ is called *right minimal* if each $\varphi \in \mathrm{End}_A(M)$ with $f\varphi = f$ is an automorphism.

**right module**

A *right module* over a ring $A$ is an additive Abelian group $M$ together with a map $M \times A \to M$ so that to every pair $(m, a)$, where $m \in M$, $a \in A$, there corresponds a uniquely determined element $ma \in M$ such that the following conditions satisfy:

1. $m(a_1 + a_2) = ma_1 + ma_2$
2. $(m_1 + m_2)a = m_1 a + m_2 a$
3. $m(a_1 a_2) = (ma_1)a_2$
4. $m \cdot 1 = m$

for any $m, m_1, m_2 \in M$ and any $a, a_1, a_2 \in A$.

**right $n$-chain ring**

A ring $A$ is called a *right n-chain ring* if for any $n + 1$ elements $x_0, x_1, \ldots, x_n \in A$ there is $i$ such that

$$x_i \in \sum_{k \neq i} x_k A.$$

**right Noetherian ring**

A ring $A$ is *right Noetherian* if the right regular module $A_A$ is Noetherian.

**right nonsingular ring**

A ring $A$ is *right nonsingular* if $\mathcal{Z}(A_A) = 0$. Here $\mathcal{Z}(A_A) = 0$ is the singular submodule of $A_A$.

**right order**

A ring $A \subseteq Q$ is called a *right order* if $Q$ is a classical right ring of fractions of the ring $A$.

**right Ore condition**

Let $A$ be a ring with a nonempty set $C_A(0)$ of all regular elements in $A$. For any element $a \in A$ and any element $r \in C_A(0)$ there exists a regular element $y \in C_A(0)$ and an element $b \in A$ such that $ay = rb$.

**right Ore domain**

A ring is a *right Ore domain* if it is both a domain and a right Ore ring.

**right Ore ring**

A ring $A$ is a *right Ore ring* if it satisfies the right Ore conditions, i.e. for any element $a \in A$ and any regular element $r \in A$ there exists a regular element $y \in A$ and an element $b \in A$ such that $ay = rb$.

**right perfect ring**

A ring $A$ with Jacobson radical $R$ is called a *right perfect ring* if $A/R$ is semisimple and $R$ is right $T$-nilpotent.

**right primitive ring**

A ring $A$ that has a faithful simple left $A$-module is called a *right primitive ring*.

### right quiver of a ring

Let $A$ be a semiperfect right Noetherian ring, $P_1, ..., P_s$ be all pairwise nonisomorphic principal right $A$-modules. Let $P(P_iR) = \overset{s}{\underset{j=1}{\oplus}} P_j^{t_{ij}}$ be a projective cover of $R_i = P_iR$ for $i = 1, ..., s$. The *right quiver* $Q(A)$ is a graph which vertices $1, .., s$ associate with principal modules $P_1, ..., P_s$ and the vertex $i$ joins with the vertex $j$ by $t_{ij}$ arrows.

### right projective global dimension

*Right projective global dimension* of a ring $A$ is supremum of projective dimensions of right $A$-modules.

### right regular element

An element $x \in A$ such that $xa \neq 0$ for any non-zero element $a \in A$ is called *right regular*.

### right regular module

*Right regular module* $A_A$ is a ring $M = A$ together with a map $\varphi : M \times A \to M$ which is the usual multiplication $\varphi(m, a) = ma \in M$.

### right Rickart ring

A ring $A$ is *right Rickart* if every one of its principal right ideals is projective.

### right self-injective ring

A ring $A$ is *right self-injective* if the right regular module $A_A$ is injective.

### right semicentral idempotent

An idempotent $e \in A$ such that $eA = eAe$ is called *right semicentral*.

### right semidistributive ring

A ring $A$ is called *right semidistributive* if the left regular module $A_A$ is semidistributive.

### right semihereditary ring

A ring $A$ is *right semihereditary* if each right finitely generated ideal of $A$ is a projective $A$-module.

### right serial ring

A ring is called *right serial* if it is a right serial module over itself.

### right singular ideal

The ideal

$$\mathcal{Z}(A_A) = \{a \in A : a\mathcal{I} = 0 \text{ for some right ideal } \mathcal{I} \subseteq_e A\}$$

is called the *right singular ideal*.

### right socular ring

A ring $A$ is *right socular* if every one of its non-zero right $A$-modules has a non-zero socle.

**right stable range one ring**

A ring has *right stable range one* whenever $a, b \in A$ and $aA + bA = A$, there exists an element $y \in A$ such that $a + by$ is invertible in $A$.

**right suitable ring**

A ring $A$ is called *right suitable* if for any element $y \in A$ there exists an idempotent $f^2 = f \in yA$ such that $1 - f \in (1 - y)A$.

**right T-nilpotent ideal**

A right ideal is *right T-nilpotent* if for any sequence $a_1, a_2, \ldots, a_n \ldots$ of elements $a_i \in \mathcal{J}$ there exists a positive integer $k$ such that $a_k a_{k-1} \ldots a_1 = 0$.

**right uniform ring**

A ring is called *right uniform* if it is uniform as a right module over itself.

**right uniserial ring**

A ring is *right uniserial* if it is a right uniserial module over itself, i.e. the lattice of right ideals is linearly ordered.

**right weak dimension**

*Right weak dimension* of a ring $A$ is supremum of the flat dimensions of right $A$-modules.

**ring**

A *ring* is a nonempty set $A$ together with two binary algebraic operations, that are denoted by $+$ and $\cdot$ and called addition and multiplication, respectively, such that, for all $a, b, c \in A$ the following axioms are satisfied:

1. $a + (b + c) = (a + b) + c$
2. $a + b = b + a$
3. There exists an element $0 \in A$, such that $a + 0 = 0 + a = a$
4. There exists an element $x \in A$, such that $a + x = 0$
5. $(a + b) \cdot c = a \cdot c + b \cdot c$
6. $a \cdot (b + c) = a \cdot b + a \cdot c$

**ring of bounded representation type**

A right Artinian ring $A$ is said to be of *bounded representation type* if there is a bound on the length of finitely generated indecomposable right $A$-modules.

**ring of finite representation type**

A ring $A$ is said to be of *finite representation type* if $A$ has up to isomorphism only a finite number of indecomposable finitely presented $A$-modules.

**ring of left bounded representation type**

A ring $A$ is called a ring of *left bounded representation type* if there is an upper bound on the number of generators required for indecomposable finitely presented left $A$-modules.

### ring of right bounded representation type
A ring $A$ is called a ring of *right bounded representation type* if there is an upper bound on the number of generators required for indecomposable finitely presented right $A$-modules.

### ring with finitely decomposable diagonal
A ring with prime radical $Pr(A)$ is called a *ring with finitely decomposable diagonal* if $A/Pr(A)$ is an FD-ring.

### semidistributive module
A module which is a direct sum of distributive modules is called *semidistributive*.

### semidistributive ring
A ring which is both a right and left semidistributive ring is called *semidistributive*.

### semihereditary ring
A ring $A$ which is both right and left semihereditary is called *semihereditary*.

### semilocal ring
A ring $A$ with Jacobson radical $R$ is called *semilocal* if $A/R$ is a right Artinian ring.

### semimaximal ring
A ring $A$ is called *semimaximal* if it is a semiperfect semiprime right Noetherian ring such that for each local idempotent $e \in A$ the ring $eAe$ is a discrete valuation ring.

### semiperfect ring
A semilocal ring $A$ is called *semiperfect* if idempotents can be lifted modulo the Jacobson radical of $A$.

### semiprimary ring
A ring $A$ with Jacobson radical $R$ is called *semiprimary* if $A/R$ is a semisimple ring and $R$ is nilpotent.

### semiprime ring
A ring which does not contain non-zero nilpotent ideals is called *semiprime*.

### semiprimitive ring
A ring is called *semiprimitive* (or *Jacobson semisimple ring*) if its Jacobson radical is equal to zero. This is a left-right symmetric notion.

### semi-reflexive module
An $A$-module $M$ is called *semi-reflexive* if

$$\delta_M : M \longrightarrow M^{**}$$

defined by

$$\delta_M(m)(f) = f(m)$$

for $m \in M$ and $f \in M^*$ is an monomorphism, where $M^* = \mathrm{Hom}_A(M, A)$.

## semisimple module
A module which can be decomposed into a direct sum of simple modules is called *semisimple*.

## semisimple ring
A ring is called *semisimple* if it is a semisimple module over itself, i.e it is a direct sum of finite number of simple right (left) ideals.

## serial module
A module which decomposes into a direct sum of uniserial submodules is called *serial*.

## serial ring
A ring which is both a right and left serial ring is called *serial*.

## simple Dynkin diagram
Dynkin diagrams of the form $A_n$, $D_n$, $E_6$, $E_7$ and $E_8$ are called *simple Dynkin diagrams*.

## simple module
A module $M$ is called *simple* if it has no non-zero proper submodules, i.e. only submodules of $M$ are the module $M$ and the zero module.

## simple ring
A ring is called *simple* if it has no two-sided ideals different from zero and the ring itself.

## simply connected species
A species is *simply connected* if the underlying graph of its quiver is a tree.

## simply laced quiver
A quiver without multiple arrows and multiple loops is called a *simply laced quiver*.

## singular element
An element $m \in M$ is called *singular* if the right ideal r.ann$_A(m)$ is essential in $A_A$.

## singular module
An $A$-module $M$ is called *singular* if $\mathcal{Z}(M) = M$.

## singular submodule
*Singular submodule* in $M$ is the submodule

$$\mathcal{Z}(M) = \{m \in M \ : \ \text{r.ann}_A(m) \subseteq_e A\} =$$
$$= \{m \in M \ : \ \exists \, I \subseteq_e A \text{ such that } mI = 0\}.$$

## skew field
A non-zero ring $A$ with identity is called a *skew field* if every non-zero element of $A$ is invertible, i.e. $U(A) = A \backslash \{0\}$.

**small submodule**

A submodule $N$ of a module $M$ is called *small* if the equality $N + X = M$ implies $X = M$ for any submodule $X$ of the module $M$. So it is a superfluous submodule.

**socle**

The *socle* of a module $M$ is the sum of all simple submodules of $M$.

**species**

A *species* is a finite family $S = (A_i, V_{ij})$, $(i, j = 1, \ldots, n)$, where the $A_i$ are prime rings and the $V_{ij}$ are $A_i$-$A_j$-bimodules.

**split ring**

A ring $A$ is called *split* if there is an $\overline{A}$-submodule $\overline{N}$ in $N$ such that $N = \overline{N} \oplus N^2$, where $N$ is the prime radical of $A$ and $\overline{A}$ is a subring in $A$ such that $A = \overline{A} \oplus N$.

**stable module**

A module $M$ is called *stable* if it has no projective summands.

**stably isomorphic modules**

Two modules $X$ and $Y$ over a ring $A$ are said to be *stably isomorphic* if there exist projective $A$-modules $P$ and $Q$ such that $X \oplus P \simeq Y \oplus Q$.

**stably finite ring**

A ring $A$ is called *stably finite* if it satisfies the following equivalent conditions for every $n \geq 1$:

1. $M_n(A)$ is a Dedekind-finite ring.
2. If $A^n \simeq A^n \oplus M$ for some $A$-module $M$ then $M = 0$.
3. Any module epimorphism $\varphi : A^n \longrightarrow A^n$ is an isomorphism.

**strict vector in $\mathbf{Z}^n$**

A vector $\mathbf{x} \in \mathbf{Z}^n$ is called *strict* if no $x_i$ is zero.

**strongly indecomposable module**

An $A$-module $M$ is called *strongly indecomposable* if the endomorphism ring $\text{End}_A(M)$ is local.

**structural matrix ring**

The *structural matrix ring* is a ring $M(B, A)$ over an associative ring $A$ associated with a reflexive and transitive Boolean matrix $B = [b_{ij}]$ defined in the following way:

$$M(B, A) = \{X = [x_{ij}] \in M_n(A) \ : \ b_{ij} = 0 \implies x_{ij} = 0\},$$

where $b_{ij} = 1$ if and only if $i \leq j$ for some pre-ordered relation $\leq$ on the finite pre-ordered set $\{1, 2, \ldots, n\}$, otherwise $b_{ij} = 0$.

**superfluous submodule**

A submodule $N$ of a module $M$ is called *superfluous* if the equality $N + X = M$ implies $X = M$ for any submodule $X$ of the module $M$. So it is a small submodule.

**suitable ring**
A ring is *suitable* if it is both right and left suitable.

**symmetrizable matrix**
An integer matrix $\mathbf{C}$ is said to be *symmetrizable* if there exists a diagonal matrix $\mathbf{F}$ with positive entries in its diagonal such that $\mathbf{CF}$ is symmetric.

**syzygy module**
A right $A$-module $M$ which is isomorphic to a submodule of a projective module is called a *syzygy*.

**T-nilpotent ideal**
An ideal which is right and left T-nilpotent is called *T-nilpotent*.

**tame algebra**
An algebra of tame representation type is called a *tame algebra*.

**tiled order**
A ring $A$ is called a *tiled order* if it is a prime Noetherian semiperfect semidistributive ring with non-zero Jacobson radical.

**torsion element**
An element $m$ of a right $A$-module $M$ is *torsion* if $\mathrm{r.ann}_A(m) \neq 0$.

**torsion module**
A module $M$ over a ring $A$ is called *torsion* if $t(M) = M$ where

$$t(M) = \{m \in M \ : \ \exists x \in A, x \neq 0, mx = 0\}.$$

**torsion submodule**
The submodule $\mathcal{Z}(M)$ of a right $A$-module $M$ over a semisimple right Goldie ring is *torsion*.

**torsion-free element**
An element $m$ of a right $A$-module $M$ such that $\mathrm{r.ann}_A(m) = 0$ is called *torsion-free*.

**torsion-free module**
A right module $M$ over a semiprime right Goldie ring $A$ is called *torsion-free* if

$$\mathcal{Z}(M) = \{m \in M \ : \ \mathrm{r.ann}_A(m) \subseteq_e A\} = 0.$$

**torsion-less module**
A semi-reflexive module is also called *torsion-less*.

**total valuation ring**
A subring $A$ of a division ring $D$ such that $x \in A$ or $x^{-1} \in A$ for each $x \in D^*$ is called a *total valuation ring*.

**transpose of a module**

Let $A$ be a semiperfect ring, and let $M$ be a right $A$-module with minimal projective resolution

$$P_1 \xrightarrow{\varphi} P_0 \longrightarrow M \longrightarrow 0.$$

Let $^* = \mathrm{Hom}_A(-, A)$. A module $\mathrm{Tr}(M) = \mathrm{Coker}(\varphi^*)$ where

$$0 \longrightarrow M^* \longrightarrow P_0^* \xrightarrow{\varphi^*} P_1^* \xrightarrow{\omega} \mathrm{Tr}(M) \longrightarrow 0$$

is called the *transpose* of $M$.

**triangular decomposition of identity**

A *triangular decomposition of identity* is a decomposition of the identity $1 = e_1 + e_2 + \cdots + e_n \in A$ into a sum of pairwise orthogonal idempotents of a ring $A$ such that $e_i A e_j = 0$ for all $i > j$.

**triangular ring**

A ring $A$ is called *triangular* if there exists a triangular decomposition of the identity of $A$.

**two-sided Peirce decomposition**

A *two-sided Peirce decomposition* of a ring $A$ is a decomposition of the following form:

$$A = \bigoplus_{i,j=1}^{n} e_i A e_j,$$

where $1 = e_1 + e_2 + \ldots + e_n$ with pairwise orthogonal idempotents $e_1, e_2,\ldots,e_n$ of a ring $A$.

**uniform dimension**

*Uniform dimension* of a module $M$ is the largest integer $n$ such that $M$ contains a direct sum $M_1 \oplus M_2 \oplus \ldots \oplus M_n$ of non-zero submodules.

**uniform module**

A non-zero module $M$ is called *uniform* if the intersection of any two non-zero submodules of $M$ is non-zero.

**uniserial module**

A module is called *uniserial* if the lattice of its submodules is a chain, i.e. the set of all its submodules is linearly ordered by inclusion.

**unit-regular ring**

A ring $A$ is called *unit-regular* if for any $a \in A$ there exists a unit $u \in \mathcal{U}(A)$ such that $a = aua$.

**valuation**

A *valuation* on a division ring $D$ is a surjective map $v : D \to G \cup \{\infty\}$ satisfying the following axioms:

  1. $v(x) \leq \infty$

2. $v(x) = \infty$ if and only if $x = 0$
3. $v(xy) = v(x) + v(y)$
4. $v(x + y) \geq \min(v(x), v(y))$

for all $x, y \in D$, where $(G, +, \geq)$ is a totally ordered group.

## valuation domain

A subring $A$ of a field $K$ is called a *valuation domain* if there is a totally ordered Abelian group $G$ and a valuation $v : K \rightarrow G \cup \{\infty\}$ on the field $K$ such that $A = \{x \in K : v(x) \geq 0\}$.

## valuation ring

An invariant valuation ring is also called a *valuation ring*.

## von Neumann-finite ring

A ring $A$ which is Dedekind-finite, i.e. if $ba = 1$ whenever $ab = 1$ for $a, b \in A$, is also called a *von Neumann-finite ring*.

## von Neumann regular ring

A ring $A$ is called a *von Neumann regular ring* if for every $a \in A$ there exists an $x \in A$ such that $a = axa$.

## weak dimension

*Weak dimension* of a ring $A$ is the common value of right and left weak dimensions of $A$.

## weakly 1-finite ring

A Dedekind-finite ring $A$ is also called a *weakly 1-finite ring*, i.e. a ring such that $ba = 1$ whenever $ab = 1$ for $a, b \in A$.

## weakly Noetherian ring

A ring $A$ is called *weakly Noetherian* if for any arbitrary ideal $I$ of $A$ and any $\overline{x} = x + I$ in $A/I$ there is an integer $n$ such that $\text{r.ann}_A(\overline{x}^n) = \text{r.ann}_A(\overline{x}^m)$, and $\text{l.ann}_A(\overline{x}^n) = \text{l.ann}_A(\overline{x}^m)$ for all $m \geq n$.

## weakly prime ring

A *weakly prime ring* $A$ is a ring for which the product of any two non-zero ideals not contained in the Jacobson radical of $A$ is non-zero.

## wild algebra

An algebra of wild representation type is called a *wild algebra*.

# Index